Receptors in Cellular Recognition and Developmental Processes

CELL BIOLOGY: A Series of Monographs

EDITORS

D. E. BUETOW
*Department of Physiology
and Biophysics
University of Illinois
Urbana, Illinois*

I. L. CAMERON
*Department of Cellular and
Structural Biology
The University of Texas
Health Science Center at San Antonio
San Antonio, Texas*

G. M. PADILLA
*Department of Physiology
Duke University Medical Center
Durham, North Carolina*

A. M. ZIMMERMAN
*Department of Zoology
University of Toronto
Toronto, Ontario, Canada*

Recently published volumes

Gary L. Whitson (editor). NUCLEAR-CYTOPLASMIC INTERACTIONS IN THE CELL CYCLE, 1980

Danton H. O'Day and Paul A. Horgen (editors). SEXUAL INTERACTIONS IN EUKARYOTIC MICROBES, 1981

Ivan L. Cameron and Thomas B. Pool (editors). THE TRANSFORMED CELL, 1981

Arthur M. Zimmerman and Arthur Forer (editors). MITOSIS CYTOKINESIS, 1981

Ian R. Brown (editor). MOLECULAR APPROACHES TO NEUROBIOLOGY, 1982

Henry C. Aldrich and John W. Daniel (editors). CELL BIOLOGY OF *PHYSARUM* AND *DIDYMIUM.* Volume I: Organisms, Nucleus, and Cell Cycle, 1982; Volume II: Differentiation, Metabolism, and Methodology, 1982

John A. Heddle (editor). MUTAGENICITY: New Horizons in Genetic Toxicology, 1982

Potu N. Rao, Robert T. Johnson, and Karl Sperling (editors). PREMATURE CHROMOSOME CONDENSA-TION: Application in Basic, Clinical, and Mutation Research, 1982

George M. Padilla and Kenneth S. McCarty, Sr. (editors). GENETIC EXPRESSION IN THE CELL CYCLE, 1982

David S. McDevitt (editor). CELL BIOLOGY OF THE EYE, 1982

P. Michael Conn (editor). CELLULAR REGULATION OF SECRETION AND RELEASE, 1982

Govindjee (editor). PHOTOSYNTHESIS, Volume I: Energy Conversion by Plants and Bacteria, 1982; Volume II: Development, Carbon Metabolism, and Plant Productivity, 1982

John Morrow. EUKARYOTIC CELL GENETICS, 1983

John F. Hartmann (editor). MECHANISM AND CONTROL OF ANIMAL FERTILIZATION, 1983

Gary S. Stein and Janet L. Stein (editors). RECOMBINANT DNA AND CELL PROLIFERATION, 1984

Prasad S. Sunkara (editor). NOVEL APPROACHES TO CANCER CHEMOTHERAPY, 1984

Burr G. Atkinson and David B. Walden (editors). CHANGES IN EUKARYOTIC GENE EXPRESSION IN RESPONSE TO ENVIRONMENTAL STRESS, 1985

Reginald M. Gorczynski (editor). RECEPTORS IN CELLULAR RECOGNITION AND DEVELOPMENTAL PROCESSES, 1986.

In preparation

Peter B. Moens (editor). MEIOSIS

Govindjee (editor). LIGHT EMISSION BY PLANTS AND BACTERIA

tm

Receptors in Cellular Recognition and Developmental Processes)

Edited by

Reginald M. Gorczynski

Ontario Cancer Institute
Toronto, Ontario, Canada

1986

ACADEMIC PRESS, INC.
Harcourt Brace Jovanovich, Publishers

Orlando   San Diego   New York   Austin
London   Montreal   Sydney   Tokyo   Toronto

ACADEMIC PRESS, INC.
Orlando, Florida 32887

United Kingdom Edition published by
ACADEMIC PRESS INC. (LONDON) LTD.
24–28 Oval Road, London NW1 7DX

Library of Congress Cataloging in Publication Data
Main entry under title:

Receptors in cellular recognition and developmental
    processes.

    (Cell biology)
    Includes bibliographies and index.
    1. Cell interaction.   2. Cell receptors.
3. Cellular recognition.   4. Developmental cytology.
I. Gorczynski, Reginald M.   II. Series.   [DNLM:
1. Cell Communication.   2. Receptors, Endogenous
Substances.   QH 603.C43 R953]
QH604.2.R43   1986   574.1'7   85-26698
ISBN 0–12–290530–X (alk. paper)

PRINTED IN THE UNITED STATES OF AMERICA

86 87 88 89   9 8 7 6 5 4 3 2 1

To Christopher and Laura with love

Contents

II (Chapters 5–8) Receptors Involved in the Regulation of Development of Multicellular Organs and Organisms

Reginald M. Gorczynski

Scott F. Gilbert

David L. Stocum

Contributors

Numbers in parentheses indicate the pages on which the authors' contributions begin.

Robert J. Bloch, Department of Physiology, University of Maryland, School of Medicine, Baltimore, Maryland 21201, (183)

Kimberly E. Dow, Department of Pediatrics (Neonatology), Queen's University, Kingston, Ontario, Canada K7L 3N6, (215)

I. J. Fidler,[1] Department of Cell Biology, The Weizmann Institute of Science, Rehovot 76100, Israel, (287)

Carl G. Gahmberg, Department of Biochemistry, University of Helsinki, 00290 Helsinki 29, Finland, (251)

Scott F. Gilbert, Department of Biology, Swarthmore College, Swarthmore, Pennsylvania 19081, (133)

Reginald M. Gorczynski, Ontario Cancer Institute, Toronto, Ontario, Canada M4X 1K9, (1, 73, 125, 245, 305)

Kimmo K. Karhi, Department of Biochemistry, University of Helsinki, 00290 Helsinki 29, Finland, (251)

David H. Katz, Department of Immunology, Medical Biology Institute, La Jolla, California 92037, (101)

Barry E. Ledford, Department of Biochemistry, Medical University of South Carolina, Charleston, South Carolina 29425, (45)

John J. Marchalonis, Department of Biochemistry, Medical University of South Carolina, Charleston, South Carolina 29425, (45)

A. Raz, Department of Cell Biology, The Weizmann Institute of Science, Rehovot 76100, Israel, (287)

Richard J. Riopelle, Department of Medicine (Neurology), Queen's University, Kingston, Ontario, Canada K7L 3N6, (215)

---

[1]Present address: Department of Cell Biology, M. D. Anderson Hospital and Tumor Institute, Texas Medical Center, Houston, Texas 77030.

J. B. Solomon, Immunology Unit, Bacteriology Department, University of Aberdeen, Aberdeen AB9 2ZD, Scotland, (9)

Joe Henry Steinbach, Department of Anesthesiology and Department of Anatomy and Neurobiology, Washington University School of Medicine, Saint Louis, Missouri 63110, (183)

David L. Stocum, Department of Genetics and Development, University of Illinois, Urbana, Illinois 61801, (165)

I. S. Trowbridge, Department of Cancer Biology, The Salk Institute, San Diego, California 92138, (267)

Gerardo R. Vasta, Department of Biochemistry, Medical University of South Carolina, Charleston, South Carolina 29425, (45)

Gregory W. Warr, Department of Biochemistry, Medical University of South Carolina, Charleston, South Carolina 29425, (45)

Preface

The aim of this book is to introduce the reader in a general way to intercellular communication and, in particular, to the evolutionary and ontogenetic role of molecules which allows cells to communicate and/or associate with one another. Communication can occur among cells across a distance as exemplified by neurons and muscle cells at the neuromuscular junction, or it can take place by actual cell contact and association as, for instance, in fertilization and differentiation. In either instance, the macromolecule used by the cells to permit this communication is designated the receptor.

In the past 20 years, rapid developments in fractionation, isolation, and biochemical characterization of both cells and subcellular tissue, coupled with an interest in the molecular biology of the process of cellular differentiation, have led to an expansion of interest in cellular receptors from an earlier relatively restricted pharmacological viewpoint as exemplified by the original studies of Langley in 1878 concerning the inhibition by atropine of the action of pilocarpine. The receptor recognizes (receives) an appropriate specific signal and transduces the information so received to provoke a specific response from the cell concerned. Signal discrimination in a mixed population of cells can be achieved by virtue of the fact that only certain cells will have a receptor capable of binding the stimulator (ligand) at an affinity sufficient to activate subsequent steps in the cascade of biological reactions. Signal transduction is generated by virtue of binding of ligand to the receptor, and generally involves alteration in the activity of some appropriate effector (enzyme, ion channel) in a manner that leads to the requisite physiological response.

If receptors express as their key function the transfer of macromolecular information through impermeable barriers, it can be anticipated that not all such receptors will necessarily be found bridging the cytoplasm of the cell with the external milieu. Thus steroid hormone receptors are found in the cellular cytoplasm, and interaction of these receptors with their specific ligand leads to translocation to the nucleus and specific activation of transcription of parts of the genome. Other receptors such as those for the thyroid hormones tri- and tetraiodothyronine are found within the nucleus itself. Nevertheless, the concern throughout this volume will be with those receptors present in the plasma membrane of cells. While recent advances have been made in exploring the biochemical mechanisms (enzymatic methylation of membrane phospholipids) whereby receptor triggering leads to signal transduction, there is an advantage to be gained in viewing intercellular communication from a more general perspective.

As organisms increase in complexity from the unicellular through the multicellular to the multiorgan state, there is a need for a concomitant increase in sophistication at the cell surface of those molecules that both recognize and signal the presence of ''self'' versus ''non-self'' and lead to the appropriate orientation and organization of the various parts of ''self'' with one another. Investigation of the phylogeny and ontogeny of receptor molecules and analysis of function of cell surface molecules may enable us to understand the forces operating to conserve receptors during the development of multicellular organisms.

How is their expression controlled (genetically/environmentally)? How do they function in the roles they play? What is the effect of modulation of their expression/function on homeostasis within the whole organism? In what follows, these and other questions will be explored with examples from many disciplines of cell biology. However, it is hoped that underlying each chapter the reader will be able to see a relevance to this guiding theme of intercellular recognition and development.

A volume of this nature could not come to fruition without the concerted effort of a number of people. I would like to thank all of the contributors, who toiled, often I am sure it seemed to them endlessly, yet eventually successfully, to meet the various deadlines I gave them for submissions and updates. My thanks, too, to my many colleagues who on a number of occasions have offered valuable advice on the organization and content of this book—to Gerald Price in particular, who has been a valued collaborator for many years. I absolve them all from any responsibility for what lies within! Without Anne Collins and Maria Boulanger I know these pages would still lie half-typed and uncollated on my desk, awash with a myriad of other unfinished work. Last, and most important of all, I would like to thank Professor Cinader, who first suggested the value of a book of this type and proceeded to support that initiative with many hours of critical review. Without his friendship, wit, and encouragement, this book would never have materialized.

Reginald Gorczynski

Commentary

This book is devoted to the steps toward the Rosetta stone for the current status and future discovery of intercellular communication.

Cell communication is dependent on a series of molecular events involving receptors and ligands that are either cell bound or secreted by one cell and taken up by surface structures—receptors—of another cell. A series of sequential events of molecular interaction at cell and organelle membranes coordinate cell metabolism within the same and between different organs. Receptors can be activated through soluble factors and, hence, at a distance. Receptor–ligand interaction can also occur between membranes of different cell types, i.e., via adhesion molecules that play a role in structural development of organs, exemplified by neural cell adhesion and embryological development under the influence of "master" cells.

Recognition and, thus, receptor–ligand interaction play a role during homing of cells in development, differentiation, and cell migration. In the immune system, macromolecules of the external world cause distortions of internal communication; the resulting change in the balance of molecular communication constitutes the immune response.

Receptor–ligand communication contributes to resistance against infectious disease. Antigen recognition by B and T cells is one component of this process; the ability of a particular parasite to attach to a cell receptor is an example of other facets. In short, interaction of the cellular milieu with external molecular changes occurs through receptors of the lymphoid system and through receptors of other cells that control the ability of parasites to attach to membranes and to reach the interior of cells.

Malfunction of a single step in cell communication results in disease and contributes to neoplastic transformation. Development of neoplastic cells and metastases depends on disappearance or blockage of receptors through which growth and metastatic spread are controlled.

Cell communication is regulated by limitation in the period during which a given stimulus can affect biochemical processes that are activated via a particular receptor. Responses, initiated by ligand–receptor combination, can be terminated by events that lead to cessation of responsiveness after messages have been received for a given time. This limitation is achieved by various processes, including endocytosis, recycling, and affinity changes in receptors, and through disassociation of micromolecular complexes with which the ligand-binding site is associated.

Factors convey signals by combination with receptors. These signals can give rise to the production of other factors and thus to the sentences of the intercellular language; the resulting intercommunication is intense and continuous. There are superfamilies of molecules, corresponding to language families, that play a role in recognition and show homologies in a wide group of animals, from vertebrates to invertebrates. The analysis of this molecular language is a major movement in the biology of the twentieth century.

B. Cinader
Department of Immunology
University of Toronto
Medical Sciences Building
Toronto, Canada M5S 1A8

Receptors in Cellular Recognition and Developmental Processes

I (CHAPTERS 1–4)

Phylogenetic Analysis of Receptors in Development of Immune Recognition

REGINALD M. GORCZYNSKI

It seems obvious that the recognition of self must be a property of all cells. In unicellular organisms this avoids iso-phagocytosis, while in pluricellular organisms self-recognition ensures cohesion and collaboration between cellular aggregates. Where the phenomenon has been studied in depth, e.g., in the vertebrate immune system, we can also state that self-recognition is an active process, in which cooperation in the recognition of and reaction to non-self between specialized cells within the same individual is often seen. In part at least, immune responses in vertebrates show evidence for linkage to the polymorphic genes of the major histocompatibility complex (MHC). The evolutionary advantage of this MHC-linked immune responsiveness is unclear, though one popular idea is that given the extensive polymorphism seen at the MHC of most species, linkage of immune-cell recognition to products of genes encoded within the MHC implies capability for increased diversity in immune recognition (Klein, 1980). Clearly a problem with this notion is to explain the adaptive advantage of species showing little MHC polymorphism (e.g., hamster) and the worry that such an interpretation seems to endow the MHC system with "Promethean foresight" (Ohno *et al.*, 1980).

Even in insects, however, the available evidence suggests that distinctions can be made between different types of foreign objects—i.e., graded discrimination is possible. While we shall see that it is by no means clear whether during

1

phylogeny recognition is always carried out by cells, or by soluble molecules in collaboration with cells, it is nevertheless possible to imagine one of two mechanisms whereby signal discrimination can occur: (1) graded responses in a recognition system using nonspecific factors, e.g., physiochemical parameters such as surface charge; (2) the development of specific factors which are superimposed upon an already existing nonspecific system.

A review of the nonimmune surface recognition of foreign material common to protozoa and to all cells of multicellular organisms is provided by Solomon in Chapter 1. A feature of such recognition in plants is the interaction between specific saccharides and glycoprotein (lectin) receptors at the cell surface—such an interaction seems to lie at the heart of the cellular adhesion process which is a feature of the agglutination of unicellular amoeba in the aggregation phase of the life cycle of the slime mold (Newell, 1977). Nevertheless, the most primitive of host immune responses in multicellular organisms, phagocytosis, is not seen in higher plants, although encapsulation can occur and phagocytosis is seen in slime molds and algae. Solomon reviews the literature concerning self-/non-self recognition (Boyden, 1962) at the cell surface (in allo- or xenotransplantation reactions) from the sponges through the annelids and mollusks to the chordates. There is convincing evidence for rejection of xeno- as well as allografts, with a growing literature on polymorphism of cell surface histocompatibility molecules in some phyla (Hildenmann et al., 1980). However, the question of whether memory (as witnessed by the phenomenon of second set rejection) exists in allo- or xenotransplants in invertebrates is still unsettled.

Far more detailed investigations have been performed on the humoral factors capable of performing specific recognition functions. It is believed that the recognition and phagocytosis of an implant in the host coelomic cavity occurs by a process similar to that associated with recognition of a transplant at the host surface, and this belief, coupled with the relative ease of experimental manipulation, has led internal phagocytosis to be the response most widely investigated. From mollusks [Helix pomatia—differential clearance of erythrocytes bearing different carbohydrate surfaces (Renwrantz, 1981)] to annelids [inhibition of selective uptake of gram-positive/gram-negative bacteria by coelomocytes in the presence of D-(+)-glucose (Fitzgerald and Ratcliffe, 1980)], we find ligand-specific cellular receptors whose recognition function is compromised by the presence of soluble sugars. Despite the fact that soluble hemagglutinins have been found in most invertebrate species studied there is no quantitative or qualitative change in these hemagglutinins following antigen stimulation, nor is there evidence for memory in invertebrate hemagglutinin-mediated anti-self recognition. In this respect then, there does not seem to be a parallel with cell surface receptors on, e.g., mammalian B lymphocytes. It is of interest that the most common reactivity seen in the hemolymph is a lectin-like hemagglutination reaction—e.g., snail lectin inhibited by methyl-DGalNAc (Hammarstrom and Kabat, 1969)—and indeed mitogen stimulation studies suggest that leukocytes of the earthworm possess a mitogen receptor for concanavalin A (Con A) which is inhibited by methyl D-mannopyranoside (Roch et al., 1975).

With respect to one of the possibilities raised above, then, it does seem that during phylogenetic development within the immune system, a highly discriminatory secondary recognition system has become superimposed upon a primordial nonspecific one. A similar conclusion is reached by Lackie (1981). Let us recognize, however, that we are not attempting here to implicate the immune system, by virtue of its capacity to react with non-self material, in the mechanism of evolution. The teleological nature of this particular argument has been forcefully attacked by Allegretti (1978).

We might now ask, in light of the above, whether there is any evidence which can be adduced for a relationship between molecules with recognition function existing within different members of a species, or a relationship during evolution between these molecules in different species? It is appropriate to investigate evidence for such a "family" of recognition molecules bearing in mind that changes in structure and/or function may occur during evolution from the primitive prototype molecules. This idea of a "superfamily" of recognition molecules showing homology within the vertebrates, chordates and invertebrates is developed further by Marchalonis et al. in Chapter 2. Comparison of amino acid composition (Marchalonis and Weltman, 1971) suggests a relatedness between recognition molecules from origins as diverse as the agglutinin of the lamprey, C-reactive protein of vertebrates (specific for phosphorylcholine), mammalian immunoglobulin molecules, and the recently described vertebrate T-lymphocyte receptor. As Marchalonis et al. stress in Chapter 2, it is "comparisons with the primitive members of the true immunoglobulin family [which] provide the strongest guidepost of homology."

In the absence of primary protein sequence data to detect sequence homology, and given the disparity in size of the molecules examined, it is probably unwise at this point in time to state the case more strongly. It may be, for instance, that this "superfamily" represents the product of convergent evolution of molecules constrained (by their very function) to evolve within certain geometric limits. However, within individual pairs of molecules where more detailed comparisons can be made (e.g., for T-cell receptors and immunoglobulin molecules) the protein sequence data (Yanagi et al., 1984; Hedrick et al., 1984), the structural resemblance (Marchalonis and Barker, 1984), and the frequent sharing of idiotypic specificities on T cells and B cells expressing a common antigen specificity (Marchalonis and Hunt, 1982) make this suggestion (evolutionary convergence, rather than direct evolutionary relatedness) less likely.

Using a variety of approaches which include the analysis of cell-surface determinants, the immune system of vertebrates has been shown to consist of a complex interacting network of lymphocytes which combine the two properties of a capacity to respond to the unexpected and the ability to remember that response, in a manner not seen in more primitive invertebrate immune systems. In general, these are functions of B and T lymphocytes in the immune system, responsible respectively for the synthesis/secretion of immunoglobulin or the expression of an array of cell-mediated responses which include the production of cytotoxic killer cells directed to virus infected targets, development of effector cells for delayed type hypersensitivity, etc. The B and T lymphocytes exhibit exquisite antigen specificity. Thus a cell with the capacity to recognize one antigen (x) is in general unable to be triggered by another antigen (y).

There are other cells of importance in the immune system which tend to exhibit less antigen specificity. The phagocytes (e.g., macrophages) of the mammal may themselves be the homologue of the primordial immune mechanism. Other less specific cells include neutrophils, granulocytes, eosinophils, and so-called natural killer cells.

Data suggesting that recognition of self from non-self by a variety of lympho-myeloid cells is mediated by lectin-like cell surface receptors is described by Gorczynski in Chapter 3. Analysis of receptors on macrophages has been performed with macrophage–tumor binding assays and presentation of macrophage-bound antigen for lymphocyte-mediated responses. Using the latter approach, a relationship has been established between the functional antigen-presenting capacity of receptors on mammalian macrophages and annelid coelomocytes. Another method of study of functionally relevant glycolipid/glycoprotein/proteogly-can molecules on immune-cell surfaces has been to assess the inhibition of effector activity in the presence of carbohydrate molecules. This method has been used to examine in some detail surface recognition receptors on natural killer cells.

A model is discussed in which macrophage–lymphocyte interaction within the mammalian immune system is mediated by a complementary interaction of glycosyltransferase and substrate saccharides on the interacting cell surfaces (Parrish et al., 1981). This model is developed to offer one possible explanation of "determinant selection" in the immune response of (responder × nonresponder)F_1 animals to complex antigens (Rosenthal et al., 1977). This mechanism proposes that an alteration in cell surface interaction between the recognition structures (a lectin on the macrophage cell surface) which make initial contact with the foreign antigen and the lymphocyte-responding cell pool forms the basis of subsequent events observable apparently only in the lymphocytes themselves. This may represent a useful approach to analysis of the apparent breakdown in immune homeostasis in chronic disease (parasite infection, tumor growth, aging, etc.).

An additional function of surface carbohydrates on immune cells seems to be their role in the appropriate homing of cells during development and differentiation (Le Douarin, 1984). In keeping with this notion, there is evidence that expression of target structures recognized by particular lectins (e.g., peanut agglutinin) changes within a maturational pathway.

Recognition seen within the immune system is but one special example of a general property of all cells, namely self-recognition. There is evidence that at least a part of this recognition is MHC-linked (e.g., evidence from auto-rosetting; autologous MLR reactivity; cell cooperation in immunity). The model described by Parrish *et al.* (1981) and discussed above (see also Chapter 3) attempts to explain Class I and Class II MHC gene-restricted immunological recognition of non-self, as well as nonimmunological self-recognition of other cells within the same organism in the context of cell surface carbohydrate determinants (and the regulation of their production by MHC-regulated glycosyltransferase genes).

One might ask whether there is any evidence that MHC molecules are indeed crucial for the survival of cells (organisms). At the single-cell level under artificial tissue culture conditions this is clearly not the case—i.e., the cell line Daudi lacks expression of HLA genes since it lacks functional genes coding for β_2-microglobulin. At the level of the whole organism the question is rather more difficult to answer, although in the "denuded lymphocyte syndrome," where lymphocytes lack surface HLA antigens, patients show a severe combined immunodeficiency disorder (Touraine *et al.*, 1978).

In an alternative situation in which nongenetically compatible cells are used to form an organism—i.e., in allophenic mice—life is again precarious. Nevertheless, one might have anticipated that successful transplantation of foreign tissue from one organism to another would not be possible, so it is pertinent to ask both why non-self recognition fails in these situations, and whether consideration of these examples provides further insight into the normal development of appropriate cell–cell recognition.

Target cells exposed to their signal ligand for prolonged periods of time may become adapted/desensitized to that ligand concentration. This process seems to occur via changes at the cell surface in the ligand receptor. There are many paths by which such desensitization can occur, e.g., (1) increased endocytosis of receptor (and hence increased internal receptor degradation) leading to receptor-down-regulation (Raff, 1976); (2) reversible inactivation of receptor by prolonged ligand binding such that receptor affinity for ligand is decreased (Lefkowitz, 1978); or (3) reversible inactivation of receptor such that a subsequent step (activation of membrane enzymes or opening of membrane ion channels) no longer occurs.

In other cases of target cell desensitization, the (reversible) change may even take place in a site internal to ligand–receptor interaction at the cell surface. Thus in morphine addiction it has been proposed that changes in adenyl cyclase activity occur such that receptor triggering by a given dose of morphine provides less enzyme activation over time; eventually cells return to their basal levels of enzyme activity. Withdrawal of morphine leads to a rapid increase in adenylate cyclase activity (perhaps responsible for many of the physiological symptoms of withdrawal seen *in vivo*) (Klee *et al.*, 1975).

Antigen-specific receptor modulation has been invoked to explain why some lymphocytes exposed to antigen at high concentrations eventually become refractory to a normally stimulating dose of the same antigen. This phenomenon may be a key to explaining tolerance to self antigens which develops during ontogeny (Zanders *et al.*, 1983).

There are other ways, however, in which expression of cell surface receptors may appropriately be regulated by events occurring at the cell surface. In Stocum's discussion of a model proposed to explain urodele limb regeneration, we will encounter the possibility that cell-surface structures (in this case of the blastema) may be modulated according to the mutual interaction of cells with their neighboring counterparts (Chapter 6). If then, the complex mammalian immune system is the analogue in microcosm of the development of the whole organism it may be anticipated that some cellular counterparts of the immune system (e.g., lymphocytes) also present cell-surface recognition structures whose expression is significantly responsive to the environment in which they develop (differentiate). This concept, to which the term adaptive differentiation (Katz, 1977) is given, is explored in detail by Katz (Chapter 4).

How is adaptive differentiation in the immune system studied? What is haplotype restriction? In the model which Katz considers, antibody production by antigen-stimulated B lymphocytes depends upon a positive cooperative interaction (help) provided by antigen-primed T lymphocytes. Primed T lymphocytes from $(X \times Y)F_1$ animals can help antigen-stimulated B lymphocytes from X or Y strain animals. But if strain X animals are lethally irradiated and reconstituted with $(X \times Y)F_1$ bone marrow, and primed F_1 lymphocytes from these chimeras are used as a source of T cells, effective help is only provided for strain X-B cells and not strain Y-B cells. This phenomenon is called haplotype restriction.

One interpretation of these data is that cellular interaction (CI) molecules exist on the surface of the communicating B and T lymphocytes, and that the expression of these CI molecules is somewhat dictated by the environment in which the lymphocytes (in this instance T lymphocytes) differentiate (adaptive differentiation) (Katz *et al.*, 1980a). However, in this particular case the restriction seen is a pseudorestriction. "Haplotype preference" is abolished if T cells are removed [from the $(X \times Y)F_1 \rightarrow X$] and adoptively primed in an irradiated

$(X \times Y)F_1$ recipient before helper cell activity is assayed! Katz refers to this pseudorestriction as "environmental restraint." A model is proposed, along with experimental data whose interpretation is consistent with the model, suggesting that nonpermissiveness in the environmental milieu is due to a dynamic process which includes the development of responses *against* the self-specific receptors for cell interaction molecules of the nonpermissive haplotype (Katz *et al.*, 1980b).

Active cell-mediated suppression as a means to explain environmentally produced regulation of intercellular communication may or may not be a mechanism restricted to the immune system. However, the phenomenological observation of a local interaction between the environment and the cell surface, which regulates the expression of those cellular receptor molecules used to communicate with other cells in the environment (local/distant), is quite general, as will become apparent from Chapter 7 by Steinbach and Bloch.

REFERENCES

Allegretti, N. (1978). *Dev. Comp. Immunol.* **2**, 15.
Boyden, S. V. (1962). *J. Theor. Biol.* **3**, 123.
Fitzgerald, S. W., and Ratcliffe, N. A. (1980). *In* "Aspects of Developmental and Comparative Immunology" (J. B. Solomon, ed.), p. 138. Pergamon Press, Oxford.
Hammarstrom, S., and Kabat, E. A. (1969). *Biochemistry* **8**, 2696.
Hedrick, S. M., Nielsen, E. A., Kavaler, J., Cohen, D. I., and Davis, M. M. (1984). *Nature (London)* **308**, 153.
Hildenmann, W. H., Bigger, C. H., Jolkiel, R. L., and Johnston, I. S. (1980). *In* "Phylogeny of Immunological Memory" (M. J. Manning, ed.), p. 9. Elsevier/North Holland Biomedical Press, Amsterdam.
Katz, D. H. (1977). *Cold Spring Harbor Symp. Quant. Biol.* **41**, 611.
Katz, D. H., Katz, L. R., Bogowitz, C. A., and Bargatze, R. F. (1980a). *J. Immunol.* **124**, 1750.
Katz, D. H., Katz, L. R., and Bogowitz, C. A. (1980b). *J. Immunol.* **125**, 1109.
Klee, W. A., Sharma, S. K., and Nirenberg, M. (1975). *Life Sci.* **16**, 1869.
Klein, J. (1980). *In* "Immunology 80" (M. Fougereau and J. Dausset, eds.), p. 139. Academic Press, London.
Lackie, A. M. (1981). *Dev. Comp. Immunol.* **5**, 191.
Le Douarin, N. M. (1984). *Cell* **19**, 537.
Lefkowitz, R. J. (1978). Regulation of β adrenergic receptors by β adrenergic agonists. *In* "Receptors and Hormone Action" (L. Birnbaumer and B. W. O'Malley, eds.), Vol. 3, p. 179. Academic Press, NewYork.
Marchalonis, J. J., and Barker, W. C. (1984). *Immunol. Today* **5**, 222.
Marchalonis, J. J., and Hunt, J. C. (1982). *Proc. Exp. Biol. Med.* **171**, 127.
Marchalonis, J. J., and Weltman, J. K. (1971). *Comp. Biochem. Physiol. [B]* **38**, 609.
Newell, P. C. (1977). *Endeavour* **1**, 63.
Ohno, S., Matsunaga, T., Epplen, J. T., and Hozumi, J. (1980). *Immunology* **80**, 577.
Parrish, C. R., O'Neill, H. C., and Higgins, T. J. (1981). *Immunol. Today* **2**, 98.

Raff, M. C. (1976). *Nature (London)* **259**, 255.

Renwrantz, L. (1981). *In* "Aspects of Developmental and Comparative Immunology" (J. B. Solomon, ed.), p. 9. Pergamon Press, Amsterdam.

Roch, P., Valembois, P., and DuPasquier, L. (1975). *Adv. Exp. Med. Biol.* **64**, 44.

Rosenthal, A. S., Barcinski, M. A., and Blake, J. T. (1977). *Nature (London)* **267**, 156.

Touraine, J. L., Betuel, H., Souillet, G., and Jeune, M. (1978). *J. Pediatr.* **93**, 47.

Yanagi, Y., Yoshikai, Y., Leggett, K., Clark, S. R., Alexsander, I., and T. W. Mak. (1984). *Nature (London)* **308**, 145.

Zanders, E. D., Lamb, J. R., Feldmann, M., Green, N., and Beverley, P. C. L. (1983). *Nature (London)* **303**, 625.

1

Invertebrate Receptors and Recognition Molecules Involved in Immunity and Determination of Self and Non-self

RECEPTORS IN CELLULAR RECOGNITION
AND DEVELOPMENTAL PROCESSES

I. INTRODUCTION

We know the first forms of life appeared about 3000 million years ago, as fossilized rod-shaped bacteria have been found in strata formed at that time. The first primitive unicellular organism to appear was probably the bluegreen algae about 800 million years later. Free oxygen probably did not form in the atmosphere until 1800 million years ago, and from 1200 to 1500 million years ago fossil cells resembling eukaryocytes appeared (Mahler and Raff, 1975). In the Cambrian period (600 million years ago) mollusks and other marine invertebrates with shells were deposited in strata which the paleontologist can study today. In the middle Cambrian period, jelly fishes and sponges emerged. It is believed from fossil evidence that insects appeared in the mid-Devonian (450 million years ago) and oysters some 225 million years later.

The phytoflagellates are at the crossroads of divergence between plants and animals. Both plants and even the single-celled protozoa are capable of recognition of self and non-self. Clarke and Knox (1979) point out that a common theme in plant recognition is the interaction between specific monosaccharides in carbohydrates of the plasma membrane with protein or glycoprotein receptors. For example, in higher plants, sugar cane leaves have a membrane receptor for the galactoside toxin produced by the fungal pathogen *Helminthosporium sacchari*. Also, the adhesion of zoospores of the pathogen *Phytophthora cinnamomi* to root surfaces of host plants seems to be mediated by carbohydrates of root slime. Despite the high concentrations of lectins in many plant seeds, Sequeira (1978) and Clarke and Knox (1979) consider their role is still unclear and the latter authors favor the arabinogalactan proteins as recognition molecules. While phagocytosis has been observed in slime molds and algae it has not been seen in higher plants, although encapsulation can occur. Compatibility and recognition in plants have been most extensively studied in pollen–stigma and host–pathogen interactions. Generally, plant autografts and allografts are accepted and most xenografts are rejected. However, Moore (1981) has recently considered how callus formation may affect the interpretation of such incompatability reactions and states that the mechanisms involved in incompatibility reactions are still largely unknown. A common hypersensitive response to infection in plants is due to production of the relatively nonspecific phytoalexins that elicit cellular incompatibility reactions (Klement and Goodman, 1967); no humoral responses have been reported (Clarke and Knox, 1979).

Carbohydrate binding, probably by lectins, has been implicated in fertilization of sea urchin and mammalian eggs, in gamete fusion in algae, in sexual compatibility in yeasts, and in cellular adhesions in slime mold morphogenesis (Callow, 1977). Probably the most fascinating example of cell–cell communication outside the animal kingdom is seen in the slime molds (Newell, 1977; Rosen and

Barondes, 1978). These slime molds resemble fungi in that they reproduce by formation of spores and are heterotrophic, but unlike fungi they lack a cell wall and ingest their food. In their nonsocial stage (the growth or vegetative stage), small solitary amoebae feed on bacteria by phagocytosis and divide by fission every 3–4 hr. When the food supply is depleted, the growth phase terminates and after an interphase of several hours the social stage commences. Guided by chemotactic signals, amoebae align and move over large distances toward discrete aggregation centers to form streams which converge in a "whirlpool" manner. Within the stream the cells are closely adherent. At each center, the cells assemble into a peglike multicellular aggregate containing thousands of cells called a pseudoplasmodium or "slug" which is surrounded by a slime sheath. Chemotaxis is mediated by $3',5'$-cAMP which is detected on the surface of amoeba by specific receptors; these tend to increase as the amoebae start to aggregate. At this time, changes in cell surface glycoproteins also occur. In parallel with the development of cell cohesiveness, lectin-mediated agglutination activities appear in the cytoplasm and are then expressed on the cell surface of the cohesive amoebae. Adhesion appears to be regulated by this appearance of lectins which interact with complementary oligosaccharide receptors on the surface of cohesive cells, as specific lectin antagonists can block cell adhesion under certain conditions (Rosen and Barrondes, 1978).

Many of these cell–cell interactions in slime molds are also found in invertebrates. For example, invertebrate hemocytes display chemotaxis toward bacteria in snails (Schmid, 1975), oysters (Cheng and Rudo, 1976), and earthworms (Marks *et al.*, 1979). Also, invertebrate lectins and their receptors probably have a role to play in recognition of self and non-self as well as defense (Section IV).

Recognition at the cell surface plays a vital role not only in preserving the integrity of "self"—that is the homogeneity of a colony—but also in the conservation of territory (along with its associated food) from allogeneic or xenogeneic aggressors.

Recognition is the first essential step in any immunological process, and for recognition to take place there must be receptors at the cell surface. A receptor could be defined as a structure which upon interaction with a ligand transmits a signal triggering a cell's function. The mere binding of a ligand to a cell surface is not indicative of the presence of a receptor, although such ligand-binding moieties may have a similar structure.

The presence or absence of cell-surface receptors for bacteria, viruses, etc., may dictate whether adherence (the first step in pathogenesis) may occur, although bacteria have their own methods of adhering, probably without specific receptors (Freter, 1980). The second line of defense is provided by recognition molecules inside the cell, which may opsonize the invasive bacteria, virus, or parasite and accelerate phagocytosis and killing. We shall examine how invertebrates can both recognize foreign cells, bacteria, viruses, and other parasites and distinguish between proteins, even though they have no lymphocytes.

The sheer complexity of the animal kingdom is a daunting prospect for the comparative immunologist. There are about 1.2 million different species of animals so far named, and about 10,000 new species are added each year. More than 90% of these species are invertebrates and at least 75% belong to the phylum Arthropoda. There are 900,000 known species of arthropods (Hickman, 1967).

The phylogeny of the invertebrates discussed in this chapter is shown in Fig. 1. This highly simplified diagram serves to show that many phyla such as Arthropoda and Mollusca have evolved quite independently of each other and on a different pathway from the echinoderms. Phylogenetic divergence is occurring all the time as species of one phyla develop quite independently of other phyla, yet convergently in terms of time. Many forms of animals are now extinct, probably due to the very severe changes of climate which occurred during this

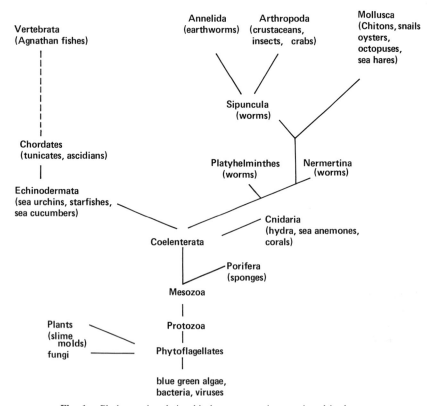

Fig. 1.   Phylogenetic relationship between species mentioned in the text.

long time period. On the other hand, some species alive today are known, from fossil evidence, to have survived virtually unchanged for hundreds of millions of years. As invertebrates appeared long before the first vertebrates—the agnathan jawless fishes (400 million years ago)—it is reasonable to suppose that one invertebrate phylum was the precursor of the vertebrates. The currently favored link is the tunicates, which have a notochord in their larval stage (Berrill, 1955).

In recent years specific cellular immunity in invertebrates has become well established. Virtually all invertebrate species studied are able to recognize "nonself" and in most cases recognition is followed by cytotoxic action towards the graft resulting in its subsequent rejection. There are now many examples of immunological memory being induced by primary xeno- or allografts, leading to accelerated "second-set" graft rejection (Section II).

Leukocytes of invertebrates may have properties very similar in terms of receptors and cell functions to the nonlymphoid leukocytes of the vertebrates. In addition, there are humoral factors in invertebrates which are also present in mammals. There is evidence that the higher invertebrates, such as Arthropoda, have rather more sophisticated immune mechanisms than lower forms, though by no means as efficient or complex as the primitive vertebrate fishes. The invertebrates' protective system would appear to vertebrate immunologists as "primitive" because of (a) relative lack of specificity, (b) absence of immunoglobulins, and (c) only relatively short-term protective immunity. However, as the invertebrates' protective systems have evolved over millions of years they must be considered successful and may be the principal basis of nonlymphoid vertebrate immunity. From such a viewpoint, studies of defense mechanisms in invertebrates may provide us with information on similar mechanisms still possessed by vertebrates but which are often obscured by the presence of lymphocytes and antibodies. On the other hand, the attitude of the invertebrate immunologist is to seek the nature of immune mechanisms involved in recognition and defense of invertebrates who do not need the sophisticated machinery of a lymphoid system for their successful survival.

In the vertebrates, much is now known about receptors on lymphocytes which are involved in recognition of "non-self." The B lymphocytes, for example, possess cell-surface receptors capable of combining with antigen which have been identified as several different classes of immunoglobulin. The nature of receptors for antigen on the surface of T lymphocytes is still uncertain. We shall see that recent research is beginning to provide information on receptors for antigens on invertebrate leukocytes.

Can we hope to find common mechanisms for recognition of self and "nonself" and common effector mechanisms involved in defense among such a huge number of species? In this chapter we examine the nature of receptors on the surface of invertebrate leukocytes and the way these cells can discriminate between "foreign" cells and molecules. In most cases the presence of receptors will be more by assumption than definition as only the subsequent effector mechanisms have been described.

The second section of this chapter deals with recognition of foreign grafts on the surface of invertebrates and the third section with the ability of phagocytic cells to recognize foreign bodies, ingest and in many cases to kill them inside the body, and the importance of opsonins in this process, which may confer a degree of selectivity. Section IV examines briefly the nature of lectins and hemagglutinins, (soluble recognition factors) and their possible role as opsonins and/or cell receptors and last, some theories of invertebrate recognition are compared.

II. RECOGNITION OF SELF AND NON-SELF IN TRANSPLANTATION REACTIONS

There is suggestive evidence for protozoan recognition from experiments on nuclear transfer in amoebae (Goldstein, 1970), as nuclear transfers between xenogeneic amoebae were far less successful than allo- or autotransfers.

A. Porifera

Self–non-self discrimination is a feature of all living cells (Boyden, 1962), so it is not surprising that even lower forms of invertebrates such as the sponges can discriminate. Sponges belong to the first metazoan diploblastic group and are very primitive multicellular organisms without a blood circulation. Despite their low level of morphological organization, sponges display species and strain specificities and even cell-type specificity.

Wilson (1907) first described species-specific reaggregation in red and yellow sponges which had been dissociated and mixed. The cells aggregated to form as whole sponges according to their color. Three distinct stages during cell recognition are recognized by Johnston and Hildemann (1982): (1) initial contact, (2) cellular adhesion, and (3) histiotypic rearrangement of cells to form new individual sponges.

Different results may be obtained in sponge cell reaggregation assays if cells are shaken or stationary. Turner (1978) found that sponge cell recognition in stationary cultures involved the motile archeocytes which preferentially collected homospecific cells, thus producing monospecific aggregates which may then reconstitute to intact sponges. The mobile archeocytes and the "grey" (mucoid) cells, which synthesize and accumulate glycogen, were necessary for species-specific aggregate formation. However, when sponge cells are subjected to agitation, the dependence on cell movement prior to recognition and sorting is eliminated. Not all sponge species can be dissociated into cells that will reaggregate under these conditions. Since very few of the sponge species that do reaggregate can form functional sponges, it is not surprising that many bispecific mixtures do not manifest specific aggregation and species-specific cell adhesion. Humphreys (1969) concluded that in most combinations initial cell adhesions are nonspecific and that subsequent species-specific cell sorting is responsible for what is usually referred to as species-specific aggregation. Aggregation is probably controlled by aggregation factors; these are described in Section IV,A.

It was once thought that sponges could only recognize orthotopic xenografts. For instance, Moscona (1968) and McClay (1974) both showed that sponge allografts were usually accepted while xenotransplants were rejected. However, observations over a longer time period on orthotopically sutured allografts and parabiosis of intact sponge finger allografts have revealed allograft rejection (Hildemann *et al.*, 1979). Also, a viscometric assay used by Curtis (1979) to measure adhesiveness of cells has revealed that non-self recognition in sponges occurs in allogeneic combinations, not just in xenogeneic situations.

Accelerated second-set allograft rejection has been reported in two species of sponges, *Hymeniacidon perleve* (Evans *et al.*, 1981) and the tropical sponge *Callyspongia diffusa* (Hildemann *et al.*, 1979). The archeocytes of sponges were implicated in the rejection process because they infiltrated into the fusion zone and are known to be phagocytic. Hildemann and co-workers (1980a) related the appearance of cytotoxicity towards allografts to progressive mobilization of arch-eocytes near the allogeneic pinacoderm interfaces. Short-term *in vitro* experiments by Van der Vyver (1980, 1983) showed no evidence for more rapid collagen layer deposition on second exposure of the fresh-water sponge *Ephydatia fluviatilis* and the marine sponge *Axinella polypoides*.

Johnston and Hildemann (1982) gave a detailed description of these and other related experiments on sponge compatibility. They proposed that sponge al-loreactivity conforms with principles of adaptive immunity found in higher animals. Their three criteria are (1) cytotoxic reactions followed by a period of sensitization, (2) selective and specific reactivity, and (3) inducible memory.

However, Bigger *et al.* (1981) found that once the allorecognition process is stimulated in sponges, the cytotoxic reaction that is triggered is nonspecific and does not discriminate between the original target and other allogeneic target tissue.

Hildemann *et al.* (1979, 1980b) proposed an extensive polymorphism of cell-surface histocompatability molecules in sponges, as no cases of graft acceptance occurred in 480 pairings of the sponge *Callyspongia diffusa* and 890 pairings of the coral *Montipora verrucosa*. On the other hand, Curtis *et al.* (1982) in studies of the marine sponge *Ectyoplasia ferox* found that sponges accepting grafts can have dissimilar plasmalemmal proteins. They proposed that graft rejection was not necessarily an adequate criterion for assessing clonal identity.

B. Cnidaria

In the class Hydrozoa, experimental transplantation of the suborder Hydra has revealed histoincompatibilities (Loeb, 1945). Hydra have no organs, blood, or lymph yet can accept autografts (self) and recognize xenografts as foreign. Early literature on grafting in Cnidaria (briefly reviewed by Campbell and Bibb, 1970) suggested that even xenografts in *Hydra* were not rejected, or only very slowly rejected, with one species or even sex eventually dominating the other. In transplants between 7 species of Hydra, Bibb and Campbell (1973) found that allografts initially attached and adhered rapidly. Some xenografts showed little initial healing after attachment and soon separated; but others healed in and after several days showed constrictions and many were rejected. However, in most early work, grafts were not observed for a long enough time, particularly when the water temperature was relatively low (Bigger and Hildemann, 1982). Ivker (1972) was the first investigator to reveal allogeneic histocompatibility reactions when different colonies of *Hydractinia echinata* were brought together.

Theodor (1970) has studied an anthozoan, gorgonacea, a species of arborescent coelenterate. Although autografts survived indefinitely, branch allografts failed to fuse. In a later study of 1479 allografts of *Eunicella stricta,* Theodor (1976) found only 0.7% of the grafts remained fused after 12 weeks. Branch xenografts of Gorgonacea showed a cytotoxic reaction within a few days as the target xeno-explant was destroyed by an induced suicidal lysis within 1–3 days after contact. The killer factor involved was preformed and diffusible (Theodor and Senelar, 1975). Bigger and Runyan (1979) have confirmed Theodor's extensive *in vitro* studies on Gorgonians by allo- and xenografting *in situ*. Whereas allografts regenerated without fusion, xenografts were rejected with necrosis.

In the natural state, these transplant reactions are probably associated with territorial protection by a single species, an essential to preserve food supply. This is clearly seen in the class Anthozoa, where Purcell (1977) has reported intra- and interspecies aggression mediated by catch tentacles of sea anemones. Lang (1971) has reported a hierarchy of aggressiveness among 52 different species of Scleractian corals. For example, individuals of *Scolymia lacera* always attacked *S. cubensis* by extracoelenteric feeding interactions and killed them. Dyrynda (1983) has found that many speices of Porifera, colonial Cnidaria, Bryozoa, and colonial Tunicata possess larvotoxic allelochemicals which may be an important defense mechanism against episettlement in the competition for space.

Allograft rejection has been described among reef-building corals *Montipara* (Hildemann *et al.,* 1975) with alloimmune memory lasting for 4 weeks and disappearing after 8–16 weeks (Hildemann *et al.,* 1977a,b, 1980b). Memory was highly specific as third-party allografts from a geographically distant source were rejected like primary allografts. Coelenterates (including corals) possess leukocyte-type cells called amoebocytes which Hildemann *et al.* (1977a) have suggested might be responsible for memory and possibly could be cytotoxic. The reader is referred to a review by Bigger and Hildemann (1982) for details.

C. Platyhelminthes

Flatworms (which have no coelom) have been studied by Lindh (1959), who showed that while certain xenografts of the same genus behave as allografts and are not rejected, most xenografts were unsuccessful and only regenerated temporarily.

D. Nemertina

In Nemertine worms (genus *Lineus*), which have a circulation but no coelomic cavity, Langlet and Bierne (1977) have demonstrated an accelerated second-set response provided the second graft was applied within 80 days after the first xenograft.

E. Annelida

Duprat (1964) first discovered that a primary xenograft between earthworms from different geographical regions was generally attacked by the recipient's phagocytes and rejected. Later, Duprat (1967) demonstrated not only a more rapid second-set response but also adoptive transfer of transplantation immunity by sensitized coelomocytes (also Cooper, 1970; Bailey *et al.*, 1971). The memory response was considered to be due to cells which had been attracted into the coelomic cavity during sensitization by a primary xenograft and were thus present in greater numbers at the time of second-set grafting (Hostetter and Cooper, 1974). For a detailed account, see Valembois *et al.* (1982).

F. Mollusca

In 1970, Cheng, reviewing earlier literature, pointed out that due to technical difficulties, transplantation of tissues to orthotopic sites had not been performed in mollusks and incompatible grafts often did not temporarily fuse to host tissues prior to rejection. Even so, Cheng and Galloway (1970) were able to show that snails can distinguish between heterotopic allografts and xenografts (digestive-gland tissues) when they are transplanted into the cephalopedal sinuses.

There are two examples of orthotopic grafting leading to rejection of xeno-grafts. Hildemann and collaborators (1974) have transplanted mantle grafts orthotopically in the pearl oyster *Pinctada margaritifera* and observed weak fusion to the host mantle. Grafts only survived for 5–23 days; necrosis of the graft was often associated with infiltration of large numbers of hemocytes into the graft bed. Later, Bayne *et al.* (1979) observed rejection of mantles of *Mya arenaria* which were placed orthotopically on *Mytilus californianus*. The xenograft reaction was assessed by infiltration of the implant by hemocytes. Although second-set xenografts were rejected faster than primary grafts and third-set grafts even faster, the authors do not claim this is evidence for memory as the primary grafts had not been rejected at the time of secondary and tertiary xenografting.

G. Insecta

Jones and Bell (1982) observed that the response of hemocytes in the American cockroach *Periplaneta americana* was more intense when xenografts from *Leucophaea maderae* were grafted compared to autografts. Cell responses were measured by counting nuclei in a transect through the mass of reacting host hemocytes. The intensity of response to xenografts increased with increasing taxonomic distance.

Thomas and Ratcliffe (1982) have shown that xenografts, but not allografts, of integument in the cockroach *Blaberus cramifer* and the giant stick insect *Extatasoma tiaratum* were rejected more or less in accordance with phylogenetic distance between donor and host. Grafts survived longer when donor and host were closely related species and were rejected more rapidly when the phylogenetic relationship was greater.

Most experiments on insects involve heterotopic implantation and these are described in Section III,C.

H. Echinodermata

Echinoderms developed convergently with the arthropods and are of particular interest as they may have an ancestor in common with the vertebrates.

Transplantation of teguments between different (colored) species of starfish in the Mediterranean Sea evinced no rejection over long periods of observation (Brusle, 1967). However, Hildemann and Dix (1972) demonstrated xenograft rejection of colored teguments between the sea cucumber *Cucumaria tricolor* and the horned sea star *Protoreaster nudosus* obtained from sites in the tropical Indo-Pacific ocean. All xenografts were slowly rejected during 129–185 days at 21/27°C, while second-set xenografts in *Cucumaria* (made while the first graft was still retained) showed accelerated rejection within 28–50 days.

Later, Karp and Hildemann (1976) observed prolonged rejection (at 14–16°C) of allografts between individual sea stars, *Dermasterias imbricata*. There was greatly accelerated rejection of second-set grafts, although in both these experiments memory was only short-term. Specificity of the memory response was demonstrated by the relatively long-term survival of third-party allografts.

Allografts exchanged between unrelated sea urchins, *Lytechinus pictus,* were rejected within about 30 days (Coffaro and Hinegardner, 1977). Memory was induced as there was an accelerated second-set rejection within about 10 days.

Bertheussen (1979) has studied mixed coelomocyte reactions in cells taken from sea urchins. With allogeneic cells, about 70% of the cell combinations induced cytotoxic reactions, and this increased to 90% when xenogeneic cells were combined. Spontaneous cytotoxicity of coelomocytes from the keyhole limpet *Megathura crenulata,* starfish *Pisaster gigantus,* and the annelid bloodworm *Glycera* Sp-C is manifest against erythrocyte and tumor-cell targets (Decker *et al.,* 1981). Effector cell activity may have been directed specifically against cell-surface glycoproteins on the target cells as a variety of mono- and disaccharides blocked killing. The ability to block killing was target-cell and effector-cell specific. This suggested that target cell recognition may be mediated by lectin-like molecules expressed on the surface of effector cells.

I. Chordata

Among the colonial tunicates (lower chordates), hindrance of fusion between two allocolonies of compound ascidians *Botryllus schlosseri* was first noted by Bancroft (1903). The same phenomenon in *Botryllus primigenus* was reported by Oka and Watanabe (1960) and has been described in detail more recently (Tanaka, 1973; Tanaka and Watenabe, 1973). The "nonfusion reaction" (NFR) between histoincompatible colonies consisted of destruction of target cells by ampullae which could dissolve or loosen the outer membrane of the test matrix. This was followed by contraction of ampullae so the incompatible colonies were separated by a necrotic area. Tanaka and Watanabe (1973) suggested this reaction was mediated by two types of factors—one was colony-specific and was located in the test matrix and blood or in granular amoebocytes, the other was released from test cells when they disintegrated and caused constriction of ampullae and necrosis. An NFR response could be induced by an injection of allogeneic blood, but not syngeneic blood (Taneda and Watenabe, 1982).

The *Botryllus* fusibility gene locus shows the same degree of polymorphism as the genes of vertebrate MHC and self-incompatibility of plants (Scofield *et al.,* 1981). This supports the hypothesis (Burnet, 1971) that vertebrate histocompatibility genes evolved from gametic self–non-self recognition systems which prevented self-fertilization in hemaphroditic organisms. So self–non-self discrimination factors in *Botryllus* may be evolutionary precursors for vertebrate MHC, although there are several major differences (Scofield *et al.*, 1981).

Fuke and Numakunai (1982) found coelomic cells from a solitary ascidian *Halocynthia roretzi* exhibited a nonphagocytic cellular reaction ("contact reaction") *in vitro* against coelomocytes of different species and other individuals of the same species.

Integumentary allografts exchanged between the solitary tunicate *Ciona intestinalis* were only rejected slowly (Reddy *et al.*, 1975). As there were problems in maintaining the tunicates after transplantation surgery no memory was found. Large numbers of lymphocyte-like cells infiltrated the allografts in contrast to the autografts.

As workers continue to perform transplantation experiments in invertebrates and techniques are developed to overcome difficulties of orthotopic grafting and the maintenance of animals after surgery, evidence that most invertebrate phyla possess means of recognition of non-self followed by active rejection mounts year by year. There is evidence for "memory" in sponges, corals, nemertines, annelids, starfish, and sea urchins. However, in many experiments "second-set" grafts have been applied before the primary grafts had been rejected. Presumably, the second-set grafts are confronted with a larger population of "activated" hemocytes sensitized by the primary graft and a more rapid rejection ensues. The criterion for immunological memory in vertebrates is that the second set of grafts must be applied after the primary graft has been rejected. The cells providing memory are lymphocytes. In invertebrates there is good evidence for "memory," but often while primary grafts are still retained (e.g., Hildemann and Dix, 1972). The "memory" is quite specific as third-party allografts have been reported to be rejected as primary grafts (e.g., Hildemann *et al.*, 1977b). Other workers (e.g., Bayne *et al.*, 1979) have not been prepared to claim memory under such conditions.

III. RECOGNITION OF NON-SELF INSIDE INVERTEBRATES

Many excellent reviews (Tripp, 1963; Salt, 1970; Sparks, 1972; Jenkin, 1976) describe how virtually all invertebrates studied are capable of phagocytizing foreign particles, heterologous proteins, or allogeneic or xenogeneic cells or tissues when implanted in the body cavity. There is very little reason to suppose that the way the host's phagocytes recognize a transplant on the surface of an animal or an implant embedded into a heterotopic site such as the coelomic cavity differ in any way. An implant in a cavity filled with fluid should allow better locomotion and better means of attachment by the phagocytes seeking to remove the non-self component compared with efforts by the same cell types to kill cells which are necessarily limited to the undersurface of a graft.

The difficulties of exchanging transplants because of the cuticle of arthropods or the calcareous shell of many mollusks means that most of our information on these phyla comes from internal phagocytosis. Responses to foreign bodies and tissue injury are very much the same in all invertebrates studied; small particles are phagocytosed, but particles greater than 10 μm diameter tend to be encapsulated rather than phagocytosed. Many higher invertebrates possess organs containing depots of phagocytes. For example, the octopus, the brine shrimp (artemia) *Crustacea anostraca* (the most primitive of living crustacea), the freshwater crayfish *Parachaeraps bicarinatus,* and the garden snail *Helix pomatia* possess fixed phagocytes mainly associated with liver-like digestive diverticula. The role of phagocytes in cellular feeding and digestion, and their importance in defense has been discussed by Jenkin (1976).

Chorney and Cheng (1980) and Ey and Jenkin (1982) have recently reviewed the specificity of phagocytic reactions in invertebrates and the role of substances which may act as opsonins. In view of the probable importance of such molecules in defense against infectious agents and their ability to confer some degree of specificity in cell recognition reactions, the relevant work is examined carefully here.

A. Mollusca

Tripp (1966) was probably the first to ascribe the role of opsonin to invertebrate hemagglutinins. When rabbit erythrocytes were pretreated with oyster hemolymph, phagocytosis by oyster hemocytes *in vitro* was enhanced. As oyster blood also contained hemgglutinins for erythrocytes for certain mammalian species, Tripp suggested the rabbit hemagglutinin might also act as an opsonin. However, opsonins are not necessarily agglutinins. For example, Stuart (1968) failed to detect agglutinins for bacteria, yeast, and human or fish erythrocytes in the blood plasma of the lesser octopus *Eledone cirrosa,* although opsonins for these cells were present. The albumin gland of mollusks may contain hemagglutinins and opsonins not found in the hemolymph. Anderson and Good (1976) observed that the hemolymph of *Otala lactea* lacked serum lectins, yet contained opsonins which stimulated phagocytosis of formalinized yeast cells but not erythrocytes *in vitro.* Yet extracts of this mollusk's albumin gland contained hemagglutinins as well as opsonins for formalinized erythrocytes. Prowse and Tait (1969) found an opsonin in the hemolymph of the snail *Helix aspersa* which appeared to be essential for phagocytosis of yeast cells and formalinized sheep erythrocytes. Renwrantz and Mohr (1978) have demonstrated that hemolymph and albumin-gland extracts can act as opsonins to facilitate elimination of human erythrocytes from the circulation of the vineyard snail *Helix pomatia.* When *Helix pomatia* were injected with colloidal carbon, chicken erythrocytes, or

Pseudomonas aeruginosa, there was a decrease in two nonrespiratory hemo-lymph proteins (probably due to their opsonic action and removal by phagocyto-sis), but respiratory proteins such as hemocyanins were unaffected (Nielsen *et al.,* 1983). Repeated injections of sublethal doses of *Pseudomonas aeruginosa* OT97 into *Helix pomatia* led to enhanced survival to challenge with lethal doses (Bayne, 1980a). Also, large doses of bacteria were cleared more rapidly in immunized snails although there were sufficient "natural opsonins" in nonim-munized snails to clear smaller doses of this bacteria (Bayne, 1980b). Bacteria, because they are small, are ingested in large numbers by each amoebocyte (blood cell) of the fresh water snail *Lymnaea stagnalis in vitro,* but they ingest fewer large cells such as erythrocytes and yeast cells (Sminia *et al.,* 1979). Preop-sonization of erythrocytes and yeast cells with snail serum greatly increased the number ingested.

Van der Knapp (1981) showed that amoebocytes (the chief phagocytic cells) of the pond snail *Lymnaea stagnalis* had a coat of acidic carbohydrates bound to the plasma membrane by proteins. Treatment of amoebocytes with proteases prevented phagocytosis and most adhesion of bacteria. Amoebocytes produced a protein with agglutinating properties which could function as an opsonin by binding to receptors on the amoebocyte.

Helix pomatia cleared the rough form of *Azotobacter vinelandii* faster than the smooth form (Renwrantz, 1981); such discrimination has often been reported in vertebrates. The snail could also discriminate between erythrocytes. Human blood group A erythrocytes were cleared twice as fast as those of group B and four times faster than rabbit erythrocytes. When the surface of yeast cells was masked with homologous *Helix* hemocyanin there was a considerable decrease in the rate of elimination compared to yeast cells coated with heterologous bovine serum albumin. *N*-Acetylgalactosamine and *N*-acetylglucosamine inhibited the opsonin-independent rate of clearance of yeast cells, indicating that clearance of yeast particles was mediated by direct binding to cell-bound carbohydrate recog-nition molecules (Renwrantz *et al.,* 1981).

Hardy and co-workers (1977) isolated and partly characterized two hemag-glutinins (one for human, the other for horse erythrocytes) from the oyster *Crassostrea gigas.* When *Escherichia coli* or *Vibrio anguillarum* were cultured with oyster hemocytes, phagocytosis was enhanced by added hemolymph. Pre-opsonization with either hemolymph or purified human hemagglutinin also en-hanced phagocytosis. However, addition of either purified human or horse hemagglutinins to the culture medium failed to enhance phagocytosis of the bacteria.

Pauley and co-workers (1971) observed that the California sea hare *Aplyisia californica* could phagocytose and rapidly clear four species of marine bacteria from the circulation but did not completely clear the terrestial bacterium *Serratia marcescens.* The rapid clearances were accompanied by a decrease in serum

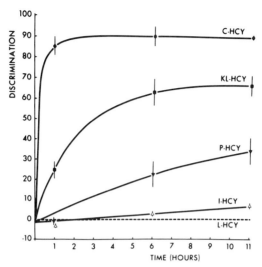

Fig. 2. Rates of clearance of different hemocyanins in the chiton *Lilophura gaimardi*. Hemocyanins from *Liolophura*, L-HCY; *Ischnoradisia*, I-HCY; *Ponerophex*, P-HCY; keyhole limpet, KL-HCY; and crayfish, C-HCY. (From Crichton and Lafferty, 1975.)

titers of a hemagglutinin and a decrease in numbers of circulating hemocytes. A type of memory response was observed, as second exposure to marine bacteria resulted in accelerated clearance.

Of great interest is the experiment of Crichton and Lafferty (1975), who have shown that the chiton *Liolophura gaimardi* can discriminate between hemocyanins on a taxonomic basis: ^{125}I-labeled hemocyanins were injected into the hemolymph and the clearance was measured during 11 hr (Fig. 2). The percentage clearance was related to the number of shared antigens with *Liolophura* hemolymph (shown by double-diffusion tests). The similar hemocyanins were cleared more slowly than the only hemocyanin (crayfish) showing no cross-reactions. Crayfish hemocyanin has a very different subunit structure compared to other mollusc hemocyanins, so major differences in size and charge as well as antigenic differences may have influenced the rate of clearance.

B. Annelida

The nonspecific uptake of gram-positive and gram-negative bacteria by coelomocytes of the lugworm *Arenicola marina* is serum-independent but becomes selective when competition for cell-surface receptors is introduced (Fitzgerald and Ratcliffe, 1981, 1982). Selectivity was differentially inhibited by preincubating coelomocytes with a bacterial lysate and D-(+)-glucose.

C. Insecta

Cameron (1934) found a varied pattern of phagocytosis, resistance and suscep-tibility of caterpillars and larvae of *Lepidoptera* toward different species of bacteria. Pneumococci, *Staphylococcus aureus,* and *Haemophilus influenzae* were actively phagocytosed and completely killed and destroyed. There was only slight phagocytosis of *Salmonella typhosa, Salmonella typhi,* and *Vibrio chol-erae* which were nevertheless all destroyed. On the other hand, *Proteus vulgaris* and *Bacillus mycoides* caused rapid death of the larvae despite active phagocyto-sis; some *Clostridium welchi* and *Staphylococcus pyogenes albus* survived active phagocytosis and caused death of larvae after several days.

In larvae of the wax moth *Galleria mellonella,* the nonpathogen *Escherichia coli* K12 was cleared faster from the hemocoel and killed, whereas the pathogen *Bacillus cereus* was more resistant to killing once phagocytosed (Walters and Ratcliffe, 1981, 1983).

Often no evidence for opsonins has been found. For example, hemocytes of the cockroach *Periplaneta americana* phagocytose yeast cells, sheep erythro-cytes, carbon and dye particles injected into the hemocoel, yet did not pha-gocytose *Bacillus thuringiensis, Escherichia coli,* or *Cornyebacterium* sp. (Ryan and Nicholas, 1972). Lysins for the first and third bacteria were present but obviously did not function as opsonins.

Scott (1971) studied adherence and phagocytosis of sheep and chicken eryth-rocytes by cockroach hemocytes *in vitro.* Sheep erythrocytes were phagocytosed but the larger chicken erythrocytes only adhered to the hemocytes. Pretreatment of the hemocytes with trypsin destroyed the receptor sites, and no adherence or phagocytosis occurred. Addition of hemolymph neither restored phagocytosis nor promoted phagocytosis by untreated hemocytes.

There is an extensive literature on recognition of foreignness in insects mea-sured by the encapsulation reaction. Salt (1963) has reviewed defense reactions of insects to metazoan parasites and lists parasites which are encapsulated and others which are not, or which can avoid eliciting any defense reactions. Salt's studies on phagocytes of caterpillars of *Ephestia kuehniella* showed they prompt-ly encapsulated alien parasites but not eggs of their habitual parasite *Nemeritis canescens* (an ichneumon fly). This protection was found to be due to a visible layer outside the chorion which was deposited on the eggs as they passed down the genital tract below the calyx (Salt, 1973). Another example of immune evasion in insects is given by *Staphylococcus aureus,* a species of bacteria pathogenic for the wax moth *Galleria mellonella,* which failed to provoke a nodular reaction by hemocytes and plasmatocytes *in vivo* (Ratcliffe and Gagen, 1976).

Carton (1976) implanted eggs of *Pimpla instigator, Drosophila melanogaster* and *Apechthis compuctor* into *Pimpla* larvae. Encapsulation occurred with greater intensity in those xenogeneic combinations with the greater taxonomic distance. Allo-implants of *Pimpla* eggs induced a slower encapsulation and none occurred with iso-implants. Lackie (1976) observed that the larvae of the tapeworm *Hymenolepis diminuta* (a hemocoelic parasite) was encapsulated in the American cockroach *Periplaneta americana* but not in the desert locust *Schistocerca gregaria*. Later, Lackie (1981a) found that cockroach serum agglutinated the larvae whereas locust serum did not. Also, the adhesive properties of cockroach hemocytes were more extensive and the range of biotic implants recognized as non-self were more wide ranging in the cockroach than in the locust. Lackie (1979) implanted allografts and xenografts from six donor species into host *Periplaneta americana*. All donor tissues were encapsulated except allografts and xenografts from the closely related *Blatta orientalis*. Lackie (1983) transplanted allogeneic and xenogeneic nymphal cuticle on to *Periplaneta americana* nymphal hosts. Her idea was that if the donor subcuticular epidermis was not recognized as foreign it would grow, fuse with the recipient's epidermal sheet, and be stimulated by the recipient's hormonal signals to produce new cuticle of donor type at the next molt. There was no evidence of recognition of foreignness. This confirmed a large amount of earlier work with tissue implants.

Lackie (1981b), reviewing the recognition process in insects, finds little evidence of the involvement of soluble molecules.

D. Crustacea

When crayfish hemocytes were cultured for a few hours with ^{51}Cr-labeled Ehrlich or Krebs II ascites tumor cells or HeLa cells, radioactivity was only released from the ascites tumor cells (Jenkin and Hardy, 1975). Crayfish hemocytes can recognize the lipopolysaccharide component of somatic O antigens of *Salmonella abortus-equi* and when HeLa cells were coated with this lipopolysaccharide many were destroyed after 4 hr incubation.

Söderhäll *et al.* (1984) observed that when fungal blastospores of *Beauveria bassiana* were coated with a hemocyte lysate from the crayfish *Astacus astacus* L. before injection into crayfish, an enhanced encapsulation reaction occurred. This demonstrated that there were opsonic factors present in the hemocytes which were absent from plasma. Phenoloxidase was one of the major proteins attached to the fungal spores, together with four proteins of the prophenoloxidase-activating system. β-1,3-Glucans derived from fungal cell walls can activate crayfish prophenoloxidase and also induce a clotting process resulting in phenoloxidase attachment to foreign surfaces (Söderhäll, 1981). Söderhäll (1982) has reviewed the possible role of the prophenoloxidase activating system and melanization in recognition in arthropods.

McKay and collaborators (1969) observed rapid blood clearance of sheep erythrocytes (SRBC) injected into the fresh-water crayfish *Parachaeraps bicarinatis* with a concomitant decrease in circulating hemocytes and a drop in titer of a hemagglutinin for sheep erythrocytes. Although no phagocytosis of SRBC by crayfish hemocytes occurred in *in vitro,* pretreatment of SRBC with crayfish hemolymph caused greater numbers to adhere to the hemocytes. McKay and Jenkin (1970) demonstrated that phagocytosis of sheep erythrocytes *in vitro* by hemocytes of *Parachaeraps bicarinatis* was proportional to the hemagglutinating activity of added crayfish serum. Both opsonic and hemagglutinating activities for certain erythrocytes were concomitantly absorbed from the serum by pretreatment with the species-specific erythrocyte. Similarly, serum hemagglutinin titres of the sea hare *Aplysia californica* decreased during early rapid blood clearance of four species of marine bacteria (Pauley *et al.,* 1971) (Section III, A). Later, Tyson and Jenkin (1973) showed that rapid clearance of *Salmonella abortus-equi* injected into the circulation of fresh water crayfish depended upon the presence of an opsonin in the hemolymph. They demonstrated the presence of an opsonin by some elegant experimentation—a second dose of *S. abortus-equi* was cleared more slowly (due to absorption of the opsonin from the circulation by the first injection). However, if the bacteria were pretreated with normal hemolymph (preopsonization), clearance was rapid. On the other hand, if the opsonin was adsorbed out of the hemolymph by pretreatment with *S. abortus-equi* and the "opsonin-less" hemolymph used to pretreat the second dose of *S. abortus-equi,* this was only cleared slowly. A lysate of crayfish hemocytes was shown to contain the opsonin.

Dosage is important in clearance experiments as the clearance rate multipled by the dose is a constant if opsonins are present. It is unfortunate that many workers have often not studied the effect of dosage on the clearance rates, as this can often demonstrate a finite amount of opsonin.

The crayfish *Procambarus clarkii* can discriminate between bovine serum albumin (BSA), human γ-globulin (HGG), and crayfish hemocyanin (Sloan *et al.,* 1975). The rates of clearance of these iodinated proteins from the circulation were widely different—the crayfish hemocyanin was cleared quite slowly, whereas HGG was cleared much more rapidly during the first 8 hr and 70% of the BSA was rapidly cleared in the first 2 hr. Both the heterologous proteins concentrated in the gills and were soon degraded. Injection of unlabeled BSA retarded the elimination of [^{125}I]BSA, suggesting a saturation of receptors, but this was not the case for HGG (possibly there were a larger number of specific receptors for this particular protein).

In similar but more elaborate experiments on the blue crab *Callinectes sapidus,* McCumber and Clem (1977) have shown that radiolabeled viruses and heterologous proteins are differentially cleared from the circulation. Whereas crab and lobster hemocyanins were cleared from the circulation very slowly, bovine γ-globulin (BGG) was cleared more rapidly. An even faster rate of clearance was shown by BSA, Limulus hemocyanin, and keyhole limpet hemocyanin, which were all cleared within 2 hr. Poliovirus and bacteriophages T2 and T4 were rapidly removed, whereas phages T3, T7, and φII were slowly cleared. There was no correlation with rate of clearance and particle size. Xenogeneic proteins and poliovirus particles were concentrated in the central axis of the gills, whereas the larger bacteriophages T2 and T4 were chiefly taken up in the hepato–pancreas. Attempts to enhance these clearance rates by injection of BSA, BGG, or T2 phage 2 weeks prior to the clearance tests were unsuccessful. The authors isolated an opsonin by affinity chromatography, and it had a molecular weight of about 120,000. Plasma of the blue crab contained a factor capable of neutralizing T2 bacteriophage *in vitro* and so could be responsible for the *in vivo* clearance (McCumber *et al.*, 1979). This factor appears to be a polymer (6–13 S) of covalently linked subunits each with a molecular weight of ~80,000. The factor was specifically reactive with the virus tail fibers (Donelly and Clem, 1982), a rare example of what appears to be "exquisite specificity."

E. Echinodermata

Hilgard and Phillips (1968) have clearly demonstrated that the purple sea urchin *Strongylocentrotus purpuratus* can discriminate between homologous and heterologous proteins. Coelomocytes cultured *in vitro* phagocytosed more than 85% of radiolabeled human serum albumin (HSA) or bovine serum albumin (BSA) in 22 hr, but less than 50% of homologous substances from the coelomic fluid of *S. purpuratus* was taken up by the coelomocytes. Further experiments by Hilgard *et al.* (1974) revealed quite extraordinary powers of discrimination by sea urchins between two heterologous albumins. Sea urchin hemocytes were cultured *in vitro* with [^{14}C]BSA at a concentration which saturated uptake by phagocytosis. Addition of chicken serum albumin (CSA) or HSA did not interfere with the uptake of [^{14}C]BSA, but addition of BSA interfered. Similar experiments with [^{14}C]CSA showed that only CSA inhibited uptake of the radiolabeled identical protein. These results are consistent with the idea that sea urchin coelomocytes possess relatively specific receptors for both BSA and CSA.

Johnson (1969) reported that cultures of coelomocytes from the sea urchins *Strongylocentrotus purpuratus* and *S. franciscanus* also show quite different responses to various species of bacteria. Gram-negative bacteria were seldom phagocytosed but encapsulated then killed or lysed. In contrast, gram-positive bacteria elicited marked phagocytosis but no other responses.

The starfishes *Patira miniata* and *Asterias forbesi* can both discriminate between xeno- and allogeneic coelomic implants of pyloric caecum (Ghiradella, 1965). Whereas allotransplants could be recovered from hosts in a healthy condition (histologically normal) from 1–5 weeks postimplantation, xeno-implants from *Asterias vulgaris* and *Henricia sanguinolenta* were eliminated through the dermal branchiae within 1 week.

Bertheussen and Seljelid (1982) detected receptors for C3 on echinoid phagocytes. When sheep erythrocyte-bound $C3_b$ was converted to the hemolytically inactive $C3_{bi}$ the opsonic effect was markedly increased. Conversion of $C3_{bi}$ to $C3_d$ caused a loss of opsonic activity. Bertheussen (1982) concluded that echinoid phagocytes display $C3_{bi}$ and possibly also $C3_b$ receptors but $C3_d$ receptors are absent.

F. Urochordata

Human erythrocytes injected into the tunicate *Ciona intestinalis* were cleared more rapidly than duck erythrocytes (Wright, 1974). Serum hemagglutinin titres for these erythrocytes were reduced in a dose-dependent fashion during the first 12–18 hr after injection, then returned to normal levels once the erythrocytes had been phagocytosed or encapsulated.

IV. MOLECULES CAPABLE OF ACTING AS RECOGNITION FACTORS

A. Soluble Factors: Lectins, Hemagglutinins

So far lectins have been obtained mainly from plants, but they also occur in bacteria and have been isolated from many invertebrates. They form one of three classes of proteins which possess specific combining sites. While various lectins combine mainly with sugars, and agglutinate cells and stimulate mitosis in lymphocytes, they are not enzymes (which combine specifically with substrates) nor are they antibodies (which combine specifically with antigens and are produced by lymphocytes) (Lis and Sharon, 1977). Most lectins interact with a single sugar structure, e.g., galactopyranose or L-fucopyranose. Some combine with closely related sugars such as mannopyranose and have a broader specificity. Other lectins only react with complex carbohydrate structures such as glycoproteins and substances on cell surfaces. In contrast to antibodies, lectin specificity is restricted to carbohydrates; another difference is that lectins are structurally quite diverse and in this latter property are more like enzymes, but do not possess catalytic activity. Lectins bind moncovalently to specific carbohydrate groups and their binding is reversible. All lectins have more than one specific carbohydrate-combining site which allows them to act as cross-linking agents.

Although lectins are all proteins, they vary in composition, molecular weight, subunit structure, and the number of sugar binding sites per molecule. Most are glycoproteins with sugar contents as high as 55%, but some have no covalently bound sugar residues (Lis and Sharon, 1977). The number of subunits may vary from 2 to 18, usually with one sugar binding site per subunit. Several lectins obtained from invertebrates have been fairly well characterized, though not in as great detail as some plant lectins such as concanavalin A.

Boyden (1966) was one of the first to suggest that invertebrate hemagglutinins may serve as recognition factors which could function in defense against infectious agents. Hemagglutinins have been detected in most, but not all, invertebrate species investigated (Cohen, 1974). They already include sponges, ascidians, earthworms, insects, snails, crabs, lobsters, oysters, mussels, octopi, and tunicates. So we may confidently look forward to many more being found in other invertebrate species.

It is tempting for the immunologist to ascribe a molecular basis for specific defense mechanisms in invertebrates to lectins which may function as opsonins, so accelerating phagocytosis. If a lectin was absent in a species we might expect the species to be more vulnerable to infection. This idea is suggested by the early taxonomic studies of Tyler and Metz (1945) on the specificity of a natural hemagglutinin in the serum of lobsters and by the following examples. Pauley and collaborators (1971) have suggested that agglutinins may be important in defense, particularly when they function as opsonins. They cite the example of spider crabs who lack agglutinin against the marine ciliate *Anophrys sarcophaga* and are killed by this protozoan. Although lobsters possess a natural agglutinin (which was isolated), they lack a serum agglutinin specific for the bacterium *Gaffkya homari,* whose growth is actually stimulated by hemolymph serum *in vitro.* Infection of lobsters with only a few *G. homari* invariably proved fatal (Cornick and Stewart, 1968).

In some species, more than one agglutinin exhibiting different specificities for erythrocytes or even a hapten have been isolated. For example, Hall and Rowlands (1974) purified two independent hemagglutinins from the hemolymph of the lobster *Homarus americanus.* One (19 S) reacted with both human and mouse erythrocytes, the other (11 S) reacted only with mouse erythrocytes. Later, two lectins were isolated: one recognized sialic acid moieties and the other *N*-acetylgalactosamine (Hartman *et al.,* 1978; Van der Wall *et al.,* 1981).

Two molecules in the hemolymph of the tunicate *Pyura stolonifera* can distinguish between DNP plus carrier versus carrier alone. The hemolymph does not possess any appreciable agglutinating activity for vertebrate erythrocytes but contains molecules that bind to glutaraldehyde-fixed SRBC (FxSRBC) as well as DNP-FxSRBC (Marchalonis and Warr, 1978). Prior absorption of hemolymph with FxSRBC still left a molecule (65,000–70,000 daltons) binding DNP-FxSRBC.

As Ey and Jenkin (1982) have recently reviewed the structure of invertebrate agglutinins, only a few will be considered here. First, the relationship of sponge aggregation factor (AF) with hemagglutinin will be considered. Moscona (1963) and Humphreys (1963) found that a water-soluble factor released by sponges specifically stimulated the aggregation of cells of the same species. The AF was shown to be a glycoprotein and contained 50% amino acids, the remainder consisting of glucosamine, fucose, mannose, galactose, and glucose. This AF exhibited the same specificity as an agglutinin on the surface of intact sponge cells (Henkart and Humphreys, 1973; Weinbaum and Burger, 1973; Turner and Burger, 1973).

The fresh-water sponge *Ephydatia fluviatilis* also possesses an AF which enhances specific cell–cell adhesion (Van de Vyver, 1975; Curtis and Van de Vyver, 1971). Aggregation-promoting activity was found in a protein with molecular weight about 50,000.

AF binds to plant lectins (Kuhns and Burger, 1971), so probably carbohydrate residues are exposed on the cell surface. Reaggregation of cells was inhibited by both glucuronic acid and the glucuronic acid disaccharide, cellobiuronic acid. As AF activity was destroyed by β-glucuronidase treatment, it appears that glucuronic acid-like residues on AF are recognized by some receptor of the dissociated sponge cell surface (Vaith *et al.,* 1979a). This receptor can be released by osmotic shock and the "hypotonic shock supernatant" has been termed "baseplate" (Turner, 1978). Aggregation factor and baseplate are thought to be together responsible for specific cell adhesion (Johnston and Hildemann, 1982).

There are also inhibitory factors which can discriminate self from non-self, which act by inhibiting movement (Galtsoff, 1964), decreasing reaggregation (Curtis and Van de Vyver, 1971), or actually lysing xenogeneic cells (Van de Vyver, 1975). It is possible these inhibitory factors may have similar activities *in vivo.*

Hemagglutinins have been isolated from several sponges. They were first described by Galtsoff (1925) and resemble AF in that they differentiate self from non-self. However, heteroagglutinins differ from AF in that they agglutinate cells recognized as non-self while AF enhance the aggregation of self. Bretting and Kabat (1976) isolated two hemagglutinins from *Axinella polypoides* which both had specificity for terminal nonreducing D-galactose (glycosidically linked β1→6). Bretting *et al.* (1976) have isolated three hemagglutinins from the sponge *Aaptos papillata* which all had binding site specificity for GlcNAc. Gold and Balding (1975) point out that there are certain similarities between sponge AFs and hemagglutinins—both are easily extracted from the sponge cell, both are responsible for mediating or promoting cell–cell interactions, their activity is dependent on the presence of calcium and transition-metal ions, and carbohydrates are involved in the specificity of the reaction.

The two main differences are that AFs have a higher sedimentation rate and the carbohydrate moiety of the AF glycoprotein is involved with specificity, whereas hemagglutinins contain very little carbohydrate but have affinity for a specific sugar residue. Sponge hemagglutinins, like plant and other invertebrate lectins, can agglutinate red blood cells by binding to cell-surface carbohydrates (Nicolson, 1974) but have not been shown to have a role in sponge aggregation since they can be separated from AF (Maclennan, 1974; Van de Vyver, 1975; Vaith *et al.*, 1979b). As these lectins are not found bound to cell membranes, they would not be expected to be involved in cell–cell recognition.

The distinct stages of cell recognition (Section II, A) may now be expanded to include these factors. Johnston and Hildemann (1982) proposed (1) species-specific self recognition mediated by "promotion" aggregation factors (PAFs), (2) allospecific non-self recognition "inhibitory" aggregation factors (IAFs), and (3) species-specific non-self recognition factors (heteroagglutinins). While these authors include agglutinins in the recognition process, currently most evidence is against this idea. The agglutinins may nevertheless be involved in defense mechanisms against microorganisms.

A detailed description of the nature of such factors is given by Turner (1978) and Ey and Jenkin (1982).

The presence of agglutinins in the albumin gland of the snail does not appear to be connected with defense but may be necessary for reproduction. Strongly reacting, highly specific agglutinins such as found in *Otala lactea* and *Helix pomatia* are very uncommon (Brown *et al.*, 1968). Snail lectin, which specifically agglutinates human blood group A constitutes 8% of the soluble protein of the albumin gland (Goldstein and Hayes, 1978). It is a glycoprotein and its hemagglutinin activity is inhibited by methyl-α-D-GalNAc (Hammarstrom and Kabat, 1969, 1971). Its structure probably consists of six identical polypeptide chains (subunits of molecular weight 26,000), each containing one intrachain disulfide bond and a carbohydrate binding site for 2-acetamido-2-deoxy-α-D-galactosyl groups (Hammarstrom *et al.*, 1972).

The lectin limulin in the hemolymph of the horseshoe crab is capable of localizing neuraminic acid residues in cell surfaces and can react with bipolymers and cells containing terminal sialic acid residues. Limulin has specificity for terminal N-acetylneuraminyl (NANA) residues (Cohen *et al.*, 1972), and Rostam-Abadi and Pistole (1982) ascribed a binding specificity for Limulin to 2-keto-3-deoxyoctonate. Limulin will agglutinate *Salmonella minnesota* and is distinct from an agglutinin-binding gram-positive bacteria (Pistole, 1978). As limulin attaches to N-acetylneuraminic acid (NANA) on erythrocyte membranes, it is interesting that NANA and sialic acids have been identified in only a few strains of bacteria, including salmonellae.

Vasta *et al.* (1982) pointed out that arthropods and some mollusks possess lectin with specificity for sialic acid in common with certain vertebrate and plant lectins. For example, *Limulus polyphemus* contains a hemagglutinin with receptors for sialic acid, 2-keto-3-deoxoctonate, and D-galactose.

Limulin has a molecular weight of 400,000 and is made up of six units each comprising three subunits of molecular weight 22,500 (Marchalonis and Edelman, 1968). Kaplan and co-workers (1977) obtained data by amino acid sequence analysis, which confirmed Marchalonis and Edelman's (1968) conclusion that the structure of limulin was unrelated to that of vertebrate immunoglobulins.

Acton and his group (1969) showed that a hemagglutinin of the oyster *Crassostrea virginica* shared a number of structural features with limulin. These included a subunit of molecular weight around 20,000, noncovalent interactions between the sub-units of the native protein, a requirement for calcium ions to maintain structural stability and biological activity, and some degree of similarity of amino acid composition. Oyster hemagglutinin resembles hemagglutinins from the snail *Helix pomatia,* the horseshoe crab *Limulus polyphemus,* and the lobster *Homarus americanus,* but homology at the level of amino acids is unlikely.

Uhlenbruck and Steinhausen (1977) have summarized the properties of high-molecular-weight antigalactose agglutinins or precipitins which constitute up to 50% of the hemolymph of the bivalve clam *Tridacna maxima.* These agglutinins are quite different from other invertebrate lectins. They are acid- and protease-sensitive molecules (molecular weight 500,000) easily split into subunits. They have anti-α-galactosyl ($1\rightarrow6$, $1\rightarrow4$) specificity similar to other invertebrate lectins (e.g., the sponge *Axinella polypoides*) as well as certain plant and vertebrate lectins. The clam agglutinin can react with galactans on algae, which are symbionts of tridacnids.

Ey and Jenkin (1982) conclude that highly purified invertebrate hemagglutinins share a considerable diversity of structure with a range of molecular weights from 79,000 to 1,000,000. The hemagglutinins are rich in aspartic and glutamic acids, and all possess a subunit structure. Most hemagglutinins are specific for carbohydrates: the most common specificity is for GlcNAc, with other species exhibiting specificity for ($\beta1\rightarrow6$)-D-Gal (sponges), AcNeu (crustacea), or Lac (*Bottrylus*).

We do not know how a lectin can function as an opsonin. After combination of the lectin with, e.g., bacteria, the molecule may undergo a molecular change (become "activated") in some way so that another part of the lectin can combine with receptors on the surface of hemocytes. This would explain why lectins are not "absorbed out" of the hemolymph under normal conditions. In vertebrates there is evidence for such activation of circulating antibody molecules after combination with antigen as part of the opsonization process.

The role of invertebrate lectins has recently been reviewed by Yeaton (1981a,b), Ey and Jenkin (1982), and Renwrantz (1983).

Renwrantz (1983) suggests three possible pathways for agglutinin-mediated recognition of foreignness: (1) direct binding to membrane-bound agglutinins,

(2) binding via an opsonin, and (3) surface agglutinins (on, e.g., bacteria) binding to sugar moieties of membrane molecules of invertebrate cells. However, Ey and Jenkin (1982) point out that most evidence that hemagglutinins function as opsonins is only circumstantial and that only the work of Renwrantz *et al.* (1981), who found that a highly purified homologous invertebrate agglutinin had opsonic activity, strongly supports this idea.

B. Lectin and Erythrocyte Receptors on Invertebrate Cells

Lectin receptors isolated from human red blood cells are glycoproteins, and this may also pertain in invertebrate cells. Hemocytes of the snail *Helix pomatia* possess carbohydrate components on their surface which have been identified by agglutination with various plant, invertebrate, and vertebrate lectins (Renwrantz and Cheng, 1977a). These carbohydrates include galactose, fucose, mannose and/or glucose, and *N*-acetylneuraminic acid. Specificity of the agglutinin reaction was shown by addition of certain sugars to the medium, which interfered with agglutination. The density of receptors may differ even between strains. Hemocytes of two different strains of the gastropod *Biomphalaria glabrata* have been shown to differ quantitatively in their arrangement of cell-surface oligosaccharides capable of binding lectins (Schoenberg and Cheng, 1980).

Yoshino (1981) used fluorescent-labeling techniques to reveal receptors for concanavalin A (Con A) on the surface of hemocytes from *Biomphalaria glabrata*. Con A binding is specific for certain carbohydrate determinants on the cell membrane and can be inhibited by α-methylmannoside or α-D-glucose. Con A binding induced a redistribution (patching and capping) of lectin–receptor complexes on "rounded" hemocytes. Yoshino (1982) observed that patching and capping is accompanied by rapid clearing of the complexes from the cell membrane by internalization. Yoshino (1983) has identified several classes of hemocyte membrane components on hemocytes of *Biomphalaria glabrata*. These consist of (1) macromolecules (primarily glycoproteins) possessing reactive sites for Con A and *Ricinus communis* agglutinin, (2) snail hemolymph determinants, (3) fibronectin-like membrane components, and (4) surface determinants with structural similarities to murine Thy-1 antigen.

Receptors for phytohemagglutinin (PHA) have been demonstrated on coelomocytes of the earthworm *Lumbricus terrestris* as the lectin induced a mitogenic response (Toupin *et al.*, 1977). Leukocytes of earthworms can also be stimulated mitogenically when cultured with Con A (Roch *et al.*, 1975). However, the magnitude of the response is much less in worm leukocytes than in vertebrate lymphocytes. The mitogenic response (measured by uptake of [^{3}H]thymidine) was also dose-dependent and selectively abolished by methyl-α-D-mannopyranoside. Only a small proportion of leukocytes were stimulated, though all cells bound Con A (again like the vertebrate lymphocyte response). Receptors for Con A have also been demonstrated on leukocytes of the earthworm *Eisenia foetida* by Roch and Valembois (1978) using fluorescent-labeled and radiolabeled Con A. Both nylon wool column adherent and nonadherent coelomocytes were bound by Con A, but these populations showed different degrees of binding. As binding of Con A was inhibited by methyl-α-D-mannose, methyl-α-D-glucose, and D-fructose, it appeared that Con A can bind to specific sites on the leukocyte surface. These sites or receptors were "free floating" in a randomly distributed manner as demonstrated by patching and capping.

Gebbinck (1980) has used fluorescein isothiocyanate-conjugated Con A to demonstrate the presence of specific receptors on free leukocytes of the sipunculid worm *Siphonosoma arcassonense* and the crab *Cancer pagurus*. However, receptors were not detected on leukocytes of certain species of sponge, crab, slug, a holotherian, and a starfish. When cells of the sponge *Halichandira panicea* were treated with trypsin, receptors for Con A were exposed. Con A receptors could also be demonstrated on erythrocytes, urn cells, and sexual cells of *Siphonosoma arcassonense* and on sexual cells of *Cancer pagurus*. Although no Con A receptors on *Asterias rubens* were found by Gebbinck (who kept the starfish in sea water), Redziniac and collaborators (1978) detected receptors for Con A on axial organ cells when they were cultured in 0.1 *M* sodium phosphate solution.

Karp and Coffaro (1982) observed mitogenic responses when coelomocytes of the sea star *Dermasterias imbricata* were stimulated with lipopolysaccharide or Con A; the response to Con A was inhibited by 0.1 *M* α-methyl-D-mannoside. No mitogenic responses to phytohemagglutinin or pokeweed mitogen were found.

Brillouet *et al.* (1981) found two cell populations isolated from the axial organ of the starfish *Asterias rubens* exhibited specific mitogenic responses. Nylon wool nonadherent cells responded to Con A, whereas adherent cells were stimulated by lipopolysaccharide. A mitogenic factor was released by these cells after stimulation with pokeweed mitogen. A soluble factor which mediates *in vivo* cell aggregation is released by the common sea star *Asterias forbesi* during its response to xenogeneic, but not allogeneic, cell challenge (Reinisch, 1974). Bayne (1982) was the first to report a response to mitogens by amoebae (Nuclearia). Con A triggered doubling of their population in 5.5 hr; lipopolysaccharide, phytohemagglutinin (PHA), and lentil lectin were all mitogenic.

Warr and colleagues (1977) showed that "lymphocyte-like" cells of the tunicate *Pyrua stolonifera* possess receptors for Con A, wheat germ agglutinin and soybean lectin (demonstrated by binding of the radioiodinated lectins), but these lectins did not induce mitogenesis. Allogeneic cells also did not stimulate mitosis.

There are a few reports of receptors for erythrocytes on invertebrate leukocytes. Some coelomocytes of the earthworm *Lumbricus terrestris* which were nonadherent to plastic formed spontaneous rosettes with sheep erythrocytes (Toupin and Lamoureux, 1976). Hemocytes from the American cockroach *Periplaneta americana* formed rosettes when cultured with sheep or chicken erythrocytes (Scott, 1971). Receptors for avian and mammalian erythrocytes on insect macrophages have also been demonstrated by rosette formation (Anderson, 1977).

When snail hemocytes were pretreated with 15 nonnative agglutinins and incubated with human erythrocytes, rosettes were formed with five of the agglutinins (Renwrantz and Cheng, 1977b). A small number of untreated hemocytes were capable of binding directly to erythrocytes of mice, rabbits, and sheep. Similar observations have previously been made with sheep erythrocytes directly binding hemocytes of crayfish (McKay and Jenkin, 1970), snail (Sminia *et al.,* 1979), and earthworm coelomocytes (Cooper, 1973).

V. RECOGNITION THEORIES

In the formation of colonies of marine invertebrates and in all associations of cells, there must be "morphological" receptors for self antigens whose function is simply to couple cells together. Docking of "self" cells at these receptors may not trigger any response; one could argue that they are not, therefore, receptors in the true sense and possibly a better term would be "docking positions" for self antigens.

Kolb (1977) has proposed a model for recognition of self and non-self. Basically, his model includes receptors for self on every phagocytic cell. When self determinants (probably carbohydrates) dock into these receptors (glycoproteins or lectins) an active process is triggered which switches *off* phagocytosis. On the other hand, foreign cells would be unable to dock into self receptors and phagocytosis can proceed unimpeded. Of course, there must be receptors for foreign cells into which they may dock (adherence) and the docking process may, or may not, trigger ingestion followed by killing and digestion.

Langman (1978) invokes a similar "switch off" mechanism following docking of self to self and emphasizes that the recognition of non-self is not only concerned with preservation of colony identity and feeding groups (Theodor, 1970; Lang, 1971; Butnet, 1971) but is perhaps more important in defense against infectious agents. One form of immune evasion by the more successful parasites is their ability to express self markers of the host. As the variety of non-self receptors increased during evolution, viruses and parasites were possibly driven toward mimicry of host self to ensure successful parasitism or infection. Some of these interesting mechanisms have been reviewed by Maramorosch and Shope (1975) and Lackie (1980).

Roseman (1970, 1974) has proposed that carbohydrates in the glycocalyx of eukaryotic cells participate in cellular adhesion. This could occur either by binding of carbohydrates on opposing cell surfaces via hydrogen bonds, or by binding of carbohydrates to the corresponding cell-surface glycosyltransferase. Roseman (1974) postulated transfer of information by a membrane messenger. The nature of the membrane messenger that might trigger DNA to initiate a specific activity is not yet clear. Obvious candidates are messenger RNA, cyclic AMP, adenyl cyclase, and even calcium ions.

Parish (1977) developed Roseman's model. He proposed that recognition factors are some of the glycosyltransferases. An invertebrate might synthesize an oligosaccharide of five sugar residues, each attached to a different glycosyltransferase that has specificity for the sugar acceptor. Random polymerization of the transferases into a recognition factor might be initiated by an additional protein that has an acceptor site on the hemocyte's surface. This protein thus confers cytophilic and opsonic properties to the recognition factor. However, this model relies heavily on glycosyltransferases, whereas it is increasingly becoming evident that lectins are the more likely molecules to be involved in recognition. If one subunit of a lectin contains a receptor for sugars on the surface of an hemocyte while other subunits possess receptors for sugars present on the cell surface of bacteria, it is easy to visualize the lectin forming a ''bridge'' between the bacteria and hemocyte, resulting in adherence, the first prerequisite of an opsonin.

Rothenberg (1978) adopts the multi-glycosyl enzymatic and recognition system (MGER) as a model that would regulate synthesis and recognition of oligosaccharides when selective processes such as infections could modulate the production of necessary receptors for phagocytosis. The MGER model proposes that (a) one germ-line gene determines specificity of glycosyltransferase, oligosaccharide, and recognition molecules (e.g., lectins), (b) diversity of structure of the receptors for ''self'' and ''non-self'' is obtained by variation in amino acid and sugar sequences, and (c) all receptors will be internally manufactured and therefore self-perpetuating.

Recognition systems in invertebrates are much more limited in their diversity compared to vertebrate lymphocytes that possess millions of different receptor structures based on the V and C genes (Cunningham, 1979). Some invertebrate receptors may be highly specific. For instance, if lectins acted as receptors on the cell surface they could exhibit high specificity for a certain sugar. It is likely that there are also structures (which may be lectins) exhibiting a broader-based specificity. In this way many foreign agents could adhere to invertebrate phagocytes via specific or quite nonspecific (yet effective) receptors which are capable of triggering phagocytosis once docking has taken place.

Generation of diversity of specificity mediated by lectin molecules has been proposed by Jenkin and Hardy (1975). If six different subunits were synthesized by the same cell and could combine in a random fashion, they could form nearly a thousand different specificities—a lower figure than calculated for the vertebrates, but perhaps sufficient for defense in the invertebrates.

Dausset (1981) proposed a role for the major histocompatibility complex (MHC) in self-recognition but emphasises that the mechanism of self-recognition is not limited to MHC products but probably involves many other molecules, probably glycoproteins. Shalev *et al.* (1981) have detected β_2-microglobulin in extracts of several invertebrates. Although extracts of *Hydra littoralis* were only weakly positive, those from the earthworm *Lumbricus terrestris* and the crayfish *Cambarus diogenes* were strongly positive for β_2-microglobulin. Roch and Cooper (1983) have detected β_2-microglobulin molecules on membranes of some leukocytes of *L. terrestris*. β_2-Microglobulin is structurally related to immunoglobulin domains and the heavy chain of HLA antigens of humans by virtue of some homology of amino acid sequences. The presence of β_2-microglobulin in invertebrates suggests that such molecules have probably evolved from one common ancestral gene.

VI. CONCLUSIONS

We have seen plenty of evidence that invertebrates possess mechanisms for discriminating between self and non-self. Even in lower vertebrates such as the sponges allogeneic rejection can occur, although this has sometimes been less easy to demonstrate in other phyla. The role of accelerated second-set responses revealed by transplant or implant rejection may be important in defense of a colony or against pathogenic microorganisms. "Memory" may need to be redefined in terms of "sensitization" of hemocytes or "activation" of these and other phagocytic cells.

Invertebrates do not all possess lectins that may act as opsonins; this is particularly the case in insects. It is difficult to understand why certain species possess lectins and others do not. In evolution, the survival of a species may have depended on its ability to produce a lectin that could combine with a certain species-specific pathogen. One example is the lack of a specific agglutinin (lectin?) in lobsters, which are highly susceptible to the pathogen *Ghaffkya homari* (Section IV,A).

The dramatic fall in numbers of circulating hemocytes following injections of foreign cells or molecules which has frequently been reported in many higher invertebrates is of interest. Circulating cells may become fixed to the body wall as a form of retention of "memory" cells or the consumption of lectins (acting as opsonins) may trigger "suicide" of hemocytes with release of their contents. In this way the level of lectins could quite rapidly be restored. There is no clonal expansion of cells producing a given lectin, and invertebrate immunity is only adaptive in the sense that the animal strives to restore its natural repertoire of humoral factors by calling forth more secreting cells as rapidly as possible.

There is some evidence that antigenic stimulation can increase humoral factors in certain invertebrates (e.g., Stein *et al.*, 1982). Also, short-term protective immunity may be induced in some higher invertebrates. For example, protection may be induced in insects by vaccines (Gingrich, 1964; Chadwick and Vilk, 1969; Karp and Rheins, 1980). As protection to soluble protein toxins can be passively transferred by immune hemolymph (Rheins *et al.*, 1981), this is good evidence that fairly specific humoral factors are produced following immunization.

Weir (1980) has pointed out how carbohydrate cell-surface antigenic determinants or markers provide a vast potential diversity of recognition factors. This poses the question of how many lectins, opsonins, or other recognition molecules may be eventually found in invertebrates. Intensive work on sponges has already revealed many such factors, although only one or two lectins have been isolated from higher invertebrates. Clearly, much more research is required on the structure of invertebrate opsonins and other soluble factors, which may even include components of the alternative pathway of complement activation (Day *et al.*, 1970). It is certain that invertebrate leukocyte receptors have a more limited range than vertebrate lymphocytes, but phagocytic cells in vertebrates may be quite similar to those of invertebrates.

REFERENCES

Acton, R. T., Bennett, J. C., Evans, E. E., and Schrohenloher, R. E. (1969). *J. Biol. Chem.* **244**, 4128.

Anderson, R. R. S. (1977). *Cell. Immunol.* **29**, 331.

Anderson, R. S., and Good, R. A. (1976). *J. Invertebr. Pathol.* **27**, 57.

Bailey, S., Miller, J., and Cooper, E. L. (1971). *Immunology* **21**, 81.

Bancroft, F. W. (1903). *Proc. California Academy of Sciences, 3rd series,* **3**, 137.

Bayne, C. J. (1980a). *Dev. Comp. Immunol.* **4**, 43.

Bayne, C. J. (1980b). *Dev. Comp. Immunol.* **4**, 215.

Bayne, C. J. (1982). *Dev. Comp. Immunol.* **6**, 369.

Bayne, C. J., Moore, M. N., Carefoot, T. H., and Thompson, R. J. (1979). *J. Invertebr. Pathol.* **34**, 1.

Berrill, N. J. (1955). "The origin of the vertebrates." Oxford University Press, London.

Bertheussen, K. (1979). *Exp. Cell Res.* **120**, 373.
Bertheussen, K. (1982). *Dev. Comp. Immunol.* **6**, 635.
Bertheussen, K., and Seljelid, R. (1982). *Dev. Comp. Immunol.* **6**, 423.
Bibb, C., and Campbell, R. D. (1973). *Tissue Cell* **5**, 199.
Bigger, C. H., and Hildemann, W. H. (1982). *In* "Phylogeny and Ontogeny of the Reticuloendothelial System" (N. Cohen and M. M. Sigel, eds.) Vol. 3, pp. 59–87. Plenum Press, New York.
Bigger, C. H., and Runyan, R. (1979). *Dev. Comp. Immunol.* **3**, 591.
Bigger, C. H., Hildemann, W. H., Jokiel, P. L., and Johnston, I. S. (1981). *Transplantation* **31**, 461.
Boyden, S. V. (1962). *J. Theor. Biol.* **3**, 123.
Boyden, S. V. (1966). *Adv. Immunol.* **5**, 1.
Bretting, H., and Kabat, E. A. (1976). *Biochemistry* **15**, 3228.
Bretting, H., Kabat, E. A., Liao, J., and Pereira, M. E. A. (1976). *Biochemistry* **15**, 5029.
Brillouet, C., Luquet, G., and Leclerc, M. (1981). *In* "Aspects of Developmental and Comparative Immunology I" (J. B. Solomon, ed.), pp. 159–170. Pergamon Press, Oxford.
Brown, R., Almodovar, L. R., Bhatia, H. M., and Boyd, W. C. (1968). *J. Immunol.* **100**, 214.
Bruslé, J. (1967). *Cah. Biol. Marine* **8**, 417.
Burnet, F. M. (1971). *Nature (London)* **232**, 230.
Callow, J. A. (1977). *Adv. Botan. Res.* **4**, 1.
Cameron, G. R. (1934). *J. Pathol. Bacteriol.* **38**, 441.
Campbell, R. D., and Bibb, C. (1970). *Transplant. Proc.* **2**, 203.
Carton, Y. (1976). *Transplantation* **21**, 17.
Chadwick, J. S., and Vilk, E. (1969). *J. Invertebr. Pathol.* **13**, 410.
Cheng, T. C. (1970). *Transplant. Proc.* **2**, 226.
Cheng, T. C., and Galloway, P. C. (1970). *J. Invertebr. Pathol.* **15**, 177.
Cheng, T. C., and Rudo, B. M. (1976). *J. Invertebr. Pathol.* **27**, 137.
Chorney, M. J., and Cheng, T. C. (1980). *Contemp. Top. Immunobiol.* **9**, 37.
Clarke, A. E., and Knox, R. B. (1979). *Dev. Comp. Immunol.* **3**, 571.
Coffaro, K. A., and Hinegardner, R. T. (1977). *Science* **197**, 1389.
Cohen, E. (1974). *Ann. N.Y. Acad. Sci.* **234**, 7.
Cohen, E., Roberts, S. C., Nordling, S., and Uhlenbruck, G. (1972). *Vox Sang.* **23**, 300.
Cooper, E. L. (1970). *Transplant. Proc.* **2**, 216.
Cooper, E. L. (1973). *In* "Non-specific factors influencing host resistance" (W. Braun and J. Ungar, eds.) pp. 11–23. Karger, Basel.
Cornick, J. W., and Stewart, J. E. (1968). *J. Fisheries Research Board, Canada* **25**, 695.
Crichton, R., and Lafferty, K. J. (1975). *Adv. Exp. Med. Biol.* **64**, 89.
Cunningham, A. J. (1978). *Dev. Comp. Immunol.* **2**, 243.
Curtis, A. S. G. (1979). *Coll. Int. Rech. Sci.* **291**, 205.
Curtis, A. S. G., and Van de Vyver, G. (1971). *J. Embryol. Exp. Morphol.* **26**, 295.
Curtis, A. S. G., Kerr, J., and Knowlton, N. (1982). *Transplantation* **33**, 127.
Dausset, J. (1981). *Dev. Comp. Immunol.* **5**, 1.
Day, N. K. B., Gewurz, H., Johannsen, R., Finstad, J., and Good, R. A. (1970). *J. Exp. Med.* **132**, 941.
Decker, J. M., Elmholt, A., and Muchmore, A. V. (1981). *Cell. Immunol.* **59**, 161.
Donelly, C., and Clem, L. W. (1982). *Dev. Comp. Immunol.* **6**, 171.
Duprat, P. (1964). *C. R. Acad. Sci. [D] (Paris)* **259**, 4177.
Duprat, P. (1967). *Ann. Inst. Pasteur, Paris* **133**, 867.
Dyrynda, P. E. J. (1983). *Dev. Comp. Immunol.* **7**, 621.

Evans, C. W., Kerr, J., and Curtis, A. S. G. (1981). In "Phylogeny of Immunological Memory" (M. J. Manning, ed) pp. 27–34. Elsevier/North Holland, Amsterdam.

Ey, P. L., and Jenkin, C. R. (1982). In "Phylogeny and Ontogeny of the Reticuloendothelial System" (N. Cohen and M. M. Sigel, eds.), Vol. 3, pp. 321–391. Plenum Press, New York.

Fitzgerald, S. W., and Ratcliffe, N. A. (1981). In "Aspects of Developmental and Comparative Immunology I" (J. B. Solomon, ed.), pp. 139–145. Pergamon Press, Oxford.

Fitzgerald, S. W., and Ratcliffe, N. A. (1982). Dev. Comp. Immunol. 6, 23.

Freter, R. (1980). In "Bacterial Adherence" (F. H. Beachey, ed.), Receptors and Recognition Series B, Vol. 6, pp. 441–458. Chapman and Hall, London.

Fuke, M. T., and Numakunai, T. (1982). Dev. Comp. Immunol. 6, 253.

Galtsoff, P. S. (1925). J. Exp. Zool. 42, 183.

Galtsoff, P. S. (1964). United States Fish and Wildlife Services, Fish Bulletin 64, p. 1.

Gebbinck, J. (1980). Dev. Comp. Immunol. 4, 33.

Ghiradella, H. T. (1965). Biol. Bull. 128, 77.

Gingrich, R. D. (1964). J. Insect Physiol. 10, 179.

Gold, E. R., and Balding, P. (1975). "Receptor-specific Proteins: Plant and Animal Lectins." Elsevier, New York.

Goldstein, I. J., and Hayes, C. F. (1978). Adv. Carbohydrate Chem. Biochem. 35, 127.

Goldstein, L. (1970). Transplant. Proc. 2, 191.

Hall, J. L., and Rowlands, D. T. (1974). Biochemistry 13, 821.

Hammarstrom, S., and Kabat, E. A. (1969). Biochemistry 8, 2696.

Hammarstrom, S., and Kabat, E. A. (1971). Biochemistry 10, 1684.

Hammarstrom, S., Westoö, A., and Bjork I. (1972). Scand. J. Immunol. 1, 295.

Hardy, S. W., Fletcher, T. C., and Olafsen, J. A. (1977). In "Developmental Immunobiology" (J. B. Solomon and J. D. Horton, eds.), pp. 59–66. Elsevier/North Holland, Amsterdam.

Hartman, A. L., Campbell, P. A., and Abel, C. A. (1978). Dev. Comp. Immunol. 2, 617.

Henkart, P., and Humphreys, J. (1973). Biochemistry 12, 3045.

Hickman, C. P. (1967). "Biology of the Invertebrates." C. V. Mosby Co., St. Louis.

Hildemann, W. H., and Dix, T. G. (1972). Transplantation 14, 624.

Hildemann, W. H., Dix, T. G., and Collins, J. D. (1974). Comtemp. Top. Immunobiol. 4, 141.

Hildemann, W. H., Linthicum, D. S., and Vann, D. C. (1975). Immunogenetics 2, 269.

Hildemann, W. H., Raison, R. L., and Hull, C. J. (1977a). In "Developmental Immunobiology" (J. B. Solomon and J. D. Horton, eds.), pp. 9–16. Elsevier/North Holland, Amsterdam.

Hildemann, W. H., Raison, R. L., Cheung, G., Hull, C. J., Akaka, L., and Okamoto, J. (1977b). Nature (London) 270, 219.

Hildemann, W. H., Johnston, I. S., and Jokiel, P. L. (1979). Science 204, 420.

Hildemann, W. H., Bigger, C. H., Jokiel, P. L., and Johnston, I. S. (1980a). In "Phylogeny of Immunological Memory" (M. J. Manning, ed.), pp. 9–14. Elsevier/North Holland, Amsterdam.

Hildemann, W. H., Jokiel, P. L., Bigger, C. H., and Johnston, I. S. (1980b). Transplantation 30, 297.

Hilgard, H. R., and Phillips, J. H. (1968). Science 161, 1243.

Hilgard, H. R., Wander, R. H., and Hinds, W. E. (1974). Contemp. Top. Immunobiol. 4, 151.

Hostetter, R. K., and Cooper, E. L. (1974). Contemp. Top. Immunobiol. 4, 91.

Humphreys, T. (1963). Dev. Biol. 8, 27.

Humphreys, T. (1969). Symp. Zoological Society (London) 25, 325.

Ivker, F. B. (1972). Biol. Bull. 143, 162.

Jenkin, C. R. (1976). In "Comparative Immunology" (J. J. Marchalonis, ed.), pp. 80–94. Blackwell, Oxford.

Jenkin, C. R., and Hardy, D. (1975). Adv. Exp. Med. Biol. 64, 55.

Johnson, P. T. (1969). *J. Invertebr. Pathol.* **13,** 42.

Johnston, I. S., and Hildemann, W. H. (1982). *In* "Phylogeny and Ontogeny of the Reticuloendothelial System" (N. Cohen and M. M. Sigel, eds.), Vol. 3, pp. 37–57. Plenum Press, New York.

Jones, S. E., and Bell, W. J. (1982). *Dev. Comp. Immunol.* **6,** 35.

Kaplan, R., Li, S. S-L., and Kehoe, J. M. (1977). *Biochemistry* **16,** 4297.

Karp, R. D., and Coffaro, K. A. (1982). *In* "Phylogeny and Ontogeny of the Reticuloendothelial System (N. Cohen and M. M. Sigel, eds.), Vol. 3, pp. 257–282. Plenum Press, New York.

Karp, R. D., and Hildemann, W. H. (1976). *Transplantation* **22,** 434.

Karp, R. D., and Rheins, L. A. (1980). *Dev. Comp. Immunol.* **4,** 629.

Klement, A., and Goodman, R. N. (1967). *Annu. Rev. Phytopathol.* **5,** 17.

Kolb, H. (1977). *Dev. Comp. Immunol.* **1,** 193.

Kuhns, W. J., and Burger, M. M. (1971). *Biol. Bull.* **141,** 393.

Lackie, A. M. (1976). *Parasitology* **73,** 97.

Lackie, A. M. (1979). *Immunology* **36,** 909.

Lackie, A. M. (1980). *Parasitology* **80,** 393.

Lackie, A. M. (1981a). *In* "Aspects of Developmental and Comparative Immunology I" (J. B. Solomon, ed.), pp. 99–104. Pergamon Press, Oxford.

Lackie, A. M. (1981b). *Dev. Comp. Immunol.* **5,** 191.

Lackie, A. M. (1983). *Dev. Comp. Immunol.* **7,** 41.

Lang, J. (1971). *Bull. Marine Sci.* **21,** 952.

Langlet, C., and Bierne, J. (1977). *In* "Developmental Immunobiology" (J. B. Solomon and J. D. Horton, eds.), pp. 17–26. Elsevier/North Holland, Amsterdam.

Langman, R. E. (1978). *Rev. Physiol. Biochem. Pharmacol.* **81,** 1.

Lindh, N. O. (1959). *Arch. Zool. Ser. 2,* **12**(14), 183.

Lis, H., and Sharon, N. (1977). *In* "The Antigens" (M. Sela, ed.), Vol. 4, pp. 429–529. Academic Press, New York.

Loeb, L. (1945). "The Biological basis of Individuality." C. C. Thomas Publications, Springfield, Illinois.

McClay, D. R. (1974). *J. Exp. Zool.* **188,** 89.

McCumber, L. J. and Clem, L. W. (1977). *Dev. Comp. Immunol.* **1,** 5.

McCumber, L. J., Hoffman, E. M., and Clem, L. W. (1979). *J. Invertebr. Pathol.* **33,** 1.

McKay, D., and Jenkin, C. R. (1970). *Aust. J. Exp. Biol. Med. Sci.* **48,** 139.

McKay, D., Jenkin, C. R., and Rowley, D. (1969). *Aust. J. Exp. Biol. Med. Sci.* **47,** 124.

MacLennan, A. P. (1974). *Arch. Biol.* **85,** 53.

Mahler, H. R., and Raff, R. A. (1975). *Int. Rev. Cytol.* **43,** 1.

Maramorosch, K., and Shope, R. E. (1975). "Invertebrate Immunity: Mechanisms of Invertebrate Vector–parasite Relations." Academic Press, New York.

Marchalonis, J., and Edelman, G. M. (1968). *J. Mol. Biol.* **32,** 453.

Marchalonis, J. J., and Warr, G. W. (1978). *Dev. Comp. Immunol.* **2,** 443.

Marks, D. H., Stein, E. A., and Cooper, E. L. (1979). *Dev. Comp. Immunol.* **3,** 277.

Moore, R. (1981). *Dev. Comp. Immunol.* **5,** 377.

Moscona, A. A. (1963). *Proc. Natl. Acad. Sci. U.S.A.* **49,** 742.

Moscona, A. A. (1968). *In* "In Vitro" (M. M. Sigel, ed.), Vol. 3, p. 13. Williams and Wilkins, Baltimore.

Newell, P. C. (1977). *Endeavour* **1,** 63.

Nicolson, G. L. (1974). *Int. Rev. Cytol.* **39,** 89.

Nielsen, H. E., Koch, C., and Drachman, O. (1983). *Dev. Comp. Immunol.* **7,** 413.

Oka, H., and Watanabe, H. (1960). *Bull. Biol. Asamushi* **10,** 153.

Parish, C. R. (1977). *Nature (London)* **267,** 711.

Pauley, G. B., Krassner, S. M., and Chapman, F. A. (1971). *J. Invertebr. Pathol.* **18,** 227.
Pistole, T. G. (1978). *Dev. Comp. Immunol.* **2,** 65.
Prowse, R. H., and Tait, N. N. (1969). *Immunology* **17,** 437.
Purcell, J. E. (1977). *Biol. Bull.* **153,** 355.
Ratcliffe, N. A., and Gagen, S. J. (1976). *J. Invertebr. Pathol.* **28,** 373.
Reddy, A. L., Bryan, B., and Hildemann, W. H. (1975). *Immunogenetics* **1,** 584.
Redziniac, G., Leclerc, M., Panijel, J., and Monsigny, M. (1978). *Biochimie* **60,** 525.
Reinisch, C. L. (1974). *Nature (London)* **250,** 349.
Renwrantz, L. (1981). *In* "Aspects of Developmental and Comparative Immunology I." (J. B. Solomon, ed.), pp. 133–138. Pergamon Press, Oxford.
Renwrantz, L. (1983). *Dev. Comp. Immunol.* **7,** 603.
Renwrantz, L., and Cheng, T. C. (1977a). *J. Invertebr. Pathol.* **29,** 88.
Renwrantz, L., and Cheng, T. C. (1977b). *J. Invertebr. Pathol.* **29,** 97.
Renwrantz, L., and Mohr, W. (1978). *J. Invertebr. Pathol.* **31,** 164.
Renwrantz, L., Schancke, W., Harm, H., Erl, H., Liebsch, H., and Gercken, J. (1981). *J. Comp. Physiol.* **141,** 477.
Rheins, L. A., Karp, R. D., and Butz, A. (1981). *Dev. Comp. Immunol.* **4,** 447.
Roch, P. G., and Cooper, E. L. (1983). *Dev. Comp. Immunol.* **7,** 633.
Roch, P., and Valembois, P. (1978). *Dev. Comp. Immunol.* **2,** 51.
Roch, P., Valembois, P., and Du Pasquier, L. (1975). *Adv. Exp. Med. Biol.* **64,** 44.
Roseman, S. (1970). *Chem. Phys. Lipids* **5,** 270.
Roseman, S. (1974). *In* "The Cell Surface in Development" (A. A. Moscona, ed.), pp. 255–272. Wiley, New York.
Rosen, S., and Barondes, S. H. (1978). *In* "Specificity of Embryological Interactions" (D. R. Garrod, ed.), Receptors and Recognition Series B, Vol. 4, pp. 235–264. Chapman and Hall, London.
Rostam-Abadi, H., and Pistole, T. G. (1982). *Dev. Comp. Immunol.* **6,** 209.
Rothenberg, B. E. (1978). *Dev. Comp. Immunol.* **2,** 23.
Ryan, M., and Nicholas, W. L. (1972). *J. Invertebr. Pathol.* **19,** 299.
Salt, G. (1963). *Parasitology* **53,** 527.
Salt, G. (1970). "The cellular defence reactions of insects," Cambridge University Monographs in Experimental Biology No. 16, p. 118. Cambridge University Press, Cambridge.
Salt, G. (1973). *Proc. R. Soc. London, Ser. B 183,* 337.
Schmid, L. S. (1975). *J. Invertebr. Pathol.* **25,** 125.
Schoenberg, D. A., and Cheng, T. C. (1980). *Dev. Comp. Immunol.* **4,** 617.
Scofield, V. L., Schlumpberger, J. M., West, L. A., and Weissman, I. L. (1981). *Nature (London)* **295,** 499.
Scott, M. T. (1971). *Immunology* **21,** 817.
Sequeira, L. (1978). *Annu. Rev. Phytopathol.* **16,** 453.
Shalev, A., Greenberg, A. H., Logdberg, L., and Brock, L. (1981). *J. Immunol.* **127,** 1186.
Sloan, B., Yocum, C., and Clem, L. W. (1975). *Nature (London)* **258,** 521.
Sminia, T., van der Knaap, W. P. W., and Edelenbosch, P. (1979). *Dev. Comp. Immunol.* **3,** 37.
Söderhäll, K. (1981). *Dev. Comp. Immunol.* **5,** 565.
Söderhäll, K. (1982). *Dev. Comp. Immunol.* **6,** 601.
Söderhäll, K., Vey, A., and Ramstedt, M. (1984). *Dev. Comp. Immunol.* **8,** 23.
Sparks, A. K. (1972). "Invertebrate Pathology." Academic Press, New York.
Stein, E. A., Wojdani, A., and Cooper, E. L. (1982). *Dev. Comp. Immunol.* **6,** 407.
Stuart, A. E. (1968). *J. Pathol. Bacteriol.* **96,** 401.
Tanaka, K. (1973). *Cell. Immunol.* **7,** 427.
Tanaka, K., and Watanabe, H. (1973). *Cell. Immunol.* **7,** 410.

Taneda, Y., and Watenabe, H. (1982). *Dev. Comp. Immunol.* **6**, 243.
Theodor, J. L. (1970). *Nature (London)* **227**, 690.
Theodor, J. L. (1976). *Zool. J. Linnean Soc.* **58**, 173.
Theodor, J. L., and Senelar, R. (1975). *Cell. Immunol.* **19**, 194.
Thomas, I. G., and Ratcliffe, N. A. (1982). *Dev. Comp. Immunol.* **6**, 643.
Toupin, J., and Lamoureux, G. (1976). *Cell. Immunol.* **26**, 127.
Toupin, J., Layva, F., and Lamoureux, G. (1977). *Ann. Immunol. [C]* **128**, 29.
Tripp, M. R. (1963). Cellular responses of mollusks. *Ann. N.Y. Acad. Sci.* **113**, 467.
Tripp, M. R. (1966). *J. Invertebr. Pathol.* **8**, 478.
Turner, R. S. (1978). *In* "Specificity of Embryological Interactions" (D. R. Garrod, ed.), Receptors and Recognition Series B, Vol. 4, pp. 201–232. Chapman and Hall, London.
Turner, R. S., and Burger, M. M. (1973). *Nature (London)* **244**, 509.
Tyler, A., and Metz, C. B. (1945). *J. Exp. Zool.* **100**, 387.
Tyson, C. J., and Jenkin, C. R. (1973). *Aust. J. Exp. Biol. Med. Sci.* **51**, 609.
Uhlenbruck, G., and Steinhausen, G. (1977). *Dev. Comp. Immunol.* **1**, 183.
Vaith, P., Müller, W. E. G., and Uhlenbruck, G. (1979a). *Dev. Comp. Immunol.* **3**, 259.
Vaith, P., Uhlenbruck, G., Müller, E. G., and Holz, G. (1979b). *Dev. Comp. Immunol.* **3**, 399.
Valembois, P., Roch, P., and Boiledieu, D. (1982). *In* "Phylogeny and Ontogeny of the Reticuloendothelial System" (N. Cohen and M. M. Sigel, eds.), Vol. 3, pp. 89–139. Plenum Press, New York.
Van der Knaap, W. P. W. (1981). *In* "Aspects of Developmental and Comparative Immunology I" (J. B. Solomon, ed.), pp. 91–97. Pergamon Press, Oxford.
Van de Vyver, G. (1975). *Curr. Top. Dev. Biol.* **10**, 123.
Van de Vyver, G. (1980). *In* "Phylogeny of Immunological Memory" (M. J. Manning, ed.), pp. 15–26. Elsevier/North Holland, Amsterdam.
Van de Vyver, G. (1983). *Dev. Comp. Immunol.* **7**, 609.
Van der Wall, J., Campbell, P. A., and Abel, C. A. (1981). *Dev. Comp. Immunol.* **5**, 679.
Vasta, G. R., Ilodi, G. H. U., Cohen, E., and Brahmi, Z. (1982). *Dev. Comp. Immunol.* **6**, 625.
Walters, J. B., and Ratcliffe, N. A. (1981). *In* "Aspects of Developmental and Comparative Immunology I." (J. B. Solomon, ed.), pp. 147–152. Pergamon Press, Oxford.
Walters, J. B., and Ratcliffe, N. A. (1983). *Dev. Comp. Immunol.* **7**, 661.
Warr, G. W., Decker, J. M., Mandel, T. E., de Luca, D., Hudson, R., and Marchalonis, J. J. (1977). *Aust. J. Exp. Biol. Med. Sci.* **55**, 151.
Weinbaum, G., and Burger, M. M. (1973). *Nature (London)* **244**, 510.
Weir, D. M. (1980). *Immunol. Today* **1**, 45.
Wilson, H. V. (1907). *J. Exp. Zool.* **5**, 245.
Wright, R. K. (1974). *J. Invertebr. Pathol.* **24**, 29.
Yeaton, R. W. (1981a). *Dev. Comp. Immunol.* **5**, 391.
Yeaton, R. W. (1981b). *Dev. Comp. Immunol.* **5**, 535.
Yoshino, T. P. (1981). *Dev. Comp. Immunol.* **5**, 229.
Yoshino, T. P. (1982). *Dev. Comp. Immunol.* **6**, 451.
Yoshino, T. P. (1983). *Dev. Comp. Immunol.* **7**, 641.

2

Molecular Recognition of Non-self Determinants: The Existence of a Superfamily of Recognition Molecules Related to Primordial Immunoglobulins

JOHN J. MARCHALONIS, GREGORY W. WARR,
GERARDO R. VASTA, AND BARRY E. LEDFORD

We review briefly our knowledge of antibodies and cellular recognition events and identify classes of molecules that have been implicated in, or hypothesized to function in, cellular recognition events. These molecules include not only immunoglobulins, particularly their variable regions, but also lectins, C-reactive proteins, T-cell receptors, complement components, products of the major histocompatibility complex, and differentiation antigens of cell surfaces. With the information at hand we have examined the hypothesis that the immunoglobulins are only one member of a family of related "recognition molecules" found in the animal kingdom. A quantitative analysis of relationships among these molecules using as an index of relatedness the parameter $S \Delta Q$ (which assesses the degree of difference between proteins on the basis of amino acid composition) revealed

45

RECEPTORS IN CELLULAR RECOGNITION
AND DEVELOPMENTAL PROCESSES

the following. Coupled with limited amino acid sequence data and circumstantial data such as molecular weights and binding specificites, this analysis has provided evidence to support the hypothesis that a "superfamily" of recognition molecules has evolved within the animal kingdom. This family includes β_2-microglobulin, C-reactive proteins, certain invertebrate, chordate and vertebrate lectins, molecules of the major histocompatibility complex, mammalian T-cell receptors, the Thy-1 differentiation antigen, and immunoglobulins. The possible relationships among these molecules are most apparent when they are compared with the primordial members of the immunoglobulin family rather than with the highly evolved chains of mammalian immunoglobulins. Furthermore, this "superfamily" does not include indiscriminately all those molecules which function in defense and recognition. A notable example of a markedly divergent molecule is the inducible bactericidal protein of the silk moth (cecropia). Evidence supporting this hypothesis, the details of phylogenetic relationships, and the implications of these conclusions are presented in this brief review.

I. INTRODUCTION

All vertebrates can produce circulating antibodies to a diverse collection of antigens. The major feature of antibody structure has apparently been conserved in evolution to the extent that all antibodies studied are multichain proteins consisting of light and heavy polypeptide chains, and these chains show a heterogeneity due to the presence of variable regions. Although definitive evidence for the existence of "classical" immunoglobulin *V* regions in lower species, particularly cyclostomes, is lacking, the circumstantial evidence is persuasive that this molecular recognition structure must have arisen early in vertebrate (or chordate?) evolution. The presence of *V* regions and the existence of a genetic mechanism for their generation is apparently a fundamental attribute of immune recognition. The primitive and universal class of antibody throughout vertebrate evolution resembles the IgM isotype as defined in mammals because all vertebrates possess heavy chains comparable to mammalian μ chains in a number of characteristic features, including size, carbohydrate content, and amino acid composition. In certain lower species such as cyclostomes, elasmobranchs, and some teleosts, IgM antibodies are the only type of immunoglobulin that can be found in the serum. Moreover, one of the two major isotypes of B-cell surface receptor immunoglobulin in all vertebrates studied appears to be a slightly modified form of the 7 S subunit $(\mu L)_2$ of the IgM class. In placoderm-derived species (elasmobranchs, teleosts, and high vertebrate orders), IgM molecules tend to exist as cyclic polymers held together by disulfide bonds, with five $(\mu L)_2$ subunits being the usual number, although teleost IgM exists as a tetrameric polymer. Moreover, 7 S IgM $(\mu L)_2$ occurs frequently within elasmobranchs, although this monomer occurs fairly rarely in the serum of higher vertebrates. By contrast, in the cyclostomes μ-like heavy chains and lower molecular weight polypeptides resembling light-chains are present, but these do not form covalently linked polymers occurring in association with each other via disulfide bonds as is the usual case in other vertebrate species (Litman and Marchalonis, 1982; Marchalonis, 1977a).

It is reasonable to infer that the basic mechanisms of cellular and humoral immune recognition are similar throughout the vertebrate subphylum and are the results of an original single evolutionary event. The observed differences between various species therefore follow from the evolutionary divergence of species and secondary effects upon the genes regulating the immune response. In this communication we plan to use the fundamental importance of variable regions in immune recognition (a) to develop a model that predicts properties of the elusive antigen specific receptor of vertebrate T cells and (b) to serve as a guidepost in ascertaining properties of the more primitive precursors of immune recognition molecules as they might exist in protochordates and in nonchordate invertebrates.

II. PHYLOGENETIC DISTRIBUTION
OF SPECIFIC RECOGNITION

Figure 1 illustrates the type of recognition responses that are carried out by a variety of cells of virtually all animal species and that are therefore presumably mediated by recognition molecules occurring on the surfaces of these cells. One interesting question, which we shall return to later, concerns whether or not these surface receptors have secreted homologous counterparts. Primordial, or nonimmune, surface recognition such as is involved in feeding, phagocytosis, and organogenesis and development is universal and is carried out both by protozoans, and to some degree by cells of all multicellular animals. It is possible that certain categories of this nonimmune recognition are carried out by lectin molecules: i.e., molecules which show the capacity to recognize various carbohydrate moieties. These molecules are certainly involved in feeding recognition in Acanthamoeba (Brown et al., 1975) and in the case of the cellular organization of the slime mould (Frazier et al., 1975). It can be argued that what we have here (Fig. 1) termed quasi-immune recognition might well be functions (Hildemann, 1974) which correlate with the mixed-lymphocyte reactions and graft-versus-host reactions mediated by T cells of vertebrates. Species (including vertebrates) that lack circulating antibodies carry out reactions functionally very similar, if not equivalent, to these kinds of reactions (Hildemann, 1974; Warr

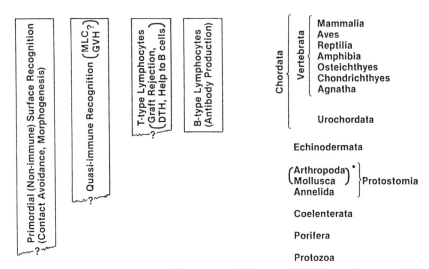

Fig. 1. Phylogenetic distribution of immune reactivities. Asterisk indicates data not available. Abbreviations: MLC, mixed-leukocyte culture response; GVH, graft-versus-host reaction; DTH, delayed-type (contact) hypersensitivity.

and Marchalonis, 1978; Warr, 1981) and, moreover, graft versus host reactions can occur in early developmental stages in the absence of clearly defined T cells (Lafferty, 1973). Both protostomate invertebrates, such as the primitive coelenterates, and deuterostomate invertebrates, including the protochordates, have been shown to carry out reactions resembling allograft reactivity (Hildemann, 1974; Cooper, 1976). The question which remains open here is whether or not these type of responses can be considered directly homologous to the more advanced category of specific T- and B-cell responses in which immunoglobulin variable regions clearly contribute to forming the combining site for antigen (Marchalonis, 1977b). From an evolutionary standpoint it would be most interesting to determine whether or not structures related to immunoglobulin variable regions or to their direct precursors are involved in quasi-immune recognition phenomena.

The next distinct response to consider is the T-lymphocyte response that is *specific* for exogenous antigen. Examples of this would be delayed-type hypersensitivity or specific helper or suppressor activity in interaction with macrophages and B cells. Putative T lymphocytes of the larva lamprey (Cooper, 1971) have the capacity to recognize haptens in a response resembling specific delayed type hypersensitivity, and this might generously be interpreted to imply the existence of variable regions specific for the particular haptens. A general conclusion to be gleaned from the functional studies of cellular responses of invertebrates and primitive chordates is that T-cell function apparently preceeds B-cell function in ontogeny and in evolution.

Molecules directly homologous to immunoglobulins (whose secretion is a strict B-cell function) have not been found in species more primitive than true vertebrates, although a tunicate, *Pyura stolonifera,* has in its hemolymph a molecule which binds the dinitrophenyl hapten, is approximately the right size to be an immunoglobulin heavy chain (70,000 daltons), and shows considerable heterogeneity in electrophoretic properties (Marchalonis and Warr, 1978). Even the most primitive of extant vertebrates, the agnathan cyclostomes, lampreys, and hagfish, can mount immune responses, albeit weakly, to various antigens and can carry out a wide range of cellular immune activities (Marchalonis, 1977a). The living vertebrates derived from placoderm stock all possess circulating antibody molecules which are clearly homologous to their counterparts in mammals. The IgM class appears universal and various other classes are believed to have arisen by gene duplication during the successive evolutionary emergence of the various vertebrate classes (Marchalonis, 1972).

It is a basic but little acknowledged fact that recognition is not carried out by cells per se but rather is mediated by molecules (receptors) that are associated with the surfaces of these cells. In some cases, for example in delayed typed hypersensitivity carried out by T cells of mammals, the recognition molecules are present on the cell surface, but very little, if any, of the corresponding recognition molecule are found in serum. In other cases, a prime example being the antigen receptors of B cells, large amounts of directly homologous immunoglobulin molecules may be found in serum (Vitetta and Uhr, 1973; Marchalonis and Cone, 1973). This generalization clearly holds for the B-cell surface IgM, which has a μ chain found in serum but which has an extra hydrophobic C-terminal stretch which facilitates association with the plasma membrane (Kehry *et al.,* 1980). IgD is also a major B-cell surface antigen receptor, but it is an immunoglobulin that is usually not found in large quantity in serum (Vitetta and Uhr, 1975). The immunoglobulin-related T-cell receptor has a variable region serologically related to immunoglobulin variable regions but has a constant region distinct from any of the classical serum immunoglobulin isotypes (Marchalonis, 1977b; Binz and Wigzell, 1978). Recent molecular biological approaches have identified and characterized putative T-cell receptor genes that are rearranged in T cells and show considerable homology to immunoglobulins in *V* and *C* regions (Yanagi *et al.,* 1984; Hedrick *et al.,* 1984) but represent translocons distinct from those specifying classical immunoglobulins (Marchalonis and Barker, 1984). If this molecule is found in serum at all, it occurs at extremely low levels.

Table I illustrates the types of molecules involved in the recognition phenomena just described. The IgM and IgD surface molecules are involved in antigen-specific recognition by B cells, and "IgT" would be involved in anti-

TABLE I

Recognition Molecules[a]

Receptor	Cell association	Specificity	Function	Molecular arrangement
IgM$_m$	B cell	Antigen	Antigen receptor	$(\lambda\mu)_2$ or $(\kappa\mu)_2$ S–S bonded tetramer
IgD$_m$	B cell	Antigen	Antigen receptor	$(\lambda\delta)_2$ or $(\kappa\delta)_2$ S–S bonded tetramer
IgT	T cell	Antigen	Antigen receptor	$(\tau)_2$ S–S bonded heavy-chain dimer; "light chains" sometimes non-covalently associated; τ chain approx. 70,000 daltons
Histocompatibility antigens	All cells	?	Cell/cell interaction	(H) (β_2M) Noncovalent dimer
β_2-Microglobulin	All cells	?	Cell/cell interaction	β_2M In noncovalent association with H
Immune response gene products (Ia)	Lymphocytes, macrophages, skin, sperm	?	Cell/cell interaction	$\alpha\beta$, Some S–S bonded dimers
Animal lectins: *Limulus* hemagglutinin (HA)	Amoebocytes ?	N-Acetylneuraminic acid	?/Opsonin	(Sub)$_{18}$ noncovalent multimer of subunits of mass 35,000 daltons; requires Ca^{2+}
Halocynthia hemagglutinin	Amoebocytes ?	N-Acetylneuraminic acid	?/Opsonin	S–S bonded multimer of subunits of mass 22,000 daltons; requires Ca^{2+}
Lamprey hemagglutinin	?	N-Acetylneuraminic acid	?	S–S bonded multimer of subunits of approx. mass 30,000 to 40,000 daltons; requires Ca^{2+}
Oyster hemagglutinin	Hemocytes	Compounds containing N-acetylneuraminic acid	Recognition	Associated with plasma membrane
C-Reactive protein	?	Phosphorylcholine (pneumococcal polysaccharide)	Opsonin	S–S bonded tetramers of subunit of mass 25,000 daltons; requires Ca^{2+}

[a] Based on Marchalonis (1980).

gen-specific T-cell recognition. Even in the case of allorecognition, Binz and Wigzell have reported that V_H-bearing molecules on T cells will recognize alloantigens (Binz and Wigzell, 1978). Lectins are widely distributed in both plants and animals and occur in the hemolymph of many, if not all, invertebrates (Gold and Balding, 1975). These are potential recognition molecules, particularly as they have been shown to act as opsonins (Jenkin, 1976). Recent work from this laboratory indicates that in the case of the oyster, an agglutinin which shows specificity for sialic acid-containing moieties also occurs in association with the plasma membrane of phagocytic hemocytes (Vasta et al., 1982a). Hemagglutinins specific for N-acetylneuraminic acid have also been found in other invertebrates, including the arachnoid Limulus polyphemus (Cohen, 1979) (horseshoe crab) and the tunicate Halocynthia pyriformis (Vasta et al., 1982b). The lamprey Petromyzon marinus, a primitive vertebrate, also possesses agglutinins specific for sialic acid in its serum (Marchalonis and Edelman, 1968). What may prove to be an interesting family of lectin-like molecules in the vertebrates is the C-reactive proteins. These molecules are specific for phosphorylcholine and represent a group of acute-phase proteins which in humans occurs at an increased level in serum and pleural fluids during pneumococcal infections or after a variety of traumas (MacLeod and Avery, 1941). Although the recognition specificity of products of the major histocompatibility complex, including allohistocompatibility antigens and immune response gene antigens (Ia), is not known, it has been proposed that these molecules are integral elements in T-cell recognition and activation (Benacerraf, 1978). β_2-Microglobulin is a small molecule about the size of a free immunoglobulin domain (molecular weight, 12,000) which is found on virtually all cells, in which sites it occurs in association with the histocompatibility antigen heavy chain (molecular weight, 45,000). Lectins have also been found on the surface of hepatocytes of vertebrate species (Sarkar et al., 1979) where they usually react with asialo glycoproteins. Lectins have also been recently found to be associated with the external membrane of T and B lymphocytes, and these express a complex specificity for oligosaccharide-bearing determinants (Decker, 1980).

The following discussion will assess various approaches used to determine whether or not these recognition molecules constitute a superfamily of recognition molecules which might be related to the immunoglobulin recognition system. Immunoglobulins are already a "superfamily" in the sense that a number of different types of immunoglobulins expressing a capacity to recognize an enormously diverse collection of antigens have arisen by apparent duplication of an ancestral polynucleotide of approximately 300 bases which encoded the precursor of both variable and constant region domains. Immunoglobulin evolution has apparently involved large numbers of tandem duplications creating distinct immunoglobulin heavy chain isotypes, various κ- and γ-chain constant regions, and an enormous diversity of variable region structures. Throughout all the duplications, the packaging information in units of about that size (100 residues) was maintained, and these are still reflected in the domain structure which is apparently a constant feature of all immunoglobulins (Marchalonis, 1977a). In a system where gene duplication and deletions are occurring, it would not be surprising to find other classes of molecules derived from the immunoglobulin family. The most notable of these is β_2-microglobulin, which is clearly homologous to a constant region domain of γ chain (Peterson *et al.*, 1972). Other molecules such as α_1 acid glycoproteins homologous to immunoglobulin domains have also been observed (Emura *et al.*, 1971).

III. THE IMMUNOGLOBULIN EXTENDED FAMILY AND THE EVOLUTION OF RECOGNITION

It is virtually a platitude to say that all cells are capable of some sort of specific recognition of other cells or molecules in their milieu, although our understanding of these multiple events at a molecular level is woefully inadequate. In many cases this recognition can be described as lectin-like because receptor molecules that are present show specificity for sugar moieties (Nowak *et al.*, 1977). Suggestive evidence has arisen from a number of sources that cell-surface molecules such as histocompatibility antigens (Strominger *et al.*, 1980; Marchalonis *et al.*, 1984) and general differentiation markers such as the Thy-1 alloantigen of T cells (Campbell *et al.*, 1979) share a not-too-distant common ancestor with immunoglobulins; moreover, it is an inevitable consequence of monophyletic theories of evolution that all serum and cell-surface proteins must have shared evolutionary history (Marchalonis, 1977a). Data supporting the postulated relationship of cell surface molecules such as histocompatibility antigens and Thy-1 to immunoglobulins comes from limited amino acid sequence analysis, and some degree of homology has been found (Strominger *et al.*, 1980; Marchalonis *et al.*, 1984). Statistical analysis of complete sequences of immunoglobulins, major histocompatibility complex (MHC) products, and Thy-1 establishes that immunoglobulins and MHC products share highly homologous domains (Ig-like domains of class I and II products), but the evidence for Thy-1 is marginal (Marchalonis *et al.*, 1984).

In order to test the validity of wider predictions about relatedness of all recognition molecules it is necessary to be able to compare, for example, the structures of lectins of invertebrate species [with particular reference to the protochordates which are related to the stream of evolution in which the immunoglobulin immune system arose (Berrill, 1955)] and immunoglobulins of lower vertebrate species which might be more closely related to the ancestral immunoglobulin types. It is to be expected that striking homologies, if they are apparent at all, would be observed between the lectins, putative cell receptors, and primitive immunoglobulin chains. At this point in time there exists very little in the way of amino acid sequence analyses of lectins and lower vertebrate immunoglobulins. A wealth of circumstantial evidence with respect to lower vertebrate immunoglobulins supports the conclusion that immunoglobulin evolution has shown a dominant conservatism (Marchalonis, 1977a; Litman *et al.*, 1971) in the sense that the existence of variable and constant regions, the presence of light and heavy polypeptide chains, the presence of domains, and the organization of molecules in terms of light/heavy pairs forming the combining site are major features of immunoglobulin structure that were shared by the most primitive antibody molecules. Such results further imply that the genetic mechanism for the generation of diversity which gave rise to heterogeneous V regions emerged very early in vertebrate or chordate evolution (Marchalonis, 1977a). Amino acid sequence data are lacking for many of the lectins, primitive immunoglobulins, and other potential candidates for membership in the putative immunoglobulin-related super-

TABLE II

Calculation of $S \ \Delta Q$ for Composition between Homologous μ Heavy Chains of Human and Stingray[a]

Amino acid	Composition (mole %)[b]		Δ	Δ^2
	Man	Stingray		
Asx	8.7	9.4	−0.7	0.49
Thr	9.5	9.4	0.1	0.01
Ser	9.7	11.5	−1.8	3.24
Glx	10.2	12.1	−1.9	3.61
Pro	7.0	6.6	0.4	0.16
Gly	7.2	6.4	0.8	0.64
Ala	7.0	6.0	1.0	1.00
Val	9.5	6.7	2.8	7.84
Met	1.2	1.4	−0.2	0.04
Ile	3.2	3.7	−0.5	0.25
Leu	8.0	7.5	0.5	0.25
Tyr	3.0	3.8	−0.8	0.64
Phe	4.0	3.9	0.1	0.01
Lys	5.2	5.3	−0.1	0.01
His	2.1	1.8	0.3	0.09
Arg	4.5	4.2	0.3	0.09
				18.37 $= S \ \Delta Q$

[a]*Daysatis centroura*. Data of Marchalonis (1972).

[b]Calculations do not include carbohydrate, cysteine, or tryptophan. Glx includes Gln plus Glu. Asx includes Asn plus Asp.

family of recognition molecules. However, amino acid composition data are more easily obtained, and a number of statistical analyses based upon quantitative comparison of amino acid compositions have been developed (Metzger *et al.*, 1968; Marchalonis and Weltman, 1971; Cornish-Bowden, 1981). It has been shown that parameters such as the $S \Delta Q$ (Marchalonis and Weltman, 1971) and a derivative the $S \Delta N$ (Cornish-Bowden, 1981) correlate with differences in amino acid sequence, and, moreover, under certain circumstances can be extremely accurate and valuable predictors of homology in amino acid sequence (Cornish-Bowden, 1980, 1981). A simple, practical method for obtaining a quantitative measure of the degree of relatedness among proteins is the application of the $S \Delta Q$ statistic first developed by Marchalonis and Weltmann (1971) and defined as follows:

$$S \Delta Q = \Sigma(X_{ij} - X_{kj})^2$$

where i and k identify the particular proteins compared and X_j is the content (mole %) of a given amino acid of type j.

TABLE III

Calculation of $S \Delta Q$ for Composition between the Unrelated Human Proteins Immunoglobulin μ Chain and Hemoglobin β Chain[a]

| Amino acid | Composition (Mole %) | | Δ | Δ² |
	μ Chain	β Chain		
Asx	8.7	9.1	−0.4	0.16
Thr	9.5	4.7	4.8	23.04
Ser	9.7	3.3	6.4	40.96
Glx	10.2	8.5	1.7	2.89
Pro	7.0	5.1	1.9	3.61
Gly	7.2	9.1	−1.9	3.61
Ala	7.0	10.6	−3.6	12.96
Val	9.5	12.5	−3.0	9.00
Met	1.2	0.8	0.4	0.16
Ile	3.2	0.0	3.2	10.24
Leu	8.0	12.7	−4.7	22.09
Tyr	3.0	2.0	1.0	1.00
Phe	4.0	5.6	−1.6	2.56
Lys	5.2	7.7	−2.5	6.25
His	2.1	6.2	−4.1	16.81
Arg	4.5	2.1	2.4	5.76
				161.10 $= S \Delta Q$

[a]From Marchalonis and Weltman (1971). *Comp. Biochem. Physiol. (B)* **38**, Relatedness among proteins: A new method of estimation and its application to immunoglobulins. Reprinted with permission. Copyright 1971 Pergamon Press, Ltd.

TABLE IV

Comparisons among Recognition Proteins by the S ΔQ Method[a,b]

	Hal HA	Lam HA	Lim HA	Lum CRP	Hu CRP	Rab CRP	Mou CRP	Lim CRP	Mar τ	Mou τ	Hu μ	Lam μ	SR μ	Hu γ	Ech γ	Ra (ARS) γ
Hal-HA	0															
Lam-HA	**24**	0														
Lim-HA	68	**39**	0													
Lum-CRP	**56**	**48**	**41**	0												
Hu-CRP	96	73	62	47	0											
Rab-CRP	**57**	**45**	**40**	**32**	**19**	0										
Mou-CRP	**56**	**43**	**40**	**35**	**17**	**3**	0									
Lim-CRP	86	59	**15**	48	83	66	69	0								
Mar-τ	66	59	70	79	107	90	69	85	0							
Mou-τ	69	55	35	35	61	50	72	44	**39**	0						
Hu-μ	84	77	69	65	30	56	45	98	99	**60**	0					
Lam-μ	**42**	**40**	**43**	34	31	31	51	57	63	**25**	**22**	0				
SR-μ	93	69	52	69	35	56	29	90	79	61	**18**	**41**	0			
Hu-γ	135	131	99	109	41	70	45	137	130	102	27	66	35	0		
Ech-γ	107	101	98	92	27	53	59	132	111	77	23	43	40	**15**	0	
R(ARS)γ	149	140	122	133	56	96	43	169	170	136	23	71	37	**20**	**31**	0
M173 γ	120	118	91	103	40	60	85	132	124	97	32	59	37	**11**	**18**	**26**
Hor γ	133	134	109	115	37	74	51	138	147	106	20	57	41	**8**	**16**	**18**
V_κ	128	105	94	112	58	73	68	140	149	130	61	85	32	68	82	57
V_λ	108	113	122	129	87	97	63	160	175	165	62	94	62	74	95	57
V_H	79	92	98	67	60	64	92	131	149	100	34	50	53	67	72	64
C_κ	134	115	126	138	74	108	66	156	161	144	54	95	48	46	55	56
C_λ	198	174	141	162	86	137	100	173	190	148	49	113	53	32	55	38
κ-MPC 321	104	82	96	111	76	90	125	146	103	121	63	92	27	71	77	73
λ-Hu Bau	138	109	116	125	83	121	74	137	181	135	40	80	57	76	80	49
DF-L	103	96	86	102	57	73	117	135	128	112	32	71	**18**	38	57	37
Lam-L	62	**54**	44	38	54	51	65	61	58	**13**	35	**10**	52	91	66	99
Mou Thy 1.2	92	78	48	53	82	79	47	66	89	56	70	59	62	97	94	125
Ra Thy 1.1	101	87	67	61	102	100	74	85	117	70	77	63	87	124	110	135
β_2M	108	93	68	88	73	72	97	96	77	78	85	73	79	80	66	111
HLA-B7	110	130	135	115	140	144	132	160	100	92	97	83	97	165	147	166
Hu C4	79	76	47	39	54	48	47	70	78	**26**	42	**28**	54	87	79	104
CIP 9	346	343	411	367	357	360	341	353	299	292	404	282	443	467	371	498

(continued)

Table IV (Continued)

	MPC 173 γ	Hor γ	V_κ	V_λ	V_H	C_κ	C_λ	κ MPC 321	λ Hu Bau	DF L	Lam L	Mou Thy 1.2	Rat Thy 1.1	β_2 M	HLA B7	Hu C4	CIP 9
Hal-HA																	
Lam-HA																	
Lim-HA																	
Lum-CRP																	
Hu-CRP																	
Rab-CRP																	
Mou-CRP																	
Lim-CRP																	
Mar-τ																	
Mou τ																	
Hu-μ																	
Lam-μ																	
SR-μ																	
Hu-γ																	
Ech-γ																	
R(ARS)γ																	
M173 γ	0																
Hor γ	17	0															
V_κ	79	75	0														
V_λ	93	74	29	0													
V_H	83	63	54	36	0												
C_κ	54	44	80	76	80	0											
C_λ	59	35	78	77	89	46	0										
κ-MPC 321	60	86	49	73	89	52	86	0									
λ-Hu Bau	88	63	77	58	68	56	39	84	0								
DF-L	40	44	26	33	41	46	51	21	57	0							
Lam-L	83	87	114	135	73	138	138	113	107	93	0						
Mou Thy 1.2	77	109	144	169	105	106	151	88	142	92	55	0					
Ra Thy 1.1	101	131	173	186	113	143	170	116	137	114	51	14	0				
β_2M	55	96	161	193	166	133	169	107	178	122	79	55	79	0			
HLA-B7	184	162	138	155	97	202	215	167	196	140	74	144	161	186	0		
Hu C4	94	90	94	114	54	156	134	122	121	82	15	75	78	108	65	0	
CIP 9	446	436	492	522	468	480	531	505	476	527	308	429	433	410	420	386	0

56

[a] The amino acid compositions of the proteins indicated were compared according to $S \Delta Q$ method of Marchalonis and Weltman (1971). The proteins are as follows: HAL-HA, *Halocynthia pyriformis* hemagglutinin, unpublished data of G. R. Vasta and J. J. Marchalonis. Lam-HA, lamprey hemagglutinin, data of Marchalonis and Edelman (1968). Lim-HA, *Limulus polyphemus* hemagglutinin, data of Marchalonis and Edelman (1968). Lum-CRP, lumpfish C-reactive protein; Hu-CRP, human C-reactive protein; Rab-CRP, rabbit C-reactive protein; Mou-CRP, mouse C-reactive protein; Lim-CRP, *Limulus polyphemus* C-reactive protein. Composition data for C-reactive proteins were taken from Fletcher *et al.* (1981). Mar-τ marmoset T-cell receptor chain, unpublished data of J. J. Marchalonis and J. Maxwell. Mou-τ mouse T cell receptor heavy chain, data of Rubin *et al.* (1980). Hu-μ, human μ chain; Lam-μ, lamprey μ chain. References to original data are given in Marchalonis and Weltman (1971). SR-μ, stingray μ chain, data given in Marchalonis (1972). Hu-γ, human γ heavy chain, source given in Marchalonis and Weltman (1971). Ech-γ, echidna (spiny anteater), monotreme mammal (*Tachyglossus aculeatus*), Atwell and Marchalonis (1977). R(ARS) γ, γ chain of affinity purified rabbit antibody to the arsonate hapten (Koshland *et al.*, 1966). M173 γ, γ chain of the murine myeloma protein MOPC 173 (Rubin *et al.*, 1980). Hor γ, horse γ chain [original data cited in Marchalonis and Weltman (1971)]. V_κ, variable region of κ light chains; V_λ, variable region of λ light chains; V_H, variable region of heavy chains; C_κ, constant region of κ light chains, C_λ, constant region of λ light chains. Data for these immunoglobulin fragments was taken from Dayhoff (1976). κ-MPC 321, murine κ myeloma protein MOPC 321 (Dayhoff, 1976). λ-Hu Bau, data for human λ chain Bau (Dayhoff, 1976). DF-L, dogfish (*Mustelus canis*, elasmobranch) light chain; Lam-L, lamprey light chain, data cited in Marchalonis and Weltman (1971). Mou Thy 1.2, mouse Thy 1.2 alloantigen (Acton *et al.*, 1979). Ra Thy 1.1, rat Thy 1.1 alloantigen (Barclay *et al.*, 1976). β₂M, β₂-microglobulin (Appella *et al.*, 1976). HLA-B7, human histocompatibility antigen HLA-B7 (calculated from data tabulated in Orr *et al.*, 1979). Hu C4, the fourth component of human complement (Gigli *et al.*, 1977). CIP9, cecropia immune protein 9A (Hultmark *et al.*, 1980).

[b] Boldfaced numerals illustrate significant comparisons.

Tables II and III illustrate sample calculations of the index for homologous proteins (Table II) such as μ chains of humans and stingray and for proteins which are clearly unrelated in sequence and function (Table III) such as the immunoglobulin μ chain and the hemoglobin β chain of humans. In the first case, μ chains of humans and stingray differ by only 18 $S \Delta Q$ units, a value which indicates a strong degree of homology (Table IV). On the other hand, the μ chain differs from the hemoglobin β chain by 161 units, indicating that the molecules are in all probability unrelated. Empirically, fewer than 2% of 720 comparison pairs of unrelated proteins were shown to produce $S \Delta Q$ values of less than 100 units.

Previous studies based on composition analysis have supported the notion of dominant conservatism in immunoglobulin evolution, indicating that μ chains ranging from lamprey and elasmobranchs to mammalian species show very similar amino acid compositions (Marchalonis, 1972), a finding leading to the conclusion that μ chains have been highly conserved in evolution. The magnitude of differences within the μ chains throughout vertebrate classes is comparable to the dispersity in composition values observed with the highly conserved cytochrome C family. Amino acid sequence analyses of μ chains throughout vertebrate classes is comparable to the dispersity in composition values observed with the highly conserved cytochrome C family. Amino acid sequence analyses of μ chains of three mammalian species—human, dog, and mouse—indicate a high degree of conservation in their constant region sequence of μ chain (Wasserman and Capra, 1978; Kehry *et al.*, 1979; Putnam *et al.*, 1973), with an overall similarity among the three μ chains of greater than 85%. Most phylogenetic comparisons are carried out in a similar way: that is, amino acid sequence analysis is carried out on corresponding proteins of a few mammalian species, and then paleontological or other data are used to estimate an approximate time of divergence between the ancestors of the species under comparison and a rate of mutation is calculated. In addition, a "prototype sequence" is often calculated based upon extrapolation of the mammalian sequence data. Within the immunoglobulin family, no such comparisons have been made on a truly large phylogenetic scale such as a comparison between shark and human or lamprey and human. In order to perform this sort of analysis in an admittedly approximate sense, we have used the $S \Delta Q$ comparison approach to ascertain the likelihood of relatedness among immunoglobulins of invertebrates and protochordates, the family of acute-phase proteins termed C-reactive proteins, immunoglobulins, T-cell receptors, and cell-surface molecules such as histocompatibility antigens, β_2-microglobulin, and the Thy-1 markers. Table IV presents $S \Delta Q$ data obtained in pairwise comparisons among the members of this putative recognition super family. In analyzing this table, it is necessary to point out that it is impossible to assign precise confidence limits for the relatedness of proteins of different length using the $S \Delta Q$ method. However, reasonable confidence values can be calculated when comparisons are between pairs of polypeptides of equal length (Cornish-Bowden, 1980). These are shown in Table V. In view of these values, and the empirical determination of over 700 $S \Delta Q$ values for proteins of known degrees of relatedness, we believe that values of 60 or less are very strong indicators of homology. Some of the more striking results to emerge from this table are as follows:

(1) The hemagglutinins isolated from *Halocynthia,* a tunicate species, the lamprey *Petromyzon marinus,* and the horseshoe crab *Limulus polyphemus* are all specific for sialic acid (Cohen, 1979; Vasta *et al.,* 1982b) and (Table VI) and, by the $S \Delta Q$ index, show strong indications of relatedness. In particular, the lamprey haemagglutinin and the *Halocynthia* hemagglutinin appear to be closely related. This is an important observation because the subunits of the molecules are similar in behavior on polyacrylamide gel eletrophoresis in sodium dodecyl sulfate (i.e., probably have similar molecular weights), have a similar range of specificities (Table VI), and further analysis might disclose the first clear-cut molecular example of direct homology between a molecule of a protochordate species and a true vertebrate. The horseshoe crab sialic acid-specific lectin is related to both, but it is clearly more divergent from the lamprey HA than is the *Halocynthia* HA.

(2) C-Reactive proteins (CRPs), which are specific for phosphorylcholine and are found in the serum of many vertebrates (Fletcher *et al.,* 1981), form a family that is clearly related and reflects expected evolutionary divergence. Human C-reactive protein, for example, is more closely related to rabbit and mouse CRPs

TABLE V

Dependence of $S \Delta Q$ Predictions on Polypeptide Chain Length

Length of polypeptide (residues per mole)	$S \Delta Q$ value[a] for		
	Relatedness	Possible relatedness	Unrelatedness
75	<112	>112	—
100	<84	84–186	>186
200	<42	42–93	>93
300	<28	28–62	>62
400	<21	21–47	>47
500	<17	17–37	>37
600	<14	14–31	>31
700	<12	12–26	>26

[a]These values are calculated from the confidence values of Cornish-Bowden (1980, 1981) for the $S \Delta Q$ index, and refer therefore to comparisons between polypeptides of equal length.

TABLE VI

Hemagglutination Profiles of Invertebrates and Lower Chordate Serum Lectins

Erythrocytes[a]		Limulus polyphemus[b]			Hadrurus arizonensis[c]			Halocynthia pyriformis[d]			Petromyzon marinus[e]		
		U	P	N	U	P	N	U	P	N	U	P	N
Human	A_1	512	1024	0	128	1024	0	0	0	0	0	0	0
	B	512	512	0	256	4096	0	0	0	0	0	0	0
	O	256	1024	0	128	1024	0	0	0	0	0	0	0
Sheep		32	128	0	32	1024	0	128	128	0	0	0	0
Duck		1024	4096	1	64	4096	0	16	64	0	0	0	0
Monkey		128	2048	0	8	1024	0	128	256	0	0	16	0
Bovine		4	32	0	16	512	0	64	128	0	0	512	0
Goat		32	128	8	8	1024	0	32	128	2	512	1024	0
Cat		256	512	128	128	256	0	256	256	0	2048	2048	0
Hamster		32	16	16	16	32	8	128	128	0	1024	2048	0
Horse		256	256	256	128	128	128	128	128	128	4096	4096	1024
Inhibitors[f]													
N-Acetylneuraminic acid (mM)		12.5			12.5			12.5			25.0		
N-Glycolylneuraminic acid (mM)		12.5			12.5			12.5			25.0		
NAN-β-Methylglycoside (mM)		6.2			25.0			—			50.0		
Sialyllactose (mM)		6.2			12.5			—			—		

2-Keto-3-deoxyoctonate (mM)	25.0	25.0	NT	50.0
N-Acetyl-D-glucosamine (mM)	—	—	—	—
N-Acetyl-D-galactosamine (mM)	100.0	—	—	—
D-Galactose (mM)	—	—	—	—
D-Glucose (mM)	—	—	—	—
D-Mannose (mM)	—	—	—	—
L-Fucose (mM)	—	—	—	—
BGF-VI[g] (% w/v)	0.003	0.003	0.25	0.125
Asialo-BGF-VI[b] (% w/v)	0.12	1.00	—	—
Fetuin (% w/v)	0.006	0.003	1.0	0.03
Asialo-fetuin (% w/v)	—	—	—	—
BSM (% w/v)	0.0007	0.001	0.01	1.0
Asialo-BSM[h] (% w/v)	0.12	0.5	—	—
Ovomucoid (% w/v)	—	—	—	—

[a] U, untreated erythrocytes; P, Pronase-treated erythrocytes; N, neuraminidase-treated erythrocytes.

[b] Data from Vasta and Cohen, (1984a).

[c] Data from Vasta and Cohen (1984b).

[d] Data from Vasta et al. (1982b).

[e] Data from Vasta and Marchalonis (1985).

[f] Figures are minimal concentrations of inhibitor required for the inhibition of two agglutination units. Bar—, no inhibition at concentrations up to 100 mM for monosaccharide and oligosaccharides or 1% (w/v) for glycoproteins. NT, not tested. *Limulus* hemagglutination-inhibition was tested with human (O); all other species were tested with horse erythrocytes.

[g] BGF-VI, bovine glycoprotein from plasma fraction VI (orosomucoid).

[h] BSM, bovine submaxillary mucin.

than it is to the corresponding proteins of the lumpfish or the horseshoe crab. The rabbit and mouse proteins are extremely close. In essence, a dendrogram consistent with the expected phylogenetic relationships can be constructed from the comparison data for C-reactive proteins cited here.

(3) Both the agglutinins and the C-reactive proteins are apparently related to one another as shown by a comparison value of approximately 56 between *Halocynthia* agglutinin and C-reactive proteins of the lumpfish, the rabbit, and the mouse.

(4) Both agglutinins and C-reactive proteins show similarity to members of the immunoglobulin family. This is particularly noticeable in comparisons with more primitive immunoglobulin chains such as those of the lamprey, although C-reactive proteins also show strong similarities to heavy chains of mammalian species.

(5) The heavy chains, both μ and γ chains illustrated here, form an extremely tightly knit family of molecules, a finding supported by amino acid sequence analysis.

(6) The T-cell receptors isolated from marmoset and mouse are proteins of approximate mass 70,000 daltons (Marchalonis *et al.*, 1980; Rubin *et al.*, 1980) that share combining-site region determinants with immunoglobulin heavy-chain variable regions but lack classical heavy-chain isotypic determinants. These molecules appear to be closely related to each other from their amino acid composition and also show the strongest similarities to the heavy and light chains of lamprey and stingray. They do not show a marked homology to heavy chains of mammals.

(7) Variable regions of κ, λ, and heavy chains are related to one another.

(8) Murine and rat Thy-1 antigens are extremely similar to one another. In addition, they show some similarities to the τ chain of T-cell receptors, C-reactive proteins, and to μ chains of lamprey and stingray, as well as to the primitive light chain of the dogfish.

(9) β_2-Microglobulin shows some relationship to γ chains and to the mouse Thy-1 alloantigen. HLA-B7 shows no striking homology with any of the other markers, although some weak homology with τ chains and primitive μ chains might be present.

(10) The fourth component of human complement shows relatedness with C-reactive proteins, the murine τ chain, μ chains, and the V_H fragment, as well as striking similarity to the lamprey light chain. It is interesting that the gene specifying C4 in the mouse lies in the MHC between the *I* region and the *D* end of the complex (Vitetta and Capra, 1978).

(11) The inducible bacteriolytic protein (Hultmark *et al.*, 1980) of the giant silk moth (*Hyalophora cecropia*) shows no similarity to any of the recognition proteins tested. Although it shows specificity in antibacterial activity, it apparently represents an unrelated family of recognition molecules.

IV. IMPLICATIONS OF PROTEIN AND GENE SEQUENCE FOR THE IDENTIFICATION OF MEMBERS OF THE EXTENDED IMMUNOGLOBULIN FAMILY

In the past year, considerable information regarding the sequence of putative T-cell receptor genes (Hedrick *et al.*, 1984; Yanagi *et al.*, 1984; Saito *et al.*, 1984) and genes specifying the poly(Ig) receptor (Mostov *et al.*, 1984) have been obtained. These support the concept that T-cell receptors are very similar to immunoglobulins and represent separate translocons from classical κ, λ, and heavy chains. The poly(Ig) receptor can be shown to be part of the immunoglobulin extended family, showing homology to variable regions (Mostov *et al.*, 1984; Williams, 1984; W. C. Barker and J. J. Marchalonis, unpublished). Detailed computer analysis of sequences of these products and MHC products and Thy-1 have allowed the following generalizations.

(1) MHC products possess an internal disulfide bonded loop or immunoglobulin-like domain which shows highly significant relatedness to immunoglobulin constant-region domains (Marchalonis *et al.*, 1984). This is illustrated in Table VII.

(2) Thy-1 is marginally related to immunoglobulin domains on the basis of

TABLE VII

Segment Comparison Scores in SD Units for Comparison of Putative Superfamily Members with Immunoglobulin Domains and β_2-Microglobulin[a]

Domain	Thy 1.1	β_2M	HLA-B7[b] (Ig-like D)	HLA-Drα[b] (Ig-like D)	HLA-DW2.2/ Dr 2.2 β[b] (Ig-like D)	α_1-AG	CRP
$C_\mu 1$	−0.2	2.2	4.1	3.7	6.7[c]	−0.8	0.2
$C_\mu 2$	−0.0	6.3[c]	6.3[c]	9.5[c]	7.9[c]	−1.2	0.7
$C_\mu 3$	−0.3	2.5	3.5	3.2	4.0	−0.1	−0.0
$C_\mu 4$	0.2	3.4	4.9[c]	6.5[c]	5.2[c]	2.0	−1.3
$C_\gamma 1$	1.1	3.4	7.8[c]	3.9	6.9[c]	−1.0	−0.5
$C_\gamma 2$	1.3	2.3	0.9	2.0	3.8	2.1	1.5
$C_\gamma 3$	1.6	4.8[c]	10.3[c]	5.0[c]	9.7[c]	1.4	0.2
β_2M	2.1	> 40.0	5.6[c]	10.8[c]	9.8[c]	−1.0	−0.2
V_H(EU)	1.0	0.3	2.7	0.5	0.2	1.0	0.2
V_H(OU)	1.8	0.9	0.4	0.3	3.8	1.0	−1.1

[a]Data of Marchalonis *et al.* (1984).

[b]Ig-like D, immunoglobulin-like domain. HLA-B7, residues 182–271; HLA-Drα, residues 110–213; HLA-Drβ residues 100–198.

[c]Value sufficiently large to conclude that the probability of its occurrence by chance is less than 1 in 10^6. Comparison was made using the program RELATE and a sequence length of 20.

quantitative statistical comparison of amino acid sequences (Marchalonis *et al.*, 1984; Marchalonis and Barker, 1984), although size, arrangement of disulfide bonds, and conformation provide ancillary indications of homology with an immunoglobulin domain (Williams, 1984). The marginal nature of the homology is indicated in Table VII.

(3) C-Reactive protein, which has a similar amino acid composition to immunoglobulins, shows no significant homology to immunoglobulins when compared on a domain basis (Table VII). It is interesting that C-reactive protein does share a functional idiotype with the myeloma protein TEPC-15 and *Limulus* hemagglutinin, both of which bind phosphorylcholine (Marchalonis *et al.*, 1984; Vasta *et al.*, 1984). Detailed comparisons of amino acid sequence among the three molecules show a short stretch of shared sequence corresponding to less than 10 residues which, on TEPC-15, maps to the juncture between the second complementarity-determining and third framework regions. This sharing of functional idiotope suggests either a convergence based upon the requirements for steric similarity in the combining site of molecules of diverse evolutionary histories but common ligand binding specificity or the sharing of minigenes corresponding to a particular site region by molecules of diverse histories.

(4) Great progress has been made in the past year with respect to the identification, isolation, and sequence analysis of genes encoding putative T-cell receptors (Hedrick *et al.*, 1984; Saito *et al.*, 1984; Yanagai *et al.*, 1984) and in the serology and molecular characterization of T-cell products corresponding to T-cell receptors for antigen (Beaman *et al.*, 1984; Mackel-Vandersteenhoven *et al.*, 1984; Marchalonis, 1984). A strong segmental homology in framework regions can be demonstrated between the putative product of at least one of these T-cell receptor genes, YT35 (Marchalonis and Barker, 1984; Marchalonis, 1984), and immunoglobulin variable regions that are known to be serologically cross-reactive with products of relevant T-cell lines (Mackel-Vandersteenhoven *et al.*, 1984). This segmental homology is illustrated in Table VIII. The overall identity between the entire V_T region of the predicted YT35 gene product and the entire V_H regions of McE and MOPC-104E are approximately 30% as determined using the computer program RELATE, but identities within the defined framework segments shown here are considerably better: FR1, 41%; FR2, 43%; FR3, 50%; J, 55% (Marchalonis, 1984). Based upon these homology results, the protein translations of the V regions of the putative T-cell receptor genes fall within the immunoglobulin family, rather than in the more extended superfamily where the degree of homology would be in the range of 10–50%. The high degree of sequence homology is consistent with numerous reports of immunoglobulin-related serological markers found on T cells and T-cell products (Cone, 1981; Marchalonis and Hunt, 1982; Warner, 1974). At this point in time two general classes of candidates for T-cell receptors have been described. The first consists of disulfide-bonded heterodimers consisting of α (molecular

TABLE VIII

Comparisons of Framework Regions of MOLT-3 Putative T-Cell Receptor Gene with Sequences of Cross-Reactive Immunoglobulin V_H

Framework 1

YT35 (V_T)	20D	A	G	V	I	Q	S	P	R	H	E	V	T	E	M	G	Q	E	V	T	L	R	C	K	P	I	S	G	H	N	S	L	F
McE	1Q	I	T	L	K	E	S	G	P	T	L	V	K	P	T	E	T	L	T	C	T	F	S	G	F	S	L	S
TEPC-15	1E	V	K	L	V	E	S	G	G	G	L	V	K	P	G	G	S	L	R	L	S	C	A	T	S	G	F	T	F	S
MOPC-104E	1E	V	Q	L	Q	Q	S	G	P	E	L	V	K	P	G	A	S	V	K	M	S	C	K	A	S	G	Y	T	F
MOPC-315	1D	V	Q	L	Q	E	S	G	P	G	L	V	K	P	S	Q	S	L	S	L	T	C	S	V	S	G	Y	S	I	T

Framework 2

YT35 (V_T)	53	W	Y	R	Q	T	M	M	R	G	L	E	L	L	I
McE	36	W	I	R	Q	P	P	G	K	A	L	E	W	L	A
TEPC-15	36	W	V	R	Q	P	P	G	K	G	L	E	W	I	A
MOPC-104E	36	W	V	K	Q	S	H	G	K	S	L	E	W	I	G
MOPC-315	36	W	I	R	Q	F	P	G	N	K	L	E	W	L	G

Framework 3

YT35 (V_T)	82	D	R	F	S	A	K	M	P	N	A	S	F	S	T	L	K	I	Q	P	S	E	P	R	D	S	A	V	Y	F	C	A
McE	66	R	L	T	G	T	K	D	T	S	S	R	N	Q	V	L	T	N	M	D	P	V	D	T	Y	F	C	A	R
TEPC-15	66	R	F	I	V	S	R	D	T	S	Q	S	I	L	Y	L	Q	M	N	A	L	R	A	E	D	T	A	I	Y	Y	C	A	R
MOPC-104E	66	K	A	T	L	T	V	D	K	S	S	S	T	A	Y	M	Q	L	N	S	L	T	S	E	D	S	A	V	Y	Y	C	A	R
MOPC-315	66	R	V	S	I	T	R	D	T	S	E	N	Q	F	F	L	K	L	D	S	V	T	A	T	Y	Y	C	A	G

Framework 4 (J)

CDR3(D)

YT35 (V_T)	121	N	Y	G	Y	T	126	F	G	S	G	T	R	L	T	V	E
McE			G	G	F	D	103	W	G	Q	G	T	L	V	T	V	S	S
TEPC-15		W	Y	F	D	V	103	W	G	A	G	T	T	V	T	V	S	S
MOPC-104E		W	Y	F	D	V	103	W	G	A	G	T	T	V	T	V	S	S
MOPC-315		L	Y	F	D	Y	103	W	G	Q	G	T	T	L	T	V	S

weight, 45,000) and β (molecular weight, 40,000) chains (Kappler *et al.*, 1983; Reinherz *et al.*, 1983). These molecules have been identified on cloned T-cell lines showing MHC-restricted function. The molecules have not been shown to be antigen specific but the T-cell response can be blocked by monoclonal antibody in a clonotype-specific fashion and the β chains show heterogeneity as judged by peptide mapping and isoelectric focusing technology. The second type of T-cell receptor molecule which has been described is antigen-specific in the sense that it will bind antigen directly (Beaman *et al.*, 1984; Marchalonis, 1984; Webb *et al.*, 1983), and this binding is not MHC-restricted. The second class of molecule also shows serological cross-reactions with immunoglobulin-related determinants, notably antigenic structures correlated with heavy-chain variable regions. This type of molecule, like the MHC-restricted heterodimers, shows clonal restriction in cell distribution and in peptide mapping and isoelectric focusing properties of the purified molecules (Mackel-Vandersteenhoven *et al.*, 1984). It is currently the practice to associate the sequences of putative T-cell receptor genes with either α or β chains, but this is not strictly proper because the gene sequences (see Table VIII) show considerable segmental homology to immunoglobulin variable regions, and thus, it is to be expected that some antisera against either idiotype or allotype or other defined V_H portions such as J_H (Mackel-Vandersteenhoven *et al.*, 1984) would react with the products of these genes. Furthermore, the α/β heterodimers do not bind antigen (Haskins *et al.*, 1984), a property which would be expected of antigen receptors and also of molecules resembling immunoglobulin so closely in sequence and presumed three-dimensional structure. As of this time, N-terminal sequence has been obtained for two putative T-cell receptors; a "β chain" of a human T-cell clone

TABLE IX

N-Terminal Sequences of a Derived T-Cell Receptor Product and Two T-Cell Products Compared with Ig-V Sequences

	1	2	3	4	5	6	7	8	9	10
YT35	D	A	G	V	I	Q	S	P	R	H
V_TM-1	D	D	?	V	I	(A)	W	(V)	(S)	H
Ti β			X	V	I	Q	S	P	R	H
V_κ(M384)	D	I	V	M	T	Q	S	P	(S)	P
V_κ(ROY)	D	I	Q	M	T	Q	S	P	(S)	—
V_λ(M104E)	Q	A	V	V	T	Q	E	S	A	—
V_λ(chicken)	A	L	T	Q	P	(A)	S	(V)	(S)	A
V_H (Horn shark)	D	V	V	L	T	Q	P	E	A	E
V_HI (EU)	Q	V	Q	L	V	Q	S	G	A	E
V_HI (M104E)	E	V	Q	L	Q	Q	S	G	P	E
V_HII(OU)	Q	V	T	L	T	E	S	G	P	A

TABLE X

Comparison of Constant-Region Sequences of T-Cell Receptors Showing Homology to
Immunoglobulin C_H1 or C_L

YT35	166	C	L	A	T	G	F	F	P	D	H	V	E	L	S	W	W	V	N	G	K
86T1	164	C	L	A	(R)	G	F	F	P	D	H	V	E	L	S	W	W	V	N	G	K
M104E	142	C	L	A	(R)	D	F	L	P	S	T	I	S	F	T	W	N	Y	Q	N	N
Eu	142	C	L	V	K	D	Y	F	P	E	P	V	T	V	S	W	N	S	G	A	L
M315	142	C	L	I	H	D	Y	F	P	G	T	M	N	V	T	W	G	K	S	G	K
M104E (λ)	134	C	T	I	T	D	F	Y	P	G	V	V	T	V	D	W	K	V	D	G	—
M41 (κ)	134	C	F	L	N	N	F	Y	P	D	I	N	—	V	K	W	K	I	D	G	—

YT35	—	E	V	H	S	G	V	S	T	D	P	Q
86T1	—	E	V	H	S	G	V	S	T	D	P	Q
M104E	T	E	V	I	Q	G	I	R	T	F	P	T
Eu	T	—	—	—	S	G	V	H	T	F	P	A
M315	—	D	—	I	T	—	T	V	N	F	P	P
M104E (λ)	T	P	V	Q	G	M	E	T	T	E	P	S
M41 (κ)	S	E	R	Q	N	G	V	L	—	D	S	K

(Acuto *et al.*, 1984) and a V_H/V_T-related molecule produced by a monoclonal
marmoset T cell leukemia line (Marchalonis, 1984). Table IX shows a comparison of the N-terminal sequences of a derived T-cell receptor product (YT35)
and these two actual T-cell products with variable region sequences of immunoglobulin heavy chains, κ chains and γ chains. It is apparent that the two T-cell
products show similarity to one another and to the product of the YT35 gene. It is
also evident that these three molecules show definite homology to immunoglobulin variable regions; in particular, the sequence QSPR/S shows similarity
to both $V_κ$ and $V_λ$ sequence and QS also appears in heavy chains. The isoleucine
(I) at position 5 is unique to the T-cell products, as is the histidine (H) at position
10. Although this stretch of sequence is extremely limited, it illustrates that the
T-cell receptors constitute a fourth translocon or immunoglobulin genetic cluster
which diverged from the ancestral immunoglobulin independently of light and
heavy chains, becasue similarities to V_H and $V_κ$ and $V_λ$ can be observed. Analyses of gene sequences in the constant region (Table X) also substantiate the
conclusion that the putative T-cell receptor genes and products characterized thus
far comprise new translocons of immunoglobulins

(5) The poly(Ig) receptor appears to be a polymer of structures related to
ancestral variable regions (Mostov *et al.*, 1984) and, although it is not as closely
related to immunoglobulin as products of putative T-cell receptor genes, the
poly(Ig) receptor has five related domains that are distantly related to immunoglobulin variable regions and can be considered a definite member of the
extended immunoglobulin family.

V. CONCLUSIONS

We have considered the putative existence of a general family of evolutionarily related recognition molecules including certain lectins, surface markers, and immunoglobulins (particularly variable regions) and have applied a quantitative statistical measure of divergence with a semiquantitative means of interpretation in order to ascertain the boundaries and possible relationships within this "superfamily" of proteins. Many of the conclusions observed here are supported by data obtained either from serological studies or direct comparison of amino acid sequences. C-Reactive protein has been shown to share an antigenic determinant related to the idiotype of the HOPC-8 antibody molecule which is specific for phosphorylcholine (Volanakis and Kearney, 1981). In addition, there is weak sequence homology between C-reactive proteins and mammalian immunoglobulins (Osmand *et al.*, 1977; Oliveira *et al.*, 1979). The Thy-1 alloantigen has been found to share limited sequence homology with immunoglobulins (Campbell *et al.*, 1979) and to share an antigenic determinant with light chains (Pillemer and Weissman, 1981), but this molecule also shows sequence stretches and antigenic determinants with actin (Dales *et al.*, 1983). β_2-Microglobulin has been found to have a weak sequence homology with mammalian γ chains (Peterson *et al.*, 1972), and the HLA-B7 molecule shows limited homology with a certain portion (around a disulfide bond) of immunoglobulin (Strominger *et al.*, 1980). The homology of the "Ig-like domain" of MHC products and immunoglobulin domains is highly significant statistically (Marchalonis *et al.*, 1984).

The amino acid composition data presented here support and extend these findings. The major point to be stressed here is that comparisons with the primitive members of the true immunoglobulin family provide the strongest guidepost of homology. We would predict that when amino acid sequence data become available on these primitive immunoglobulin chains and T-cell receptors and agglutinins, direct evidence of homologies will be obtained. Of the three variable regions studied, V_H shows the best indication of similarity with primitive immunoglobulins, C-reactive proteins, and lectins. It is possible that antigenic determinants, and correlated sequences, related to V_H structures are shared by lectins, C-reactive proteins, and T-cell receptors either because of the direct evolutionary relatedness among the combining sites of diverse recognition molecules or because of convergent evolution in the sense that all specific receptors for a given molecule such as phosphorylcholine would have to have approximately the same geometry and therefore would have similar antigenic determinants. At this point in time insufficient data are available to test this proposition. In the case of the T-cell receptor, however, it is unlikely that the observed result is merely an example of a convergence, because more than 15 distinct idiotypes have been found on T cells which are shared with the V_H-dominant portion of antibodies directed against the corresponding antigen (Marchalonis and Hunt, 1982), and strong segmental homology to V-region framework regions is apparent (Marchalonis and Barker, 1984; Marchalonis, 1984). It appears unlikely that such a large number of similarities would appear by chance, although it is not unlikely that once or twice molecules directed against similar substrates might have similar antigenic configurations in their combining site regions.

ACKNOWLEDGMENTS

This work was supported in part by grants NIH-RO1 CA42049 and GM 30672 NSFDCB-84 19880 to John J. Marchalonis, NSF-PCM 81-08872 to Gregory W. Warr, and NIH-RO1 CA30151 to Barry E. Ledford. Gerardo R. Vasta is the recipient of a Fogerty International Fellowship of the National Insititutes of Health. We thank Lesley Secker for typing the manuscript.

REFERENCES

Acton, R., Barstad, P., and Zwerner, R. (1979). *Methods Enzymol.* **58**, 211–221.
Acuto, O., Fabbi, M., Smart, J., Poole, C. B., Protentis, J., Royer, H. D., Schlossman, S. F., and Reinherz, E. L. (1984). *Proc. Natl. Acad. Sci. U.S.A.* **81**, 3851–3855.
Appella, E., Tanigaki, M., Natori, T., and Pressman, D. (1976). *Biochem. Biophys. Res. Commun.* **70**, 425–430.
Atwell, J. L., and Marchalonis, J. J. (1977). *J. Immunogenetics* **4**, 73–80.
Barclay, A., Letarte-Muirhead, M., Williams, A. F., and Faulkes, R. A. (1976). *Nature (London)* **263**, 563–567.
Beaman, K. D., Ruddle, N. H., Bothwell, A. L. M., and Cone, R. E. (1984). *Proc. Natl. Acad. Sci. U.S.A.* **81**, 1524–1528.
Benacerraf, B. (1978). *J. Immunol.* **120**, 1809–1812.
Berrill, N. J. (1955). "The Origin of Vertebrates." Oxford University Press, New York.
Binz, H., and Wigzell, H. (1978). *Cont. Top. Immunobiol.* **7**, 113–177.
Brown, R. C., Bass, H., and Coombs, J. P. (1975). *Nature (London)* **254**, 434–435.
Campbell, D. G., Williams, A. F., Bayley, P. M., and Reid, K. B. M. (1979). *Nature (London)* **282**, 341–342.
Cohen, E. (1979). "Biomedical Applications of the Horseshoe Crab (Limulidae)." A. R. Liss, New York.
Cone, R. E. (1981). *Prog. Allergy* **29**, 182–221.
Cooper, A. J. (1971). *In* "Proc. Fourth Annual Leucocyte Culture Conf." (O. R. McIntyre, ed.), pp. 137–147. Appleton-Century-Crofts, New York.
Cooper, E. L. (1976). "Comparative Immunology." Prentice Hall, Englewood Cliffs, N.J.
Cornish-Bowden, A. (1980). *Biochem. J.* **191**, 349–354.
Cornish-Bowden, A. (1981). *Trends Biochem. Sci.* **6**, 217–219.
Dales, S., Fujinami, R. S., and Oldstone, M. B. A. (1983). *J. Immunol.* **131**, 1332–1338.
Dayhoff, M. O. (1976). "Atlas of Protein Structure and Sequence," Vol. 5, Suppl. 2. National Biomedical Research Foundation, Washington, D.C.

Decker, J. M.(1980). *Mol. Immunol.* **17**, 803–808.

Emura, J., Ikenaka, T., Collins, J. H., and Schmid, K. (1971). *J. Biol. Chem.* **246**, 7821–7823.

Fletcher, T. C., White, A., Youngson, A., Pusztai, A., and Baldo, B. A. (1981). *Biochim. Biophys. Acta* **671**, 44–49.

Frazier, W. A., Rosen, S. D., Reitherman, R. W., and Barondes, S. A. (1975). *J. Biol. Chem.* **250**, 7714–7721.

Gigli, I., von Zabern, I., and Porter, R. R. (1977). *Biochem. J.* **165**, 439–446.

Gold, E. R., and Balding, P. (1975). "Receptor-Specific Proteins." Excerpta Medica, Amsterdam.

Haskins, H., Kappler, J., and Marrack, P. (1984). *Ann. Rev. Immunol.* **2**, 51–66.

Hedrick, S. M., Nielsen, E. A., Kavaler, J., Cohen, D. I., and Davis, M. M. (1984). *Nature (London)* **308**, 153–158.

Hildemann, W. H. (1974). *Nature (London)* **250**, 116–120.

Hultmark, D., Steiner, H., Rasmuson, T., and Boman, G. (1980). *Eur. J. Biochem.* **106**, 7–16.

Jenkin, C. R. (1976). *In* "Comparative Immunology" (J. J. Marchalonis, ed.), pp. 80–97. Blackwell, Oxford.

Kappler, J., Kubo, R., Haskins, K., Hannum, C., Marrack, P., Pigeon, M., McIntyre, B., Allison, J., and Trowbridge, I. (1983). *Cell* **35**, 295–302.

Kehry, M., Sibley, C., Fuhrman, J., Schilling, J., and Hood, L. E. (1979). *Proc. Natl. Acad. Sci. U.S.A.* **76**, 2932–2936.

Kehry, M., Ewald, S., Douglas, R., Sibley, C., Raschke, W., Fambrough, D., and Hood, L. (1980). *Cell* **21**, 393–406.

Koshland, M. E., Englberger, F. M., and Shapanka, R. (1966). *Biochemistry* **5**, 641–651.

Lafferty, K. J. (1973). *In* "The Biochemistry of Gene Expression in Higher Organisms" (J. K. Pollak and J. W. Lee, eds.), pp. 593–605. Australia and New Zealand Book Company, Sidney.

Litman, G. W., and Marchalonis, J. J. (1982). *In* "Biological Significance of Immune Regulation" (E. Gershwin and L. N. Ruben, eds.), pp. 29–60. Marcel Dekker, New York.

Litman, G. W., Rosenberg, A., Frommel, D., Pollara, B., Finstad, J., and Good, R. A. (1971). *Int. Arch. Allergy Appl. Immunol.* **40**, 551–575.

Mackel-Vandersteenhoven, A., Moseley, J. M., and Marchalonis, J. J. (1984). *Cell. Immunol.* **88**, 147–161.

MacLeod, C. M., and Avery, O. T. (1941). *J. Exp. Med.* **73**, 191–200.

Marchalonis, J. J. (1972). *Nature [New Biol.]* **236**, 84–86.

Marchalonis, J. J. (1977a). "Immunity in Evolution." Harvard University Press, Cambridge.

Marchalonis, J. J. (1977b). *Cont. Top. Mol. Immunol.* **5**, 125–160.

Marchalonis, J. J. (1980). *Cont. Top. Immunobiol.* **9**, 255–288.

Marchalonis, J. J. (1985). *Scand. J. Immunol.* **21**, 99–107.

Marchalonis, J. J., and Barker, W. C. (1984). *Immunol. Today* **5**, 222–223.

Marchalonis, J. J., and Cone, R. E. (1973). *Transplant. Rev.* **14**, 3–49.

Marchalonis, J. J., and Edelman, G. M. (1968a). *J. Exp. Med.* **127**, 891–914.

Marchalonis, J. J., and Edelman, G. M. (1968b). *J. Mol. Biol.* **32**, 453–465.

Marchalonis, J. J., and Hunt, J. C. (1982). *Proc. Exp. Biol. Med.* **171**, 127–145.

Marchalonis, J. J., and Warr, G. W. (1978). *J. Dev. Comp. Immunol.* **2**, 443–460.

Marchalonis, J. J., and Weltman, J. K. (1971). *Comp. Biochem. Physiol. [B]* **38**, 609–625.

Marchalonis, J. J., Warr, G. W., Rodwell, J. D., and Karush, F. (1980). *Proc. Natl. Acad. Sci. U.S.A.* **77**, 3625–3629.

Marchalonis, J. J., Vasta, G. R., Warr, G. W., and Barker, W. C. (1984). *Immunol. Today* **5**, 133–142.

Metzger, H., Shapiro, M. B., Mosimann, J. E., and Vinton, J. E. (1968). *Nature (London)* **219**, 1166–1168.

Mostov, K. E., Friedlander, M., and Blobel, G. (1984). *Nature (London)* **308**, 37–43.

Nowak, T. P., Kobiler, D., Roel, L. E., and Barondes, S. H. (1977). *J. Biol. Chem.* **252**, 6026–6030.

Oliveira, E. B., Gotschlich, E. C., and Liu, T. Y. (1979). *J. Biol. Chem.* **254**, 489–502.

Orr, H. T., Lopez De Castro, J. A., Parham, P., Ploegh, H. L., and Strominger, J. L. (1979). *Proc. Natl. Acad. Sci. U.S.A.* **76**, 4395–4399.

Osmand, A. P., Gewurz, H., and Friedenson, B. (1977). *Proc. Natl. Acad. Sci. U.S.A.* **74**, 1214–1218.

Peterson, P. A., Cunningham, B. A., Berggard, I., and Edelman, G. M. (1972). *Proc. Natl. Acad. Sci. U.S.A.* **69**, 1697–1701.

Pillemer, E., and Weissman, I. L. (1981). *J. Exp. Med.* **153**, 1068–1079.

Putnam, F. W., Florent, G., Paul, C., Schinoda, T., and Shimizu, A. (1973). *Science* **182**, 287–291.

Reinherz, E. L., Meuer, S. C., and Schlossman, S. F. (1983). *Immunol. Today* **4**, 5–8.

Rubin, B., Suzan, M., Kahn-Perles, B., Boyer, C., Schiff, C., and Bourgois, A. (1980). *Bull. Inst. Pasteur* **78**, 305–346.

Saito, H., Kranz, D. M., Takagaki, Y., Hayday, A. C., Eisen, H. N., and Tonegawa, S. (1984). *Nature (London)* **309**, 757–762.

Sarkar, M., Liao, J., Kabat, E. A., Tanabe, T., and Ashwell, G. (1979). *J. Biol. Chem.* **254**, 3170–3174.

Strominger, J. L., Orr, H. T., Parham, P., Ploegh, H. L., Mann, D. L., Bilofsky, H., Saroff, H. A., Wu, T. T., and Kabat, E. A. (1980). *Scand. J. Immunol.* **11**, 573–592.

Vasta, G. R., and Cohen, E. (1984a). Sialic acid-binding lectins in the "whip scorpion" (Mastigoproctus giganteus) serum. *J. Invertebr. Pathol.* **43**, 333–342.

Vasta, G. R., and Cohen, E. (1984b). *Comp. Biochem. Physiol.* **77**, 721–727.

Vasta, G. R., Sullivan, J. T., Cheng, T. C., Marchalonis, J. J., and Warr, G. W. (1982a). *J. Invert. Pathol.* **40**, 367–377.

Vasta, G. R., Warr, G. W., and Marchalonis, J. J. (1982b) Tunicate lectins: Distribution and specificity. *Comp. Biochem. Physiol.* **73B**: 887–900.

Vasta, G. R., Marchalonis, J. J., and Kohler, H. (1984). *J. Exp. Med.* **159**, 1270–1276.

Vasta, G. R., and Marchalonis, J. J. (1985). Serum and egg lectins from the sea lamprey *Petromyzon marinus,* submitted.

Vitetta, E. S., and Capra, J. D. (1978). *Adv. Immunol.* **26**, 147–193.

Vitetta, E. S., and Uhr, J. W. (1973). *Transplant. Rev.* **14**, 50–75.

Vitetta, E. S., and Uhr, J. W. (1975). *Science* **189**, 964–969.

Volanakis, J. E.,and Kearney, J. F. (1981). *J. Exp. Med.* **153**, 1604–1614.

Warner, N. L. (1974). *Adv. Immunol.* **19**, 67–216.

Warr, G. W. (1981). *J. Invert. Pathol.* **38**, 311–314.

Warr, G. W., and Marchalonis, J. J. (1978). *Q. Rev. Biol.* **53**, 225–241.

Wasserman, R. L., and Capra, J. D. (1978). *Science* **200**, 1159–1161.

Webb, D. R., Kapp, J. A., and Pierce, C. W. (1983). *Ann. Rev. Immunol.* **1**, 423–438.

Williams, A. F. (1984). *Immunol. Today* **5**, 219–221.

Yanagi, Y., Yoshikai, Y., Leggett, K., Clark, S. P., Aleksander, I., and Mak, T. W. (1984). *Nature (London)* **308**, 145–149.

3

Self–Non-self Discrimination and Cell-Surface Carbohydrate Receptors in the Immune System

REGINALD M. GORCZYNSKI

RECEPTORS IN CELLULAR RECOGNITION
AND DEVELOPMENTAL PROCESSES

I. INTRODUCTION

Cell-surface carbohydrates are believed to mediate recognition and inter-cellular communication in a variety of biological species and for an array of phenomena. Thus, lectins have been implicated in the cohesiveness which characterizes the developmental stage of the slime mold *Dictyostelium discoideum* when deprived of its food source (Rosen *et al.*, 1973); in the mating systems of plants (Goldstein *et al.*, 1980); and, in higher organisms, in retinotectal recognition in the developing central nervous system of the chick (Rutishauser *et al.*, 1976).

Much of the contemporary immunology is devoted to a study of the biochemistry of lymphocyte receptor (recognition) molecules and an analysis of structure–function relationship in such antigen-specific molecules. There is some evidence that carbohydrate structures may be of particular relevance. Thus, specialized cells of the vertebrate immune system recognize and communicate with each other using cell-surface glycoproteins encoded by the major histocompatibility complex (MHC). Monosaccharides have been reported to inhibit a variety of functions of cells in the immune system (Gorczynski *et al.*, 1984a; Koszinowski and Kramer, 1981; Muchmore *et al.*, 1980; Pimlott and Miller, 1984).

There is evidence from studies both of the phylogeny and the ontogeny of immune recognition to suggest that the phagocytic macrophage (or its probable invertebrate equivalent, e.g., hemocytes) represents the earliest developing cell capable of self–non-self discrimination (see for instance, Solomon, Chapter 1, this volume). Initially the focus in what follows will thus primarily be concerned with the evidence that macrophage antigen discrimination is a function of cell-surface moieties with carbohydrate specificity. Thereafter, evidence of functional cell surface recognition receptors on other cells of the immune spleen will be considered.

We are probably more familiar with the extensive information content of codes built upon nucleic acids and amino acids. It should nevertheless be realized that given the variety of spatial configurations open to carbohydrates (i.e., their linkage in α- or β-forms; by O- or N-glycosidic bonds; in linear or branched arrays) and the number of commonly used subunits to build such glycosidic chains (glucose, galactose, mannose, fucose, arabinose, xylose, sialic acid, *N*-acetylglucosamine, *N*-acetylgalactosamine), the potential informational content available to a repertoire built only of cell-surface carbohydrates is enormous.

Indeed, accessible surface moieties of both eukaryotes and prokaryotes are characteristically carbohydrate (often in association with protein or lipid).

II. EVIDENCE FOR LECTINS AS RECEPTOR STRUCTURES ON THE SURFACE OF MAMMALIAN PHAGOCYTIC MACROPHAGES AND HEPATOCYTES

A. Methods of Analysis

Biochemical analysis of cell-surface receptors for ligands has developed along fairly conventional lines, using subcellular fractionation and physicochemical separation procedures, assaying for material which binds to radiolabeled ligand or for which such specific binding is inhibited by known soluble hapten (sugar) competitors.

Analysis of carbohydrate specific receptors on isolated cells has preceded essentially by one of two methods: (1) by direct visualization of the binding of the carbohydrate antigen bearing structures (e.g., bacteria) and its inhibition by soluble sugar; and (2) by direct analysis of the binding with subsequent radioimmunoassay or bioassay of the bound material as assessed by the induction of specific immune responses to the bound antigen. For a recent summary of the former approach, the reader is referred to a review by Weir (1980).

A galactose-specific lectin on the surface of hepatocytes was described by Morell et al. (1968). Similar lectins were subsequently shown on spleen and peritoneal macrophages with a capacity to bind sialidase-treated erythrocytes (with exposed galactose residues). This binding reaction may have functional relevance in the observed rapid clearance of sialidase-treated erythrocytes from the blood stream (Kolb-Bachofen et al., 1982). While the asialoglycoprotein receptor on hepatocytes is the most widely studied receptor system for recognition and binding of specific nonreducing carbohydrate termini, at least three others are known to be operative in the liver. These are a mannose/N-acetylglucosamine receptor; a fucose receptor; and a phosphomannosyl receptor (Ashwell and Harford, 1982).

Other studies have documented lectins specific for mannose and N-acetylglucosamine on macrophages (Sharon, 1984). Sung et al. (1983) were able to inhibit binding and phagocytosis of Saccharomyces cerevisiae zymosan by mannan prepared from Klebsiella at concentrations (50% inhibition at 10^{-2} mg/ml), having no effect on phagocytosis of IgG-coated erythrocytes. Preincubation of macrophages to coverslips bearing cross-linked mannan also specifically blocked subsequent ingestion of zymosan but not IgG-coated erythrocytes. These data suggest independent receptors on macrophages for zymosan- and opsonin-coated particles. Similar findings have been reported for mannose/N-acetylglucosamine specific lectin by Stahl *et al.* (1978).

The work in this laboratory has centered on the ability to dissect mammalian macrophage heterogeneity. The method used to study this heterogeneity has been to assess the ability of antigen pulsed cells to ''present'' antigen in an immunogenic form to a murine B-lymphocyte pool such that B lymphocytes are triggered to differentiate to antibody-producing cells (Gorczynski *et al.*, 1979). These are in turn enumerated in a standard hemolytic plaque assay 4 days later. In principle, analysis of receptor binding by these two methods can be depicted as shown in Fig. 1.

Studies by Friemer *et al.* (1978) on the binding of bacteria to peritoneal exudate cell monolayers on flying coverslips and the inhibition of the same by soluble monosaccharides (at concentrations of the order of 10 mM) argue in favor of glucose and galactose specificities being recognized quite regularly on different bacterial organisms. Similar findings were reported by Weir *et al.* (1979), using mouse fibrosarcoma or embryo fibroblast cells as alternative targets for analysis.

We have used a series of glucose-derived carbohydrates (Dextran B_{1355}, Dextran B_{1299}, Dextran B_{512}) which differ in the degree of branching and the type of sugar linkage ($\alpha1\rightarrow6$, $\alpha1\rightarrow3$, etc.) as well as Levan (a fructose polymer) and Ficoll (a synthetic sucrose polymer). We reported on the isolation of independent subsets of macrophages within individuals of inbred strains of mice which could apparently discriminate between the distinct specificities of these carbohydrates. Our initial studies were interpreted in terms of a minimum of three subpopulations of cells showing preference for an $\alpha1\rightarrow6$ glucosidic linkage, $\alpha1\rightarrow3$ glucosidic linkage, or having a broad specificity (Gorczynski *et al.*, 1979). By

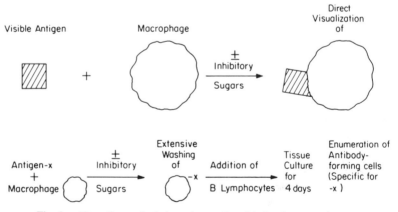

Fig. 1. Alternative methods to analyze antigen binding by macrophages.

investigating the binding specificities of mouse macrophages grown from bone marrow cells as colonies in methylcellulose, we have shown that this heterogeneity is a reproducible function possibly associated with different stages in monocyte differentiation (Gorczynski et al., 1980).

B. Biochemistry of Receptor

The biochemistry of carbohydrate specific receptors is probably most well understood in the liver. This topic has been reviewed in detail elsewhere (Ashwell and Harford, 1982) and is summarized below. The hepatic receptor for asialoglycoproteins is probably the best characterized at the physicochemical and/or biochemical level. Its existence was first inferred from the reduced survival times of desialylated serum glycoproteins (e.g., ceruloplasmin), which in turn could be prolonged by treatment of asialoceruloplasmin with β-galactosidase (see review by Ashwell and Morell, 1974). Analysis of amino acid and carbohydrate composition of affinity-purified receptor from a variety of species suggests similarity between the molecules prepared from rat, human, and rabbit liver (Kawasaki and Ashwell, 1976; Tanabe et al., 1979; Baenziger and Maynard, 1980). A characteristic protein with subunits varying in molecular weight in different species from 40,000 to 50,000 has been described in all three cases, with evidence for association of those proteins into the functional receptor in the rat [105,000 kD for receptor based on radiation inactivation (Steer et al., 1981)] and rabbit [260,000 kD (Kawasaki and Ashwell, 1976)].

Using the binding protein immobilized on Sepharose, specificity for galactose (or more especially N-acetylgalactosamine) has been described using hapten inhibition of binding of [^{125}I]asialoorosomucoid, or using as probes proteins to which carbohydrates have been chemically attached (neoglycoproteins) (Krantz et al., 1976). Interestingly, adhesion of hepatocytes to ligand-bearing substratum leads to patching of the surface receptor, indicating lateral mobility of the receptor in the plasma membrane (Weigel, 1980). Moreover, using antibody to soluble receptor, it has been documented that the molecule spans the lipid bilayer of the plasma membrane (Harford and Ashwell, 1981). Since clathrin, the major coated pit protein, is on the internal side of the plasma membrane, (Pearse, 1976) and ligand receptor interaction occurs in coated pits (Kolb-Bachofen, 1981) (which in turn have been implicated in endocytosis of ligands in a number of cell types), it seems likely that the portion of the receptor leading to endocytosis is that located on the cytoplasmic side of the plasma membrane. The intracellular conditions for, and site of, ligand receptor dissociation are still to be unequivocally shown. Both microtubules and microfilaments are involved in intracellular movement of asialoglycoproteins (Kolset et al., 1979). Moreover, the inhibition of ligand degradation by antilysosomal agents (ammonium chloride, chloroquine) imply that lysosomes are important in asialoglycoprotein metabolism (Berg and Tolleshaug, 1980).

Receptor reutilization has been described in culture rat hepatocytes (Tolleshaug and Berg, 1979; Regoeczi *et al.*, 1978). The mechanism whereby sparing of receptor and destruction of ligand occurs is not clear. The receptor may reorient in the lysosomal membrane to avoid proteolytic digestion (Tanabe *et al.*, 1979), or alternatively the ligand receptor complex may dissociate before it reaches the lysosome (Wall *et al.*, 1980).

Stockert *et al.* (1976) examined the fate of desialylated orosomucoid in rats and found that removal of galactose (to expose terminal N-acetylglucosamine) was accompanied by hepatic uptake which was not inhibited by asialoorosomucoid. A separate recognition system specific for N-acetylglucosamine was inferred. Achord *et al.* (1977) then described the inhibition of uptake of agalactoorosomucoid by infused mannans, implying that the presumptive receptor could recognize both mannose and N-acetylglucosamine termini. Subsequent isolation and purification of this receptor from rat liver revealed it to be a distinct entity from the galactose-specific asialoglycoprotein receptor already discussed. The purified material has a molecular weight on sodium dodecyl sulfate (SDS) gel electrophoresis of 31,000 with specificity for N-acetylglucosamine, mannose, and N-acetylmannoseamine (Kawasaki *et al.*, 1978). Mannan oligosaccharides linked $\alpha 1 \rightarrow 6$ were superior inhibitors to those linked $\alpha 1 \rightarrow 2$ or $\alpha 1 \rightarrow 4$. While initially it had been thought that only liver Kupffer cells and not hepatocytes contained this receptor, there are now data which contradict this hypothesis. Proof that hepatocytes contain this as a cell surface rather than intracellular receptor is still lacking, however.

Lactoferrin (from human milk) is rapidly cleared from the circulation in mice, while transferrin or asialotransferrin is not. The former contains fucous residues linked $\alpha 1 \rightarrow 3$ to the N-acetylglucosamine adjacent to galactose (terminal sugars of transferrin are sialic acid, galactose, and N-acetylglucosamine) (Prieels *et al.*, 1978). Periodate oxidation of lactoferrin or treatment with glycosidases delayed clearance. Fucosylation ($\alpha 1 \rightarrow 3$ link) of asialotransferrin led to its rapid clearance. Mannan and derivatives of orosomucoid (which would block binding to receptors for galactose, N-acetylglucosamine, or mannose) had no inhibitory effect on clearance of fucosylated asialotransferrin. Subsequent analysis of homegenates of rat liver revealed a distinct molecular entity (67,000–73,000) with fucose-binding specificity. Antibody to this molecule did not abolish activity to the mannose N-acetylglucosamine receptor.

Finally, extending work initially begun with human fibroblasts in culture which suggested evidence for a cell-surface structure that could recognize a carbohydrate marker on different forms of lysosomal glycosidase for which mannose 6-phosphate was a structural analogue, a similar discrete mannose 6-phosphate-specific receptor has also recently been described in bovine liver (Sahagian *et al.*, 1981). A subunit molecular weight of 215,000 has been reported for this receptor.

We have explored in a preliminary fashion the biochemical nature of the receptor structures on macrophages which discriminate between different immunogens by using enzyme digestion procedures. All enzyme digestion was for 30 min at 37°C, enzymes being used at the following concentrations: trypsin, 5 mg/ml; pronase, 25 mg/ml; phospholipase-C, 50 μg/ml. In all cases the ability to bind Dextran B_{1355} and Dextran B_{1299} in an immunogenic form was lost immediately after enzyme treatment, but recovered after incubation for 4 hr at 37°. Antigen binding in all cases was dependent upon the presence of calcium but not magnesium in the medium (Table I). Similar data from studies of the binding of *Corynebacterium parvum* to mouse macrophages have been reported by Weir (1980). Here a loss of binding was observed after treatment of macrophages with periodate, but binding was restored after borohydrate reduction of the oxidized terminal hydroxy groups of the sugar residue on the macrophage cell surface. Both sets of data are consistent with the notion that hydroxyl groups on terminal sugars of the macrophage receptor attach to their target in a noncovalent manner involving a Ca^{2+} bridge, as suggested by Weir (1980).

C. Universality of Lectin-Like Receptors on Phagocytic Cells

The mammalian hepatic receptors specific for different carbohydrate groups were already discussed. An early observation was made that α acid glycoproteins isolated from avian species were rapidly cleared from the circulation in rabbits (Regoeczi *et al.*, 1975). Such a feature suggests undersialylation of avian glycoproteins and the absence of a hepatic receptor for asialoglycoproteins. Both predictions have been upheld (Lumney and Ashwell, 1976), though an avian liver receptor (subunit size 26,000) with specificity for mannose and *N*-acetylglucosamine has now been described (Kuhlenschmidt and Lee, 1980). Other lectin receptors with different specificities have been reported from the avian/reptilian liver, the expression of which may be developmentally regulated (Barondes, 1981). The significance of this latter observation is unclear.

In order to examine the universality of the ability of macrophages and macrophage-like cells to discriminate between distinct carbohydrate specificities, we have compared the inhibition of binding (for subsequent immune induction) of saccharide antigens caused by addition of an excess of soluble sugars using different sources of antigen-binding cells. For this experiment the two sources of antigen binding cells used were murine peritoneal exudate cells and co-

TABLE I

Nature of Recognition Site on Macrophage Cell Membrane and Dependence of Antigen Binding on Divalent Ions

Enzyme treatment	Incubation after enzyme treatment (4 hr/37°C)	Antigen used for pulse[a]	Antibody-forming cells (day 4)[b]
None	−	$TNP\text{-}Dex_{1355}$	310 ± 36
		$TNP\text{-}Dex_{1299}$	258 ± 41
	+	$TNP\text{-}Dex_{1355}$	329 ± 39
		$TNP\text{-}Dex_{1299}$	306 ± 36
Trypsin	−	$TNP\text{-}Dex_{1355}$	61 ± 21
		$TNP\text{-}Dex_{1299}$	40 ± 18
	+	$TNP\text{-}Dex_{1355}$	216 ± 24
		$TNP\text{-}Dex_{1299}$	259 ± 33
Pronase	−	$TNP\text{-}Dex_{1355}$	36 ± 11
		$TNP\text{-}Dex_{1299}$	51 ± 19
	+	$TNP\text{-}Dex_{1355}$	263 ± 42
		$TNP\text{-}Dex_{1299}$	219 ± 31
Phospholipase C	−	$TNP\text{-}Dex_{1355}$	19 ± 8
		$TNP\text{-}Dex_{1299}$	32 ± 11
	+	$TNP\text{-}Dex_{1355}$	259 ± 62
		$TNP\text{-}Dex_{1299}$	269 ± 39
None	M^{2+} Cation-free PBS	$TNP\text{-}Dex_{1355}$	42 ± 11
		$TNP\text{-}Dex_{1299}$	18 ± 8
	PBS + Ca^{2+} only	$TNP\text{-}Dex_{1355}$	259 ± 33
		$TNP\text{-}Dex_{1299}$	262 ± 41
	PBS + Mg^{2+} only	$TNP\text{-}Dex_{1355}$	84 ± 21
		$TNP\text{-}Dex_{1299}$	65 ± 14
None	None	No antigen	38 ± 11

[a]All CBA macrophage populations (3×10^5 freshly washed, anti-Thy 1.2 treated, 2000 R, peritoneal exudate cells) were pulsed for 90 min at 37°C with 10^{-1} μg/ml antigen. After exhaustive washing, 1×10^5 macrophages were cultured in triplicate with 3×10^6 spleen cells of CBA mice.
[b]Anti-TNP plaque-forming cells were enumerated at day 4. Data shown are arithmetic mean ± SEM of triplicate cultures.

elomocytes harvested from a common earthworm (*Lumbricus terestris*). In addition, binding of antigen to mouse peritoneal cells was evaluated by two different methods. In the first, that also used with the coelomocyte population, we assessed antigen presentation for antibody production by antigen-pulsed cells in a fashion similar to that shown in Table I. As an alternative independent analysis of recognition by mouse macrophages, we studied the quantitative binding of radiolabeled spontaneous mammary adenocarcinoma tumor target cells. The logic behind such an investigation comes from the evidence that macrophages *in vivo* play an important role in tumor surveillance, though whether this is via an effector-cell function or regulation of an alternative effector mechanism [natural killer cells (NK), or natural anti-tumor antibody] is unclear. A similar series of studies have been performed by Weir *et al.* (1979). Typical data from such a study are shown in Table II.

It seems clear from the data of this simple study that the primitive recognition system of the earthworm coelomocytes also has the capacity to distinguish TNP-

TABLE II

Inhibition of Antigen Binding by Mouse Macrophages Caused by Soluble Sugars

		Percentage inhibition of binding using			
		Mouse peritoneal cells		Coelomocytes	
Sugar used for binding inhibition (50 mM) (main linkage group)[a]	Binding of tumor target to macrophage monolayer, no sugar[b]	TNP-Dex$_{1355}$[c]	TNP-Lev[c]	TNP-Dex$_{1355}$[c]	TNP-Lev[c]
None (control response)	$1.1 \pm 0.2 \times 10^4$ per 10^5 macrophages	328 ± 46	251 ± 44	304 ± 38	244 ± 46
D-Mannose	65**	45**	40*	8	12
N-Acetyl-D-galactosamine	75**	46*	52*	42*	36*
N-Acetyl-D-glucosamine	24	10	9	6	11
Sialic acid	10	18	12	9	6
Glucosamine	60*	36*	19	10	12
Galactosamine	55*	31*	40*	9	8
Glucose	35*	8	8	12	8
Galactose	35*	10	6	36*	26
Lactose ($\alpha 1 \rightarrow 3$)	68**	50*	21	46*	24
Sucrose ($\alpha 1 \rightarrow 2$)	26	35*	72**	16	59*
Fructose ($\beta 2 \rightarrow 6$)	21	35*	42*	8	21
ATH anti-ATL (anti-Ia)	84**	71**	88**	59**	70**

[a]Where sugars were used as competitive inhibitors of saccharide binding they were included in the binding reaction mixture for 90 min at 20°C. ATH anti-ATL antiserum was used at a 1/20 dilution. Significant differences shown by asterisks: *$p < 0.05$, **$p < 0.01$.

[b]Macrophage monolayers were seeded into 96-well flat-bottomed Falcon microtiter plates with 10^5 peritoneal exudate cells harvested freshly from normal CBA mice and pre-treated with anti-Thy 1.2 and complement. After 3 hr at 37°C, ^{51}Cr labeled tumor target cells (10^5/well) were added and binding was allowed to proceed for a further 90 min at 37°C. Thereafter the layers were washed gently and bound target cells were evaluated by release of radioactive label from the wells with 5% acetic acid.

[c]Day 4 antibody response when 3×10^5 irradiated (2000 R) mouse macrophages or coelomocytes were incubated in 0.2 ml medium with 10^{-1} μm/ml TNP-Dextran$_{1355}$ or TNP-Levan with or without the various sugars shown. Thereafter the cells were washed exhaustively, and 1×10^5 cells were mixed with 3×10^6 mouse spleen B lymphoctyes and incubated for 4 days before enumeration of antibody-producing cells by standard techniques.

Dextran$_{1355}$ and TNP-Levan via receptors which are competitively inhibited by different sugars. In addition, these receptors do not necessarily show the same specificity as those on murine macrophages (viz., the inhibition seen with D-mannose and galactosamine), though considerable overlap is seen and perhaps to be expected (inhibition with N-acetyl-D-galactosamine, lactose, and sucrose). Most interesting of all perhaps is the evidence that a mouse anti-Ia antiserum inhibits binding in all situations studied, suggesting that structures cross-reactive with mouse I-region antigens are involved in the binding of these complex carbohydrates to the cell-surface.

D. Role of Cell-Surface Oligosaccharides (in Glycoproteins or Glycolipids) in Macrophage Antigen Recognition

An alternative means by which carbohydrate lectin interactions could form a basis for recognition in macrophages would involve not macrophage surface lectins and sugars on target cells but the converse, i.e., sugars on the surface of macrophages and lectins either on the target surface, or acting as an exogenous bridging molecule to sugars on the target cells. A number of lectins have been demonstrated that bind to phagocytic cells in a manner which is inhibited by addition of soluble sugars for which those lectins are specific. Binding of concanavalin A (Con A) to mannose residues, of wheat germ agglutinin (WGA) to sialic acid and N-acetylglucosamine, and of peanut agglutinin (PNA) or soybean agglutinin (SYA) to galactose residues results in characteristic changes in the cells. Large vacuoles appear at the cell surface and are later pinocytosed (Goldman et al., 1976).

Exogenous lectin, being multivalent with respect to sugar binding, may indeed form a bridge between the macrophage and the target (e.g., bacteria, tumor cells, red cells). Bar-Shavit and Goldman (1976) showed that yeast cells (with a high surface mannan content) coated with Con A bound in large numbers to macrophages and were subsequently phagocytosed. These effects were inhibited by methyl-α-mannoside. WGA, Con A, and phytohemagglutinin (PHA) also enhance binding of mouse macrophages to tumor cells (Kurisu et al., 1980). WGA induced killing of syngeneic and allogeneic tumor target cells was inhibitable by N-acetylglucosamine.

Lectin-mediated binding to macrophages may also occur via target cell lectins. Thus, gram-negative bacteria adhere to eukaryotic cells in a mannose-inhibitable fashion via bacterial surface lectins (Beachy, 1981). Similar bacterial lectins with specificity for L-fucose, sialic acid, galactose, etc., have been described (Sharon et al., 1981). As a result of binding to bacterial lectins, a burst of metabolic activity is elicited from the macrophages (Magnon and Snyder, 1979), followed by ingestion and destruction of the bacterial target (Ohmann et al.,

1982). These reactions do not take place in the presence of mannose or methyl-α-mannoside, yet these same sugars have no effect on immune phagocytosis (Bar-Shavit et al., 1980).

E. Role of Hepatic Cell-Surface Receptors in Glycoprotein Homeostasis

The correlation already noted between the level of serum asialoglycoproteins and functional activity in the liver in health and disease implies that the asialoglycoprotein receptor at least plays a role in regulation of serum glycoprotein homeostasis or in the clearance of bioactive glycoproteins (e.g., immune complexes) after they have served a physiological function. However, the majority of the receptor is found intracellularly, consistent with an alternative hypothesis that envisages a subcellular recognition role, e.g., in regulating flow of endocytosed vesicles to the lysosome.

III. A POSSIBLE ROLE FOR LECTINS IN DICTATING INTERCELLULAR COMMUNICATION WITHIN THE IMMUNE SYSTEM

A. Genetic Restriction of Immune Responses to (Recognition of) Foreign Antigens—Concept of Determinant Selection

Effective macrophage–lymphocyte interaction occurs only when the cells share genes encoded within the major histocompatibility complex (MHC) (Benacerraf and Germain, 1978). The T-lymphocyte recognition of foreign antigen is apparently quite unlike B-lymphocyte recognition. The latter cells seem to see nominal antigen alone. The T lymphocytes in contrast see antigen only in the context of self MHC-encoded molecules. Independent subsets of T cells see antigen either in the context of so-called class I MHC molecules (H2 in the mouse; HLA in humans) (Zinkernagel and Doherty, 1979) or class II MHC molecules (I-A or I-E region associated in the mouse; HLA-DR, -SB in humans).

Thus consider an antigen-induced T-cell proliferative response which is under genetic control, such that responder (R) and nonresponder (NR) animals exist. It has been found that such immune response (IR genes) map within the MHC to the I region. In F_1 hybrid mice (responder phenotype dominant) an effective response is induced only by antigen-pulsed F_1 or R macrophages, not by NR macrophages (Rosenthal et al., 1977).

Two models have been invoked to explain this dual specificity for T-cell recognition. The first, the two-receptor model, proposes that the T lymphocyte has two discrete recognition sites, respectively specific for the foreign cell-surface antigen and the MHC-restricting antigen. The second, the one-receptor or altered-self model, suggests that T cells recognize an antigen–MHC complex via receptors specific for this unique complex of antigen with the appropriate MHC determinants (Blanden, 1980; Matzinger, 1981; Cohn and Langman, 1982).

The debate continues concerning the relative merits of the one- and two-receptor models to explain restriction of T-lymphocyte recognition. Ishi *et al.* (1982) have looked at MHC-encoded restriction molecules involved in presentation of antigen by allogeneic macrophages to induce proliferation in T cells. They concluded that nonresponsiveness and selective restriction is most likely due to elimination of self-reactive clones from the T-cell repertoire and is not due to failure of certain I-encoded molecules to form immunogenic complexes with certain antigens. A similar conclusion was reached by Hedrick *et al.* from analysis of cross-reactivity in T-cell hybridoma clones specific for pigeon cytochrome *c* (Hedrick *et al.*, 1982). While T cells from B10.S(9R) mice were not stimulated by tuna cytochrome *c* fragment 81-103 in the context of $IA^s_\beta:IE^k_\alpha$, responder T cells with this specificity were present in a cloned population from the B10.A mouse. The dose-response curve of this clone to tuna cytochrome *c* fragment 81-103 in the context of $IA^s_\beta:IE^k_\alpha$ (low responder) was indistinguishable from the response in the context of $IA^k_\beta:IE^k_\alpha$ (high responder). Thus there does not seem to be an intrinsic inability of this antigen to associate with Ia molecules of nonresponder strains. Nor indeed was there any evidence in this study that only high-affinity T cell clones (which compensated for a normally poor association between tuna cytochrome *c* 81-103 and $IA^s_\beta:IE^k_\alpha$) were studied.

Despite these studies, other evidence from Hedrick's work suggested that the source of the Ia molecule couplexed with antigen did in fact dictate to a degree the specificity of stimulation for a given T-cell clone. In addition, experiments in Rosenthal's laboratory have provided evidence that one function of the MHC-linked immune response gene mapping in the I region in experimental situations involves some process occurring within the macrophage whereby only discrete regions of complex antigens become available in appropriate form for presentation to responding lymphocytes—so-called determinant selection (Rosenthal *et al.*, 1977). Strain 13 or strain 2 guinea pigs respond to distinct determinants on foreign insulin molecules, with strain 13 T cells reacting with an immunogen encoded in the B chain while T cells from strain 2 animals respond to a conformatorial determinant in the A-chain loop. The F_1 hybrid animals stimulated with insulin-pulsed strain B antigen-presenting cells (macrophages) respond only to the B-chain determinants. Similarly, stimulation of F_1 animals with insulin-pulsed strain 2 antigen-presenting cells (macrophages) leads to an immune response only to the A-chain loop determinants.

B. MHC-Determined Cell-Surface Glycosyltransferases as a Mechanism to Explain MHC-Restricted Lymphocyte–Lymphocyte Interaction

Recently, Parish *et al.* (1981) have proposed a detailed model which, developing upon the theme of both protein and carbohydrate antigens controlled by K, D/L, and I regions of the mouse MHC, postulates that these regions of the MHC actually code for glycosyltransferases (or regulate the expression of structural genes for glycosyltransferases mapping elsewhere in the genome). In a simplistic form, this model proposes (for T cells) three alternative forms of recognition/activation, as shown in Fig. 2. The fundamental variables forming the structure of this model result from binding of one or the other or both of the postulated glycosyltransferase and antigen-specific receptor (with clonally expressed V_H gene sequences).

According to this model, in part (i), where T cells (T_A) contact self MHC-bearing cells in the absence of simultaneous binding of antigen to their antigen specific receptors, an MHC-controlled glycosyltransferase GT_A catalyses the transfer of a sugar residue (from a UDP-sugar) intermediate to the cell surface; cell contact is broken, and no cell activation occurs. In contrast, in part (iii), contact of T_A with a cell bearing both of the self MHC antigens and an antigen recognized simultaneously by the antigen-specific receptor on the T-cell surface leads to an interaction (allosteric?) between receptors on the T-cell surface (antigen-specific receptor and GT_A) such that an alteration in the specificity of GT_A

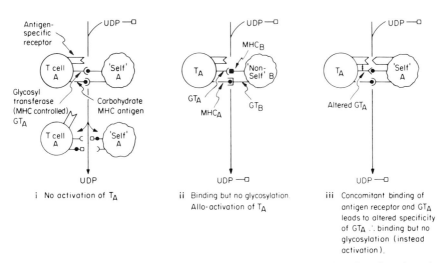

Fig. 2. T-Cell recognition of self and non-self determinants regulated by cell-surface glycosyltransferases (see also Parish *et al.*, 1981).

occurs. The latter now recognizes, but does not catalyze glycosylation of, the contacting cell, and the stabilized binding ultimately leads to T-cell activation. As an adjunct, in part (ii) the model explains alloactivation of T cells simply as the binding by GT_A of non-self MHC (shown as MHC_B in the diagram) without the glycosylation [which occurs when GT_A binds the self antigen MHC_A, as in part (i)]. The net stable and persistent binding of T_A and stimulator cell is thus analogous to that shown for MHC-restricted antigen recognition in part (iii).

   While the reader is left to several reviews by Roseman (1970), Rothenberg (1978), and Parish et al. (1981) to explore evidence for and against such a model, there are two key features that are particularly pertinent to the present discussion. First, in this model, MHC restriction in the T-lymphocyte recognition repertoire is proposed to depend upon somatic selection for transferases which bind and glycosylate carbohydrate MHC antigens expressed within the host thymus during the period of lymphocyte differentiation (Sia and Parrish, 1981). Evidence in support of this notion would include the finding that in bone marrow radiation chimeras, MHC antigens of the recipient influence the MHC antigens subsequently demonstrable on donor cells, and that in H-2 mutant mice simultaneous expression of modified protein and carbohydrate antigens are seen. As will be discussed further, this somatic selection does not seem to occur for macrophages. Rather, an inherent genetic determination of recognition specificities seems to be the case; in consequence, the repertoire of recognition structures is postulated to be different from (more limited than) that for T lymphocytes (see also Fig. 3). Second, activation of lymphocytes by antigen follows from an interaction of the receptor for antigen and the cell surface glycosyltransferase such that the latter no longer glycosylates the "self" target carrier structure [part (iii) in Fig. 2].

   Such glycosyltransferase recognition systems have been proposed by others as potential mediators of cellular recognition in chick neural retinal cells (Rutishauser et al., 1976); of growth control of cultured mouse cells; of the regulation of embryonic development in early chick embryos (see review by Le Douarin, 1984); and of self–non-self discrimination in invertebrates (Rothenberg, 1978). Indeed other loci in the mouse (and their equivalent in humans), such as the T/t locus involved in intercellular communication during development, have also been postulated to encode families of glycosyltransferases (Cheng and Bennett, 1980).

C. Macrophage Cell-Surface Glycosyltransferases Produce Determinant Selection with Complex Antigens

   If in fact we assume that such a recognition system is involved in the antigen-discrimination system (immune response gene linked) demonstrated in mouse macrophages (as already discussed), the activation of T lymphocytes by antigen [shown as part (iii) in the earlier model, Fig. 2] could be represented as shown in Fig. 3. Note that the glycosyltransferase of macrophages, GT'_A, is not proposed to be necessarily equivalent to that on T lymphocytes, since the selection pressure on self recognition by the two classes of cells is probably quite different (e.g., host and/or genetic influences on the receptor recognition repertoire; see earlier discussion and Table III). In particular, for reasons which will become more apparent below, it is suggested that the binding affinity of receptors on macrophages is greater than for T cells, such that the primary binding reaction involves the macrophage glycosyltransferase GT'_A and the T-lymphocyte self antigen (MHC_A).

However, a more interesting model which could incorporate the notion of antigen-determinant selection by I-region (MHC) products and the blocking of antigen binding by saccharides (and anti-Ia antibody) would suggest that antigen actually binds to the postulated glycosyltransferase on macrophages. Such a model is shown in Fig. 4.

In this model, macrophages of responder/nonresponder mice are proposed to express receptors with differential binding affinity for unique determinants on a given antigen (here only two determinants, the first indicated by a rectangle and the second by a triangle, are shown for sake of clarity). Binding at high affinity to a glycosyltransferase on "B" macrophages leads to an interaction with free GT'_B sites on the membrane such that these continue to show specificity for the MHC antigen of "B" but are now unable to glycosylate it. Multipoint binding to T_B cells (with antigen-specific receptor for first type of determinant) occurs. In turn, the binding of the antigen receptor causes modification (via conformational change?) of the T-cell glycosyltransferase (GT_B in Fig. 2 terminology) such that it now binds to (but does not glycosylate) the MHC antigen on the macrophage. By a similar argument, the weaker binding of the first type of determinant to the macrophage glycosyltransferase is proposed to leave the specificity of free GT_B sites on the macrophage unchanged—subsequent attachment of these antigen-pulsed cells to antigen-specific T cells bearing the receptor for the second type of

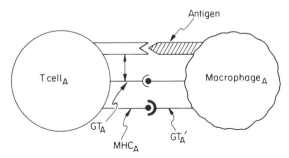

Fig. 3. Activation of T cells by syngeneic antigen-pulsed macrophages.

determinant will lead merely to glycosylation and no activation. Thus strain B shows a T-lymphocyte response to the first type of determinant, and conversely strain A to the second type of determinant. This model would predict that the optimum conditions for demonstrating differences in antigen binding by subpopulations of macrophages would exist when saccharide antigens were used. In fact, when we investigated potential heterogeneity in antigen recognition by

TABLE III

Alteration in Antigen Recognition Repertoire of T Lymphocytes and Macrophages Caused by Growth in Semi-Allogeneic Environments

Source of responding cells[a]	Percentage specific cytotoxicity from splenic T-lymphocytes[b]			Sugar used to inhibit macrophage binding	Percentage inhibition of binding of tumor targets by macrophages[c]
	$\bar{v}$ C3H-TNP	$\bar{v}$ C57BL-TNP	$\bar{v}$ DBA		
C3H → C3H				None	[1.6×10^4/ 2×10^5 macrophages]
	28 ± 4.7			Mannose	31 ± 8*
		1.3 ± 0.4		Galactose	38 ± 7*
			36 ± 4.7	N-Acetylgalactosamine	68 ± 11**
C3B6F$_1$ → C3B6F$_1$				None	[1.7×10^4/ 2×10^5 macrophages]
	32 ± 4.2			Mannose	10 ± 5
		25 ± 3.1		Galactose	7 ± 5
			32 ± 5.2	N-Acetylgalactosamine	36 ± 9*
C3H → C3B6F$_1$				None	[1.6×10^4/ 2×10^5 macrophages]
	29 ± 3.8			Mannose	35 ± 7*
		26 ± 4.2		Galactose	33 ± 7*
			31 ± 4.6	N-Acetylgalactosamine	66 ± 12*

[a]C3H marrow was obtained from a pool of six donors each of C3H or C3B6F$_1$ origin. Cells were treated with a monoclonal anti-Thy 1.2 and complement and transplanted to 1000-R irradiated 10-week-old recipients (six per group). These recipients were sacrificed for use in the experiment described 50 days after transplantation; a minimum of four per group were available for test and spleen cells were pooled within the three groups. Significant differences are marked with asterisks: *$p < 0.05$, **$p < 0.01$.

[b]See text for details: 2×10^5 spleen cells were stimulated in six replicate cultures for 5 days with 2×10^5 2000-R irradiated stimulator cells (DBA spleen or TNP-modified C3H or C57BL/6 spleen cells). After 5 days, 5×10^3 ^{51}Cr-labeled targets of the appropriate specificity were added to each culture and specific cytolysis of the added targets measured by analyzing ^{51}Cr released at 4 hr. Data shown are the arithmetic mean ($\pm$ SEM) of the six replicate samples ($\bar{v}$).

[c]As in Table II.

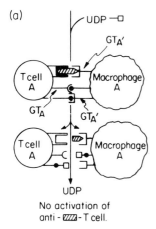

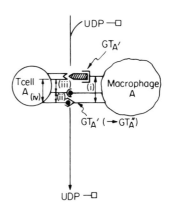

(a)

No activation of
anti - ▨ - T cell.

Binding of antigen to GT$_A$′ causes change
in specificity of GT$_A$′ (to GT$_A$″) ∴ binding,
but no glycosylation of MHC$_A$ on T$_A$ occurs.
Subsequent interaction ((ii),(iii) or (iv), or
all of these) changes the specifities of GT$_A$
such that no glycosylation of macrophage
MHC antigens occurs ∴ T cell activation
of anti – △ -T cells takes place.

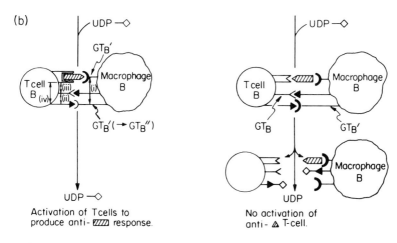

(b)

Activation of T cells to
produce anti- ▨ response.

No activation of
anti- △ T-cell.

(Presentation of antigen on Macrophage$_B$ to T$_A$ (or vice-versa), leads to
activation as in (a)).

Fig. 4. Binding of antigen by a macrophage cell-surface glycosyltransferase can explain determinant selection.

macrophage subpopulations we found no evidence for subpopulation of cells within a given strain preferentially able to present a protein antigen in immunogenic form but did find heterogeneity to saccharide antigens (Gorczynski *et al.*, 1979). More pertinent to an experimental test of this model would be a search for the putative changed glycosyltransferase specificities of antigen-binding macrophages compared to non-antigen-binding cells.

IV. GENETIC AND ENVIRONMENTAL EFFECTS DICTATING THE CELL-SURFACE RECOGNITION REPERTOIRE IN LYMPHOCYTES AND MACROPHAGES

There is a further important distinction drawn in the above model between recognition by T lymphocytes and macrophages. While the specificity repertoire of T lymphocytes is proposed to be generated (somatically) such as to exclude those glycosyltransferases which do not glycosylate MHC antigens expressed within the thymus environment in which differentiation occurs, no such restriction need be placed on the repertoire for macrophages (Zinkernagel *et al.*, 1978). Any differences which exist between the repertoire of strain A and B macrophages could be inherent in the mechanism whereby genetically determined MHC antigens (which are in turn expressed on the macrophages in question) are proposed to regulate the expression of glycosyltransferase activity. We would thus expect that in allogeneic bone marrow radiation chimeras the receptor repertoire of these two cell types (macrophages and T lymphocytes) relative to the repertoire in the syngeneic radiation chimera would be quite different. We have examined this question using C3H or (C3H $\times$ C57BL/6)F$_1$ bone marrow radiation chimeras. The recipient mice were given 1000 R 2 hr before transfer of 2×10^7 C3H bone-marrow cells. Seven weeks post reconstitution, spleen lymphocyte suspensions were prepared from these animals, the cells were treated with (anti-H2^b + C') to remove surviving host cells in the F$_1$ radiation chimeras, and the following assays were performed on the cells.

1. MHC restriction of T cells was assessed in cytotoxic assays, stimulating 2 $\times$ 10^5 spleen cells in culture with 2×10^5 irradiated C3H-TNP, C57BL/6-TNP, or DBA/2J spleen cells. Cytotoxicity was assessed in these cultures at day 5 using ^{51}Cr-labeled Con A-stimulated blast cell targets (a 5:1 ratio for effector:target was used for assay). Where C3H-TNP or C57BL/6-TNP targets were used, a 10-fold excess of nonradioactive, non-TNP-labeled targets were added to ensure that the cytotoxicity measured was directed to TNP-modified determinants.

2. Binding of 2×10^5 adherent spleen macrophages (anti-Thy 1.2 + C' treated) for 1×10^5 ^{51}Cr-labeled C3H adenocarcinoma tumor target cells was measured in the presence or absence of various competing sugars (see Table I).

Data for this experiment are shown in Table III.

It is clear from this study that while the repertoire of T-lymphocyte receptors is altered (restriction now being learned for the "new" antigens in the F_1 hybrid thymic environment), no such alteration of the macrophage repertoire is evident. Thus, as already suggested, the macrophage receptor repertoire seems to be intrinsically determined and is not apparently influenced by the environment of differentiation (see also Gorczynski et al., 1984a).

V. EVIDENCE FOR TARGET RECOGNITION BY LECTIN-LIKE MOLECULES ON OTHER CELLS IN THE IMMUNE SYSTEM

A. Subsets of Lymphomyeloid Cells Defined by Lectin Binding

PNA has been found to bind to hemopoietic stem cells, cortical thymocytes, in vitro T cell lines, acute lymphoblastic leukemia cells with T-cell markers, and murine pre-T cells (Reisner et al., 1978; Newman and Delia, 1983; Roelants et al., 1979). Mature peripheral T cells and medullary thymocytes are PNA$^-$ (Reimer et al., 1979). In addition, murine bone marrow pre-B cells, germinal-center B cells, and B-cell follicular lymphomas are PNA$^+$, as are CFU-C (myeloid stem cells assayed in vitro) and some myeloid leukemias (Rose et al., 1981; Saveriano et al., 1981; Nicola et al., 1980; Newman and Boss, 1980). Thus while not characterizing a particular lineage the expression of targets binding to PNA does seem to describe a maturational and or a functional phase (resting or activated) within a developmental pathway.

SYA binds to hemopoietic stem cells and B-cell subsets (Sharon, 1980). Helix pomatia A hemagglutinin binds to T and B cell subsets and so-called natural killer (NK) cells (Axelsson et al., 1978). The usefulness of documenting these phenomena is brought out by a recent study by Reimann et al. (1984) which compared lectin binding to lymphopoietic and myelopoietic cells in mouse bone marrow by flow cytometry. Normal mice as well as athymic (nu/nu mice) and animals with a severe combined immunodeficiency (SCID) were used in the analysis. The data showed that SCID mice, which show a deficiency in lympho-poiesis, lacked a marrow-cell subset of small-size cells which stained weakly with Con A and pokeweed mitogen (PWM) but intensely with PNA. Such developmental stage-specific lectins may be important in appropriate cell migra-tion (to organ sites of cell differentiation) (Le Douarin, 1984).

Mature T cells within the lymphoid system are believed to recognize and communicate with other cells using cell surface glycoproteins encoded by genes of the major histocompatibility complex (Benacerraf, 1981), "seeing" those glycoproteins either alone or in association with other foreign determinants

(Zinkernagel and Doherty, 1974; Shearer, 1974). Nevertheless, there are few reports which document the inhibition of T-cell function by purified MHC-encoded molecules (Lemonnier et al., 1978). Monosaccharides have been reported to inhibit T-suppressor cells (Koszinowski and Kramer, 1981) and to inhibit antigen-induced proliferation of human lymphocytes to their specific target cells (Muchmore et al., 1980). More recently, Pimlott and Miller (1984) reported that a glycoprotein prepared from P815 tumor cells can inhibit binding of antigen-specific cytotoxic T lymphocytes. If a lectin-like molecule on CTL is genuinely involved in recognition, and not merely in stabilizing effector:target conjugates, this implies haplotype-specific patterns of glycosylation. Some private class II antigenic specificities have been reported to be determined by sugar residues (McKenzie et al., 1977).

B. Inhibition of Function of Immune Cells by Soluble Sugars

Mention was made above of so-called natural killer (NK) cells. These cells lyse tumor target cells and some undifferentiated normal cells in the absence of deliberate immunologic priming. Genetic and or experimental NK deficiency leads to increased lymphoreticular malignancies in humans and increased susceptibility to transplantable tumors in experimental animals (Roder et al., 1980; Sullivan et al., 1980; Talmadge et al., 1980; Kasai et al., 1981). Susceptibility to radiation-induced thymic leukemias in NK-deficient beige (bg/bg) mice is decreased by restoration of NK activity (Warner and Dennert, 1982). NK activity declines with advancing malignant disease in a heterogeneous group of cancer patients studied at a point in time, and as function of time in ovarian cancer patients followed longitudinally (Pross and Baines, 1976, 1980).

The nature of the target antigen recognized by NK cells is still controversial. Roder (1980) and, independently, Obexer (1983) have isolated from susceptible targets membrane proteins with the capacity to inhibit NK activity/conjugate formation. Target antigens exist on normal bone marrow and thymus cells (Riccardi et al., 1981). In vivo passaged targets tend to show a lower sensitivity to lysis than in vitro grown targets, suggesting some selection (by NK) for resistance in vivo (Brooks et al., 1981). There is a heterogeneity in target antigens, with some antigens being more important than others. Thus for instance, the human target K562 competes most effectively not only with itself but with other targets as well (MacDougall et al., 1983). Recent studies have implicated a role for transferrin receptors on susceptible target cells as a prototypic NK target antigen (Vodenlich et al., 1983). While this concept too is controversial (Dokhelar et al., 1984), there are data consistent with the idea (Baines, 1983). Excess ferritransferrin has been reported to inhibit NK activity, and a role for hydrodoxyl radicals (perhaps produced via the Haber Weiss pathway, using iron as a cofactor) in lysis by NK cells may be important.

Independent studies have suggested that glycolipids and/or glycoproteins are the target structures recognized by NK cells (Roder *et al.*, 1979; Young *et al.*, 1981). Changes in cell-surface sialylation of galactosyl and *N*-acetylgalac-tosaminyl residues in tumor metastasis have been reported (Yogeeswaran and Salk, 1981). Given the role NK cells are believed to play in the regulation of metastasis (Talmadge *et al.*, 1980), these changes are also consistent with NK recognition of carbohydrate moieties on target cells.

As reported initially by Stutman *et al.* (1980) and subsequently by Brunda *et al.* (1983), we have shown that both simple and complex sugars can independently inhibit NK activity from murine (Gorczynski *et al.*, 1984a) and human (R. M. Gorczynski, unpublished) NK cells. A pattern of inhibition characteristic of discrete subpopulations of NK cells can be described which varies with the genetic background of the mice under study. When bone marrow cells are transplanted to lethally irradiated recipient mice, the profile of NK recognition seen in the chimaera is characteristic of the donor, not of the recipient. This is in contrast to the recognition profile of alloreactive T cells (Gorczynski *et al.*, 1982), which changes in accordance with alteration in the host differentiative (thymus) environment (but is reminiscent of antigen discrimination by macrophages as already discussed).

A criticism of the interpretation offered for inhibition of NK lysis by sugars concerns the idea that the inhibition seen is at a postrecognition event. Wright *et al.* (1983) reported that soluble sugars can inhibit the activity of cytolytic lymphokines released by human/murine NK cells. Vose *et al.* (1983) also postulated a role for sugars in a postrecognition event of the lytic process. Using a long-term assay (72 hr) we confirmed that the function of NK-produced cytotoxins is inhibited by sugars. However, of three sugars tested (Table IV), only Glc-NAc-($\beta 1\rightarrow 2$)Man-α-OCH$_3$ caused significant inhibition of lysis by the NK-derived cell supernatant. In contrast, whether sugars were added throughout the 4-hr assay or only in the first 20 min (of NK : target conjugate formation), all three sugars used caused inhibition of lysis derived from NK cells themselves, with the order of activity for the 3 sugars the same in each assay (see Table IV). These data are consistent with the idea that NK cell activity recorded in short-term assays may detect a different cellular function from that detected in long-term assays. A recent report claimed that data analogous to that described by Wright and Bonavida could only be seen using mycoplasma-infected targets (Wayner and Brooks, 1984).

We have also shown that a family of monoclonal antibodies directed against the stage-specific embryonic antigens (SSEA-1) expressed by restricted tissue types of embryonic and normal adult tissue can inhibit human NK effector cells in a manner which correlates well with the expression of SSEA-1 by those targets (Fox *et al.*, 1983; Harris *et al.*, 1984). [The immunodominant regions of SSEA-1 are likely to be carbohydrate in nature (Gooi *et al.*, 1981).] These studies also support the concept of heterogeneity in human NK effector populations. Using

TABLE IV

Comparison of Inhibition of Tumor Target Lysis by Carbohydrates in Different Assays of
NK Cell-Mediated Activity

Sugar inhibitor used (concentration)	NK activity[a] from poly(1-C)-stimulated spleen cells		
	4-hr ^{51}Cr with cells		16-hr ^{51}Cr with cell supernatant
	Sugar throughout assay	Sugar during conjugate formation only	
N-Acetyl-D-glycosamine (50 × 10^{-3} M)	7.1 ± 0.9*	7.9 ± 0.8*	19 ± 2.5
Glc-NAC-β1 → 2-Man α-OCH$_3$ (2.5 × 10^{-5} M)	6.2 ± 1.3*	7.2 ± 0.8*	13 ± 3.6*
Man-α1 → 6-Man-α-OCH$_3$ (2.5 × 10^{-5} M)	5.1 ± 0.6*	6.0 ± 1.1*	18 ± 1.7
None	16 ± 3.0	11 ± 2.1	23 ± 4.1

[a]NK activity is expressed as lytic units/10^6 effector cells or per milliliter supernatant. I.L.U. gives 20% specific lysis in 4 hr (first two columns) or 16 hr (last column). Significantly different data, *$p < 0.05$.

four different ^{51}Cr-labeled tumor target cells, the relative cytolytic activity of 21 individual donor tested for those targets was not constant; nor indeed was the ability of anti-SSEA-1 Mab to inhibit NK activity for those targets constant within the 21 donors tested (Harris *et al.*, 1984).

VI. IMPLICATIONS OF PROPOSED MODEL FOR IMMUNE RESPONSE IN DISEASE

A. Ability of Intracellular Parasites to Evade Immunosurveillance

It is worthwhile to consider some of the more general features suggested by the model of Fig. 4. It might be anticipated, for instance, that where the macrophage per se still retains an important host protective function, it would be profitable to search for evidence that there is a role for cell-surface glycosyltransferases in this protection. In cutaneous parasitic disease, as exemplified by *Leishmania* parasites, there is now a well documented role for innate genetic resistance to infection in mice by given strains of the parasite (reviewed in Gorczynski, 1982). Recently, Handman *et al.* (1979) have reported that in the highly susceptible (to

L. tropica) BALB/c strain, infected macrophages showed a marked diminution in expression of self MHC antigen following parasite infection—within the model of this chapter, this could be viewed as both a cause and effect of altered cell-surface glycosyltransferases, which in turn would then markedly alter the subsequent immunostimulation of T lymphocytes (a concomitant feature of *L. tropica* infection in BALB/c).

Interpretation of defective T-cell stimulation by *L. tropica*-infected BALB/c macrophages within this model would thus predict that if *L. tropica* antigen were presented without the concomitant alteration in macrophage cell surface recognition molecules (GT'_A) there would perhaps be no defective T-lymphocyte stimulation. We and others recently analyzed T-cell stimulation by BALB/c macrophages exposed to solubilized inert parasite antigen (rather than the live parasite) or host resistance after vaccination with irradiated parasites and found just this result (Gorczynski and MacRae, 1982; Howard *et al.*, 1982). Indeed, the recognition repertoire of T lymphocytes in BALB/c mice is such that, triggered under the appropriate circumstances, these T cells are as capable of responding to produce protective immunity to *L. tropica* as those of the "naturally resistant" CBA mouse.

Even more interesting given the discussion above (see also Table III) is the report from Howard *et al.* (1980) analyzing the susceptibility to *L. tropica* infection in reciprocal bone marrow radiation chimeras between susceptible and resistant congenic strains of mice. The susceptibility phenotype of the transplanted animal was shown to be typical of the (macrophage) donor strain.

B. Altered Phagocytosis in Tumor-Bearing Individuals

In mice bearing spontaneously appearing tumors, we have described an alteration in the recognition repertoire of resident macrophages. Typical data (using TNP-carbohydrate pulsed antigens as immunogen for stimulation of mouse B lymphocytes to antibody production) are shown in Table V.

Here, too, there is suggestive evidence that alteration of the macrophage cell-surface recognition receptors occurs alongside development of a major pathological condition and is reversed when overt disease is controlled.

The fact that phagocytic cells in the liver also bear cell-surface-specific lectin molecules may also be of use in the targeting of specific therapeutic agents to the liver, using bioactive molecules covalently linked to, e.g., galactose saccharides. Since it is believed that such liver cell-surface lectins are involved in glycoprotein homeostasis, an analysis of the serum levels of asialoglycoproteins has also been used to follow liver disease (cirrhosis, hepatitis, primary liver cancer). Sobue and Kosoka (1980) showed a correlation between tumor size and the ratio of intact to desialylated transferrin in such cases.

TABLE V

Alteration of Carbohydrate Receptor Repertoire in Splenic Macrophages
of Tumor-Bearing Mice

Source of macrophage[a]	Anti-TNP PFC after pulsing, with[b]				
	None	TNP-T$_4$	TNP-Dextran$_{1355}$	TNP-Dextran$_{1299}$	TNP-Levan
Normal	11 ± 5	219 ± 38	316 ± 51	268 ± 39	251 ± 46
Tumor bearer	28 ± 5	186 ± 29	68 ± 31	43 ± 12	164 ± 33
Tumor resected	17 ± 6	201 ± 35	184 ± 46	162 ± 39	219 ± 38

[a]Macrophages were obtained by velocity sedimentation (cells sedimenting in the region 4.0–7.0 mm/hr) of (anti-Thy 1.2 + C′)-treated irradiated (2000 R) spleen cells. Three donors were used per group. Tumor-bearer and tumor-resected mice were given 1×10^6 cells of a primary spontaneous adenocarcinoma subcutaneously 20 or 40 days earlier (tumors were removed under ether anesthesia at 20 days; approximate size 1.5 cm^3); 3×10^5 cells were pulsed with antigen (10^{-1} μg/ml) for 90 min at 37°C and washed exhaustively before culturing with fresh normal spleen cells.

[b]Arithmetic mean (± SEM) of anti-TNP PFC (plaque-forming cells) on day 4 in cultures (three per group) receiving 3×10^6 normal spleen cells (from a pool of three 8-week donor C3H mice) and 1×10^5 antigen-pulsed cells (first column).

Finally, given that extracellular lectins may be important in macrophage–target interaction (e.g., the mannose/N-acetylglucosamine lectin in rabbit serum), it has been suggested that we may even consider use of lectins in the treatment of infected wounds.

If NK cells are critically involved in the regulation of tumor growth *in vivo,* and particularly of tumor metastases, it may prove possible to manipulate NK activity in a specific (clonotypic) manner, once we have defined the recognition structure on antigen-specific discrete subpopulations of NK cells. This sort of approach has been used to manipulate tumor growth *in vivo* of those tumors where growth control *in vivo* is mediated by immune lymphocytes (Miller *et al.,* 1982; Tilfkin *et al.,* 1981; Flood *et al.,* 1980; Gorczynski *et al.,* 1984b). A framework for understanding such immunoregulation is given by Jerne's network model (1974) for cell–cell interaction in the immune system. Antireceptor immunity may then be considered as anti-idiotypic in nature.

VII. SUMMARY

There is now abundant evidence to indicate that invertebrate mononuclear cells recognize foreign targets by virtue of the presence of receptors with exquisite carbohydrate specificity and that mammalian mononuclear macrophages and related cells show similar recognition structures. This phylogenetic preservation of sugar-specific recognition suggests an essential role for this mechanism in self–non-self discrimination. Within the liver, where the biochemistry of such specific lectin cell-surface receptors has been studied in greater depth than for other cells and tissues, several molecularly distinct lectins with discrete recognition specificity have been described.

A model consistent with a role for the conservation of sugar-specific recognition in the MHC restriction phenomena observed for lymphocyte and lymphocyte–macrophage cooperation in mammalian immune induction has also been presented. As discussed in this chapter, minimal expansion of this model can usefully explain determinant selection by macrophages for immune responses to complex determinants. An understanding of this determinant selection may be critical to comprehend phenomena such as the susceptibility to parasite, bacterial, and viral infections in humans, as well as to understand the mechanism of MHC-related immunological abnormalities (autoimmune disease, allergy, immune deficiency, etc.). We have ourselves produced evidence that an alteration in functional macrophage carbohydrate recognition may occur along with tumor growth in mice receiving a transplantable tumor, and that this alteration is reversed on tumor resection. A further complication to this potential role for macrophages in immunosurveillance can be found in the growing evidence to suggest that tumor cells may express alien MHC specificities (Schirrmacher *et al.*, 1980). It should be apparent to the reader (see Fig. 4) that depending upon the new MHC determinants expressed, a significant perturbation in cell–cell recognition (by macrophages, NK cells, and T cells) and homeostasis will be predicted as a result of the expression of such alien MHC antigens.

ACKNOWLEDGMENTS

The author would like to thank A. Collins and M. Boulanger for their excellent assistance during the preparation of this manuscript. This work was supported by the NCI and MRC (#MA440) of Canada and by a grant from the UNDP/World Bank/WHO Special Program for Research and Training in Tropical Diseases.

REFERENCES

Achord, D. T., Brot, F. E., and Sly, W. S. (1977). *Biochem. Biophys. Res. Commun.* **77,** 409.
Ashwell, G., and Harford, J. (1982). *Ann. Rev. Biochem.* **51,** 531.
Ashwell, G., and Morell, A. G. (1974). *Adv. Enzymol.* **41,** 99.
Axelsson, B., Kimura, A., Hammarstrom, S., Wigzell, H., Nilsson, K., and Mellstadt, H. (1978). *Eur. J. Immunol.* **8,** 757.
Baenziger, J. U., and Maynard, Y. (1980). Human hepatic lectin. *J. Biol. Chem.* **255,** 4607.

Baines, M. S. (1983). *Immunol. Lett.* **7**, 51.

Barondes, S. H. (1981). *Ann. Rev. Biochem.* **50**, 207.

Bar-Shavit, Z., and Goldman, R. (1976). *Exp. Cell. Res.* **99**, 221.

Bar-Shavit, Z., Goldman, R., Ofek, I., Sharon, N., and Mirelman, D. (1980). *Infect. Immunol.* **29**, 417.

Beachy, E. H. (1981). *J. Infect. Dis.* **143**, 325.

Benacerraf, B. (1981). *Science* **212**, 1229.

Benacerraf, B., and Germain, R. (1978). *Immunol. Rev.* **38**, 70.

Berg, T., and Tolleshaug, H. (1980). *Biochem. Pharmacol.* **29**, 917.

Blanden, R. V. (1980). *Immunol. Today* **1**, 33.

Brooks, C. G. (1981). *Int. J. Cancer* **28**, 191.

Brunda, M. J., Wiltrout, R. H., Holden, H. T., and Varesio, L. (1983). *Int. J. Cancer* **31**, 373.

Cheng, C. C., and Bennett, D. (1980). *Cell* **19**, 537.

Cohn, M., and Langman, R. (1982). *Behring Inst. Mitteilungen* **70**, 219.

Dokhelar, M-C., Garson, D., Testa, U., and Turaz, T. (1984). *Europ. J. Immunol.* **14**, 340.

Flood, P. M., Kripke, M., Rowley, D. A., and Schreiber, H. T. (1980). *Proc. Natl. Acad. Sci. U.S.A.* **77**, 2209.

Fox, N., Damjanov, I., Knowles, B. B., and Solter, D. (1983). *Cancer Res.* **43**, 669.

Friemer, N. B., Ogmundsdottir, H. M., Blackwell, C. G., Sutherland, I. W., Grahame, L., and Weir, D. M. (1978). *Acta Pathol. Microbiol. Scand. [B]* **86**, 53.

Goldman, R., Sharon, N., and Loton, R. (1976). *Exp. Cell. Res.* **99**, 408.

Goldstein, I. J., Hughes, R. C., Monsigny, M., O'Sawa, M., and Sharon, N. (1980). *Nature (London)* **285**, 66.

Gooi, H. C., Feizi, T., Kapadia, A., Knowles, B. B., Solter, D., and Evans, M. J. (1981). *Nature (London)* **292**, 156.

Gorczynski, R. M. (1982). *Dev. Comp. Immunol.* **6**, 199.

Gorczynski, R. M., and MacRae, S. (1982). *Cell. Immunol.* **67**, 74.

Gorczynski, R. M., MacRae, S., and Jennings, J. J. (1979). *Cell Immunol.* **45**, 276.

Gorczynski, R. M., Benzing, K., MacRae, S., and Price, G. B. (1980). *Immunopharmacology* **2**, 327.

Gorczynski, R. M., MacRae, S., and Kennedy, M. (1982). *Cell. Immunol.* **73**, 44.

Gorczynski, R. M., Harris, J. F., Kennedy, M., MacRae, S., and Chang, M. P. (1984a). *Immunology* (in press).

Gorczynski, R. M., Kennedy, M., Polidoulis, I., and Price, G. B. (1984b). *Cancer Res.* **44**, 3291.

Handman, E., Cerdig, R., and Mitchell, S. F. (1979). *Aust. J. Exp. Biol. Med. Sci.* **57**, 9.

Harford, J., and Ashwell, G. (1981). *Proc. Natl. Acad. Sci. U.S.A.* **78**, 1557.

Harris, J. F., Chin, J., Jewett, M. A. S., Kennedy, M., and Gorczynski, R. M. (1984). *J. Immunol.* **132**, 2502.

Hedrick, S. M., Matis, L. A., Hecht, T. T., Samelson, L. E., Longo, D. L., Heber-Katz, E., and Schwartz, R. H. (1982). *Cell* **30**, 141.

Howard, J. G., Hole, C., and Liew, F. C. (1980). *Nature (London)* **288**, 161.

Howard, J. G., Nicklin, S., Hale, C., and Liew, F. Y. (1982). *J. Immunol.* **129**, 2206.

Ishii, N., Nagy, Z. A., and Klein, J. (1982). *J. Exp. Med.* **156**, 622.

Jerne, N. K. (1974). *Ann. Immunol. (Inst. Pasteur)* **125C**, 373.

Kasai, M., Yoneda, T., Habu, S., Maruyama, Y., Okumura, K., Tokunaga, T. (1981). *Nature (London)* **291**, 334.

Kawasaki, J., Etoh, R., and Yamashina, I. (1978). *Biochem. Biophys. Res. Commun.* **81**, 1018.

Kawasaki, T., and Ashwell, G. (1976). *J. Biol. Chem.* **251**, 1296.

Kolb-Bachofen, V. (1981). *Biochem. Biophys. Acta* **645**, 293.

Kolb-Bachofen, V., Schlepper-Schafer, J., Vogell, J., and Kolb, H. (1982). *Cell* **29**, 859.

Kolset, S. O., Tolleshaug, H., and Berg, T. (1979). *Exp. Cell Res.* **122,** 159.

Koszinowski, U. H., and Kramer, M. (1981). *Nature (London)* **289,** 181.

Krantz, M. J., Holtzman, N. A., Stockwell, C. P., and Lee, Y. C. (1976). *Biochemistry* **15,** 3963.

Kuhlenschmidt, T. B., and Lee, Y. C. (1980). *Fed. Proc.* **39,** 1968.

Kurisu, M., Yamazaki, M., and Mizumo, D. (1980). *Cancer Res.* **40,** 3798.

Le Douarin, N. M. (1984). *Cell* **360,** 353–360.

Lemonnier, F., Mescher, M. F., Sherman, L., and Burakoff, S. (1978). *J. Immunol.* **120,** 114.

Lumney, J. K., and Ashwell, G. (1976). *Proc. Natl. Acad. Sci. U.S.A.* **73,** 341.

MacDougall, S. L., Shustik, C., and Sullivan, A. K. (1983). *Cell. Immunol.* **76,** 39.

McKenzie, I. F. C., Clarke, A., and Parrish, C. R. (1977). *J. Exp. Med.* **145,** 1039.

Magnon, D. F., and Snyder, J. S. (1979). *J. Infect. Immunol.* **26,** 1014.

Matzinger, P. (1981). *Nature (London)* **292,** 497.

Miller, R. A., Maloney, D. G., Warnmke, R., and Levy, R. (1982). *N. Engl. J. Med.* **306,** 517.

Morell, A. G., Irvine, R. A., Steinlieb, I., Scheimberg, I. H., and Ashwell, G. (1968). *J. Biol. Chem.* **243,** 155.

Muchmore, A. V., Decker, J. M., and Blaese, R. M. (1980). *J. Immunol.* **125,** 1306.

Newman, B. A., and Delia, D. (1983). *Immunology* **49,** 147.

Newman, R. A., and Boss, M. A. (1980). *Immunology* **40,** 193.

Nicola, N. A., Burgess, A. W., Staber, F. G., Johnson, G. R., Metcalf, D., and Battye, F. L. (1980). *J. Cell Physiol.* **103,** 217.

Obexer, G. (1983). *Immunobiology* **165,** 15.

Ohmann, L., Hed, J., and Standahl, O. (1982). *J. Infect. Dis.* **146,** 751.

Parish, C. R., O'Neill, H. C., and Higgins, T. J. (1981). *Immunol. Today* **2,** 98.

Pearse, B. M. F. (1976). *Proc. Natl. Acad. Sci. U.S.A.* **73,** 1255.

Pimlott, N. J. G., and Miller, R. G. (1984). *J. Immunol.* **133,** 1763.

Prieels, J. P., Pizzo, S. V., Glasgow, L. R., Paulson, J. C., and Hill, R. L. (1978). *Proc. Natl. Acad. Sci. U.S.A.* **75,** 2215.

Pross, H. F., and Baines, M. G. (1976). *Int. J. Cancer* **18,** 593.

Pross, H. F., and Baines, M. G. (1980). In ''Natural Cell Mediated Immunity Against Tumors'' (R. Herberman, ed.), p. 1063. Academic Press, New York.

Regoeczi, E., Hatton, M. W. C., and Charlwood, P. A. (1975). *Nature (London)* **254,** 699.

Regoeczi, E., Debanne, M. T., Hatton, M. W. C., and Koj, A. (1978). *Biochem. Biophys. Acta* **54,** 372.

Reimann, J., Ehman, D., and Miller, R. G. (1984). *Cytometry* **5,** 194.

Reimer, Y., Binyaminoy, M., Rosenthal, E., Sharon, N., and Lamot, B. (1979). *Proc. Natl. Acad. Sci. U.S.A.* **76,** 447.

Reisner, Y., Itzicovitch, L., Maschover, A., and Sharon, N. (1978). *Proc. Natl. Acad. Sci. U.S.A.* **75,** 2933.

Riccardi, C., Santoni, A., Barlozzari, T., Herberman, R. B. (1981). *Cell. Immunol.* **60,** 136.

Roder, J. L. (1980). In ''Natural Cell Mediated Immunity Against Tumors'' (R. Herberman, ed.), p. 939. Academic Press, New York.

Roder, J. C., Ahrland-Richter, L., Jondal, M. (1979). *J. Exp. Med.* **150,** 471.

Roder, J. C., Haliotis, T., and Klein, M. (1980). *Nature (London)* **284,** 553.

Roelants, G. E., London, J., Mayor-Withey, K. S., and Serrano, B. (1979). *Eur. J. Immunol.* **9,** 139.

Rose, M. L., Habeshaw, J., Kennedy, R., Sloane, J., Wiltshaw, E., and Davies, A. J. S. (1981). *Br. J. Cancer* **44,** 68.

Roseman, S. (1970). *Chem. Phys. Lipids* **5,** 270.

Rosen, S. D., Kafka, J. A., Simpson, D. L., and Barondes, S. M. (1973). *Proc. Natl. Acad. Sci. U.S.A.* **70,** 2554.

Rosenthal, A. S., Barcinski, M. A., and Blake, J. T. (1977). *Nature (London)* **267,** 156.

Rothenberg, R. E. (1978). *Dev. Comp. Immunol.* **2,** 23.

Rutishauser, U., Thiery, J.-P., Brackenbury, R., Sela, B.-A., and Edelman, G. M. (1976). *Proc. Natl. Acad. Sci. U.S.A.* **73,** 577.

Sahagian, G. G., Distler, J., and Jourdain, G. W. (1981). *Proc. Natl. Acad. Sci. U.S.A.* **78,** 4289.

Saveriano, M., Prinnan, M., Sauter, V., and Osmond, D. G. (1981). *Exp. J. Immunol.* **11,** 870.

Schirrmacher, V., Hiibsch, D., and Garrido, F. (1980). *Proc. Natl. Acad. Sci. U.S.A.* **77,** 5409.

Sharon, N. (1980). *In* "Immunology 80" (Progress in Immunology IV) (M. Fougereau and J. Dausset, eds.), p. 254. Academic Press, New York.

Sharon, N. (1984). *Immunol. Today* **5,** 143.

Sharon, N., Eshdat, Y., Silverbatt, F. J., and Ofek, I. (1981). "Bacterial Adherence to T Cell Surface Sugars," Ciba Foundation Symposium, Vol. 80, p. 119. Churchill, London.

Shearer, G. M. (1974). *Eur. J. Immunol.* **4,** 527.

Sia, D. Y., and Parrish, C. R. (1981). *Immunogenetics* **1,** 12.

Sobue, G., and Kosaka, A. (1980). *Hepatogastroenterology* **27,** 200.

Stahl, P. D., Rodman, J. S., Miller, M. J., and Schlesinger, P. H. (1978). *Proc. Natl. Acad. Sci. U.S.A.* **75,** 1399.

Steer, C. J., Kempner, E. J., and Ashwell, G. (1981). *J. Biol. Chem.* **256,** 5851.

Stockert, R. J., Morell, A. G., and Scheinberg, I. H. (1976). *Biochem. Biophys. Res. Commun.* **68,** 988.

Stutman, O., Dien, P., Wisun, R. E., and Lattime, E. C. (1980). *Proc. Natl. Acad. Sci. U.S.A.* **77,** 2895.

Sullivan, J. L., Byron, K. S., Brewster, F. E., and Purtilo, D. T. (1980). *Science* **210,** 543.

Sung, S. J., Nelson, R. S., and Silverstein, S. C. (1983). *J. Cell Biol.* **96,** 160.

Talmadge, J. E., Meyers, K. M., Prieur, D. J., and Starkey, J. R. (1980). *Nature (London)* **284,** 622.

Tanabe, T., Pricer, W. E., Jr., and Ashwell, G. (1979). *J. Biol. Chem.* **254,** 1038–1043.

Tilfkin, A. F., Shaaf-Lafontaine, N., Van Acker, A., Boccadoro, M., and Urbain, J. (1981). *Proc. Natl. Acad. Sci. U.S.A.* **78,** 1809.

Tolleshaug, H., and Berg, T. (1979). *Biochem. Pharmacol.* **28,** 2919.

Vodenlich, L., Sutherland, R., Schneider, C., Newman, R., and Greaves, M. (1983). *Proc. Natl. Acad. Sci. U.S.A.* **80,** 835.

Vose, B. M., Harding, M., White, W., Moore, M., and Gallagher, J. (1983). *Clin. Exp. Immunol.* **51,** 517.

Wall, D. A., Wilson, G., and Hubbard, A. L. (1980). *Cell* **21,** 79.

Warner, J. F., and Dennert, G. (1982). *Nature (London)* **300,** 31.

Wayner, E. A., and Brooks, C. G. (1984). *J. Immunol.* **132,** 2135–2139.

Weigel, P. H. (1980). *J. Cell Biol.* **87,** 855.

Weir, D. M. (1980). *Immunol. Today* **1,** 45.

Weir, D. M., Grahame, L. M., and Ogmundsdottir, H. M. (1979). *J. Clin. Lab. Immunol.* **2,** 51.

Wright, S. C., Weitzen, M. L., Kahle, R., Granger, G. A., and Bonavida, B. (1983). *J. Immunol.* **130,** 2479.

Yogeeswaran, G., and Salk, P. L. (1981). *Science* **212,** 1514.

Young, W. W. Jr., Durdik, J. E., Urdal, D., Hakomori, S., and Henney, C. S. (1981). *J. Immunol.* **126,** 1.

Zinkernagel, R. M., and Doherty, D. C. (1974). *Nature (London)* **251,** 547.

Zinkernagel, R. M., and Doherty, P. C. (1979). *Adv. Immunol.* **27,** 51.

Zinkernagel, R. M., Callahan, G. N., Klein, J., and Dennert, G. (1978). *Nature (London)* **271,** 251.

4

Homeostatic Control in the Immune System: Involvement of Self-Recognition Receptors and Cell Interaction Molecules

DAVID H. KATZ

I. INTRODUCTION

Somatic cell differentiation has been a subject of extensive interest and investigation for many years, because a clearer understanding of the mechanisms governing such processes should unlock many of the unsolved problems and questions concerning embryogenesis, ontogeny, and neoplastic transformation. There has been a vast amount of information accumulated, particularly in recent years, about the correlations of cell-surface membrane changes with various stages of differentiation. This approach has been extremely informative in terms of the consequences of translation of information from the genome as it pertains to control of phenotypic programming and has opened an avenue to possibly

RECEPTORS IN CELLULAR RECOGNITION
AND DEVELOPMENTAL PROCESSES

understanding the manner by which alterations in the program may occur and thereby result in neoplastic changes.

Nonetheless, there is still a vast gap in our understanding of the various inductive and selective forces which govern the translational and/or transcriptional events at the genetic level, which, in turn, control the program(s) for differentiation. In this chapter, we will analyze such questions from a somewhat different point of view—namely, to examine to what extent cell differentiation is influenced by cell-surface molecules on the differentiating cells as well as those on neighboring cells and structures with which they may be interacting. The basis for this approach stems from studies performed in our laboratory, and related observations from other laboratories, that have been concerned with the role of histocompatibility gene products in the control of cellular interactions between T lymphocytes and B lymphocytes and macrophages in the development of humoral immune responses. Since many reviews have already been written on this subject, the main emphasis of this discussion will be on recent studies from our own laboratory which, we believe, directly bear on the relationships between receptors involved in intercellular communication and the phenotype of cell–cell interactions ultimately displayed by a given lymphoid cell population.

II. GENETIC RESTRICTIONS ON IMMUNOCOMPETENT CELL INTERACTIONS

The discovery of major histocompatibility complex (MHC)-linked genetic restrictions on cell–cell interactions represented a significant advance in our understanding of the fine specificity of such intercellular communication events and has broadened our view of the biological significance of MHC genes and their products. The initial demonstrations of MHC restrictions on T-cell–B-cell interactions in the mouse (Katz et al., 1973a,b; Kindred and Shreffler, 1972) and on macrophage–T-lymphocyte interactions in the guinea pig (Rosenthal and Shevach, 1973; Shevach and Rosenthal, 1973) were followed by demonstrations of the involvement of MHC gene products in controlling the ability of cytotoxic T lymphocytes (CTL) to effectively lyse target cells (Bevan, 1975; Blanden et al., 1975; Gordon et al., 1975; Koszinowski and Ertl, 1974; Schmitt-Verhulst and Shearer, 1975; Shearer, 1974; Zinkernagel and Doherty, 1974a). In fact, MHC-linked genetic restrictions have by now been identified in virtually every conceivable type of cell–cell interaction involving, directly or indirectly, cells of lymphohematopoietic origin. Although the principal MHC genetic loci involved may vary from one type of interaction to another (i.e., I region genes control lymphocyte–lymphocyte and macrophage–lymphocyte interactions, whereas K/D genes govern CTL–target cell interaction), the basic phenomenology is the same, namely, that the most efficient cell–cell interactions are transacted when the two interacting partner cells share identical *I* or *K*/*D* region genes, as the case may be. Essentially, two bodies of thought have arisen as possible explanations for such genetic restrictions. One line of thinking considers that genetic restrictions are manifestations of the preferable, or even necessary, perception by lymphocytes of antigen in some type of molecular association with MHC determinants on the surface membranes of partner cells with which they interact (i.e., altered-self, complex antigenic determinants) (Zinkernagel and Doherty, 1974b; Benacerraf, 1978; Rosenthal, 1978). The second line of thought considers genetic restriction as a reflection of a distinct cell–cell recognition system involving cell interactions (CI) molecules, at least some of which are encoded by MHC genes, which determine the specificity with which interacting partner cells can effectively communicate (Katz *et al.,* 1973a,b; Katz and Benacerraf, 1974; Katz and Benacerraf, 1975). To date, both of these possibilities are still viable and will require much additional experimentation to sort out.

The interesting paradox regarding the original discovery of MHC restrictions is that the first concepts that linked immunocompetent cell communication events to self-recognition of MHC gene products evolved from a phenomenon which depended on recognition of ''not-self''—namely, the allogeneic effect (Katz, 1972). The allogeneic effect described that phenomenon in which introduction of histoincompatible T cells to previously immunized recipients circumvented the normal requirement for antigen-specific helper T cells in secondary antibody responses; this resulted from the development of an active graft-versus-host reaction in recipient lymphoid organs. The very fact that the allogeneic effect stimulated target T or B cells as a result of interaction at their cell-surface MHC molecules prompted the consideration that perhaps precisely the same pathway was involved in syngeneic interactions, occurring perhaps by similar molecular mechanisms. Indeed, experiments designed to address this question demonstrated that physiologic *in vivo* T–B-cell interactions in the mouse were genetically restricted by MHC-linked genes. As summarized in Fig. 1, the basic observation was that antigen-specific T cells, primed to keyhole limpet hemocyanin (KLH), were capable of providing specific helper function for B cells, primed to the 2,4-dinitrophenyl (DNP) hapten, of semihistocompatible or histocompatible, but *not* histoincompatible, donor origin in secondary antibody responses of the IgG class (Katz *et al.,* 1973a,b). At about the same time, others demonstrated a requirement of *H-2* identity for successful thymus reconstitution of nude mice (Kindred and Shreffler, 1972) and the existence of MHC-linked genetic restrictions in macrophage–T-cell interactions in *in vitro* proliferation assays (Rosenthal and Shevach, 1973; Shevach and Rosenthal, 1973).

Genetic mapping studies established linkage of such genetic restrictions on T–B cell interactions to the *I* region of *H-2* (Katz and Benacerraf, 1975; Katz *et al.,* 1975). Since such experiments had been designed to specifically circumvent

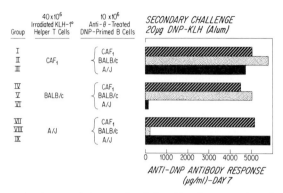

Fig. 1. Conventional parental, but not F_1, helper T cells are restricted in providing helper activity to partner B cells of isologous parental or F_1 type. Unirradiated CAF_1 recipient mice were injected with 40×10^6 KLH-primed spleen cells from CAF_1, BALB/c, or A/J donors. At 24 hr later, all recipients were irradiated with 650 rads and then injected with T cell-depleted, DNP-primed B cells from either CAF_1, BALB/c, or A/J donors, as indicated. All cell transfers were performed by the intravenous route. Shortly after the transfer of B cells, all recipients were challenged with 20 μg of DNP-KLH adsorbed on alum. The data are presented as geometric mean levels of serum anti-DNP antibodies in individual mice (5 mice/group) bled on day 7 after secondary challenge.

potential defects in macrophage–lymphocyte interactions and specific or non-specific suppressive effects (Katz *et al.*, 1973a,b,c; Katz *et al.*, 1974), the original interpretation was that genetic identity between the T cell and the B cell was necessary for the relevant T-cell surface molecules, distinct from antigen-specific receptors, to bind to the corresponding B-cell molecule (termed "acceptor" sites) for effective interactions to occur (Katz *et al.*, 1973a,b). The respective molecules were defined as cell interaction (CI) molecules and the *I* region genes encoding them as *CI* genes (Katz and Benacerraf, 1975).

Subsequently, the involvement of MHC gene products in controlling the ability of cytotoxic T lymphocytes (CTL) to effectively lyse the virus-infected, chemically modified or minor H antigen-bearing target cells was found (Bevan, 1975; Blanden *et al.*, 1975; Gordon *et al.*, 1975; Koszinowski and Ertl, 1974; Schmitt-Verhulst and Shearer, 1975; Shearer, 1974; Zinkernagel and Doherty, 1974a,b). These observations demonstrated that CTL are most efficient in lysing target cells derived from a similar MHC genotype (Katz *et al.*, 1978; Zinkernagel *et al.*, 1978b), with the critical genetic loci involved mapping to the *K* and *D* regions of the MHC, thus differing from the *I* region location of the MHC genes involved in T-cell–B-cell–macrophage interaction.

In view of the substantial experimental and theoretical attention that has been accorded to this subject, one could validly question whether any evidence exists to indicate that such genetic restrictions, which are, by necessity, identified and demonstrated under experimentally contrived circumstances, are physiologically relevant. There are at least four pieces of information which support the belief that MHC restrictions portray the actual physiology of cell–cell communication in the lymphohematopoietic system. First, the allogeneic effect demonstrates unequivocally that specific interaction at cell-surface MHC molecules induces a discrete and measureable biological response by the target cell in such interactions (Katz, 1972), thus proving that MHC molecules can play a role in cell triggering. Second, the existence of MHC-linked immune response (*Ir*) genes, which map in precisely the same genetic locations as *CI* genes, determine the immune response phenotype of an individual to various specific antigens, thus linking indisputably the MHC to functional responsiveness (Benacerraf and Katz, 1975; Benacerraf and McDevitt, 1972; McDevitt *et al.*, 1972). Third, the fact that *CI* genes determine the effectiveness of cell–cell interactions necessary for nonlymphoid hematopoietic stem-cell differentiation (Lengerova *et al.*, 1973; Sharkis *et al.*, 1979) provides evidence for a biological significance for such restrictions that extends beyond the immune system. (Parenthetically, such genetic restrictions on cell–cell interactions which do not involve, in any obvious way, specific immunologic responses provide evidence for a recognition system independent of that employed for recognition of the antigenic universe.) Finally, it is now established that the self-recognition repertoire by which interacting cells perceive themselves most efficiently is influenced significantly by elements in the environmental milieu to which developing cells are exposed during their early differentiation, i.e., the process of adaptive differentiation (Katz, 1976, 1977; Katz and Benacerraf, 1976; Katz *et al.*, 1976).

Adaptive differentiation describes the process by which differentiating stem cells adapt their functionally expressed self-recognition repertoire, and hence their ultimate cooperating phenotype, as a result of exposure to the MHC phenotype of the environment in which they differentiate (Katz, 1976, 1977; Katz and Benacerraf, 1976; Katz *et al.*, 1976). This has been substantiated largely by experimentation with irradiation bone marrow chimeras, most notably the results obtained with chimeric lymphocytes of $F_1 \rightarrow$ parent type [lethally irradiated parental hosts, of A or B type, repopulated with $(A \times B)F_1$ bone marrow stem cells] which no longer display the indiscriminate interacting phenotype (i.e., for either parent) typical of conventional F_1 lymphocytes, but rather interact preferentially with partner cells of host parental type or F_1 type (Bevan, 1977; Hodes *et al.*, 1980; Kappler and Marrack, 1978; Katz *et al.*, 1978; Singer *et al.*, 1979; Sprent, 1978a,b; Waldmann, 1978; Waldmann *et al.*, 1978; Zinkernagel *et al.*, 1978a,b). Still to be determined are the ground rules of adaptive differentiation and the underlying mechanisms by which the self-recognition repertoire is sculpted by elements in the environment.

III. STUDIES ON ADAPTIVE DIFFERENTIATION OF
LYMPHOCYTES IN BONE MARROW CHIMERAS

As an example of such studies (Katz *et al.*, 1978), we tested the capacities of helper T lymphocytes and hapten-specific B lymphocytes primed in the environments of various combinations of bone marrow chimeras prepared between two parental strains (i.e., A/J and BALB/c) and their corresponding F_1 hybrid (CAF_1) to interact with primed B and T lymphocytes derived from conventional parent and F_1 donors as well as all of the corresponding bone marrow chimera combinations. Our results demonstrated clearly that (a) $F_1 \rightarrow F_1$ chimeric lymphocytes displayed no restrictions in terms of cooperative activity with all of the various partner-cell combinations; (b) parent$\rightarrow F_1$ chimeric lymphocytes manifested effective cooperative activity only for partner cells from F_1 or parental donors corresponding to the haplotype of the original bone marrow donor, thereby behaving phenotypically just like conventional parental lymphocytes; and (c) as shown in Fig. 2, $F_1 \rightarrow$parent chimeric lymphocytes displayed restricted haplotype preference in cooperating best with partner lymphocytes sharing the H-2 haplotype, either entirely or codominantly, of the parental chimeric host. Suitable control studies ruled out the existence of either nonspecific or specific

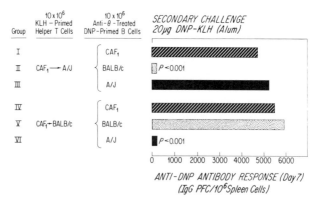

Fig. 2. Spleen cells from KLH-primed $CAF_1 \rightarrow A/J$ (groups I–III) and $CAF_1 \rightarrow BALB/c$ (groups IV–VI) were cotransferred with T cell-depleted B cells from DNP-*Ascaris*-primed conventional CAF_1, BALB/c, or A/J donors into 650-rad irradiated CAF_1 recipients. All recipients were challenged with 20 μg of DNP-KLH in alum shortly after cell transfer. The data are presented as geometric mean levels of IgG plaque-forming cells/10^6 spleen cells of groups of four recipients each assayed on day 7 after cell transfer and challenge. Statistically significant differences, as measured by Student's *t* test, are indicated adjacent to the pertinent horizontal bar. The cells provide helper activity for *in vivo* adoptive secondary responses, displaying the restricted cooperating phenotype of the parental host. (Adapted from Katz *et al.*, 1978.)

suppression mechanisms as possible explanations for the restricted partner-cell preference of $F_1 \rightarrow$ parent chimeric lymphocytes.

Since similar observations with bone marrow chimeras in the CTL systems were interpreted as evidence for a central role of the thymus in dictating the self-recognition repertoire to precursor T lymphocytes, we conducted experiments in which the cooperating preference of helper T cells originating from F_1 bone marrow, but differentiating in adult thymectomized, lethally irradiated F_1 recipients, reconstituted with either F_1 or homozygous parental thymus grafts was investigated (Katz *et al.*, 1979a). Cooperating preference was assayed by determining the levels of helper activity provided by antigen-primed T cells derived from such thymic chimeras for hapten-primed B lymphocytes obtained from conventional F_1 or parental donors in adoptive secondary antibody responses *in vivo*. The results of these analyses revealed a tendency for helper T cells derived from parental thymic chimeras to provide better help for B cells of the same parental type corresponding to the origin of the thymus graft than for the opposite parent. Such preference was, however, only marginal and only rarely statistically significant. Moreover, this marginal preference, when observed, pertained only to responses of the IgG class; no concordant preference in helper activity for IgE antibody responses was observed even with the same populations of thymic chimera helper T cells. Finally, in no instance was there any evidence of "restriction" in the classical sense of presence versus absence of help, as routinely observed in studies concerning genetic restrictions of T–B-cell cooperative interactions using conventional lymphoid cell populations.

Thus, our studies with thymectomized $F_1 \rightarrow F_1$ bone marrow chimeras reconstituted with parental thymuses demonstrated that the thymic microenvironment exerts relatively little influence on the cooperative phenotype of helper T cells generated in such thymic chimeras. The next chimera study was conducted to analyze further the sites of dominant influence on lymphocyte maturation with regard to the self-recognition capabilities normally displayed by regulatory helper T cells (Katz *et al.*, 1980b). This was accomplished by utilizing lymphocytes obtained from a variety of differentiating environments including (1) $F_1 \rightarrow$ parent and (2) intact parental mice rendered tolerant as neonates to the MHC determinants of a second parental strain. Lymphocytes were removed from these environments and adoptively primed to KLH in irradiated, thymectomized F_1 recipients. The resulting helper T cells were then analyzed for their partner-cell preferences when mixed with conventional DNP-primed B lymphocytes of either parental or F_1 origin in adoptive secondary responses in irradiated F_1 recipients. As shown in Fig. 3, irrespective of their initial environmental origins, T cells of such types could be adoptively primed to develop totally unrestricted helper cell activity for B lymphocytes of both parental types as well as B cells of F_1 type. Such results indicate that the dominant influence on cooperative capabilities of

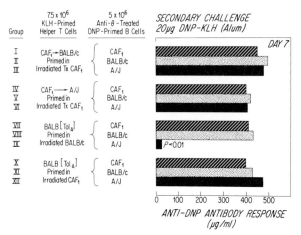

Fig. 3. Spleen cells from unprimed $F_1 \rightarrow$ parent chimeras (groups I–VI) were adoptively primed to KLH in irradiated thymectomized conventional CAF_1 recipients. Spleen cells from BALB/c mice, rendered tolerant to A/J MHC determinants as neonates (by injection of 50×10^6 irradiated CAF_1 spleen cells within the first 24 hr after birth) were adoptively primed to KLH in either irradiated BALB/c recipients (groups VII–IX) or irradiated CAF_1 recipients (groups X–XII). All adoptive priming consisted of injecting 50×10^6 donor spleen cells into irradiated recipients which were then immunized with 20 μg of KLH in CFA immediately after transfer. The adoptively primed KLH-specific helper cells were recovered 7 days later and cotransferred with T cell-depleted B cells from DNP–*Ascaris*-primed conventional CAF_1, BALB/c, or A/J donors into 650-rad irradiated CAF_1 recipients. All recipients were secondarily challenged with 20 μg of DNP-KLH in alum shortly after cell transfer. The data are presented as geometric mean levels of serum anti-DNP antibodies in groups of four recipients each. Statistically significant differences are indicated adjacent to the pertinent horizontal bar. The cells display unrestricted helper activity for parental B cells following adoptive priming in irradiated F_1 recipients. (Adapted from Katz *et al.*, 1980b.)

helper T cells is exerted by the extrathymic microenvironment in which such cells undergo their early differentiation. Moreover, they demonstrate that the haplotype restriction displayed by helper T cells primed in, and taken directly from, $F_1 \rightarrow$ single parent chimeras is actually a pseudorestriction, since helper T cells with unrestricted cooperating phenotypes can be induced in such $F_1 \rightarrow$ single parent chimeric populations when adoptively primed in irradiated F_1 recipients. This pseudorestriction in cooperative capabilities was explained by a new concept termed environmental restraint.

Environmental restraint describes the process by which the environmental milieu can exert nonpermissive influences on the development of functional interacting partner cells corresponding to one of the possible (and actually existing) CI phenotypes inherent in a given lymphoid cell population. In other words, despite the fact that the F_1 lymphoid cells residing in an $F_1 \rightarrow$ parent chimera consist of self-recognizing subpopulations corresponding to each of the two inherited parental CI types, the parental host environment is permissive for expression (in that environment) of only that subpopulation corresponding to the CI phenotype of the parental host; that same environment is nonpermissive for emergence of the second parental type subpopulation for reasons that have yet to be delineated.

Thus, our current working hypothesis is that adaptive differentiation is a dynamic, rather than a static process and that the self-recognition repertoire within a given species enjoys a certain degree of plasticity. Moreover, we feel that the plasticity of the self-recognition repertoire is determined by the occurrence of responses against self-specific receptors for CI molecules (i.e., αCI) and these, in turn, determine the immune response phenotype for a given individual.

Perhaps the most important lesson from the $F_1 \rightarrow$parent chimera experiments has been the realization that the answers to all of the mysteries pertaining to self-recognition and adaptive differentiation are present in conventional heterozygous F_1 individuals. Consequently, experimental analysis of lymphoid cells from F_1 hybrids under various circumstances should allow us to unravel such mysteries. As shown in Fig. 4, it can be viewed that an $(A \times B)F_1$ individual contains a minimum of three subsets of self-specific interacting partner cells, one each corresponding to the two respective parental types (A and B) and the third corresponding to an F_1-specific subset (A/B). Each respective subset carries specific CI molecules (CI_A, CI_B, and $CI_{A/B}$), for which there are corresponding αCI receptors (αCI_A, αCI_B, $\alpha CI_{A/B}$). One need only envisage the possibility that responses can be generated against such αCI receptors (i.e., anti-αCI) under certain circumstances to realize that the cooperating phenotype can exhibit considerable plasticity.

The occurrence of such anti-αCI responses was suggested by experiments demonstrating that the cooperating phenotypes of conventional F_1 lymphocytes could be orchestrated by certain experimental manipulations, including (1) parental cell-induced allogeneic effects during priming of either T or B lymphocytes (Katz *et al.,* 1979b, 1980a) and (2) incorporation of lymphoid cells of parent B-type into cooperative interactions between F_1 hybrid T cells and B cells of parent A-type, and vice versa (Katz *et al.,* 1981a; also see later in this chapter). In both types of experiments, appropriate controls ruled out allosuppression phenomena or blocks in effective macrophage–lymphocyte interactions. Most importantly, the effects observed were exquisitely haplotype specific. The development of anti-αCI responses could explain the permissiveness of expression of one subpopulation of self-recognizing cells in the face of nonpermissiveness of expression of the second subpopulation of self-recognizing cells. Likewise, such anti-αCI responses provide a suitable explanation for manifestations of environmental restraint within an $F_1 \rightarrow$parent chimera, as already discussed.

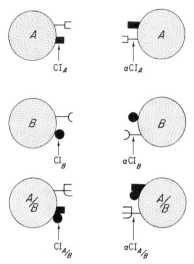

Fig. 4. Depicted are the three minimal subsets of potential self-specific interacting partner cells in heterozygous (A × B)F$_1$ individuals. Subsets A and B correspond to the inherited CI specificities of the respective parental A and B donor mice, while subset A/B represents a unique F$_1$-specific subset of interacting cells. The corresponding CI$_A$, CI$_B$, and CI$_{A/B}$ target molecules and the corresponding receptors for such molecules (αCI$_A$, αCI$_b$, and αCI$_{A/B}$) are depicted.

IV. EXPERIMENTAL MANIPULATIONS OF THE COOPERATING PHENOTYPES OF CONVENTIONAL F$_1$ HYBRID HELPER CELLS

A series of experiments were conducted to demonstrate means by which to maneuver the cooperating phenotypes of conventional F$_1$ hybrid lymphocytes. The first of such studies demonstrated circumstances in which restrictions in F$_1$–parent partner-cell interactions determined by *Ir* genes could be willfully directed by induction of parental cell-mediated allogeneic effects during priming of the F$_1$ helper T-cell population to the antigen governed by the relevant *Ir* genes (Katz *et al.*, 1979b). Responses to the synthetic terpolymer L-glutamic acid, L-lysine, L-tyrosine (GLT) in the mouse are controlled by the *H-2*-linked *Ir-GLT* genes. The (responder × nonresponder) F$_1$ hybrid mice, themselves phenotypic responders, can be primed with GLT to develop specific helper cells capable of interacting with DNP-primed F$_1$ B cells in response to DNP-GLT. Unlike the indiscriminate ability of F$_1$ helper T cells for conventional antigens (i.e., not *Ir* gene-controlled), which can help B cells of either parental type (as well as F$_1$) equally well, GLT-primed F$_1$ T cells can only provide help under normal circumstances for B lymphocytes of responder parent origin (Katz *et al.*, 1973c); they are unable to communicate effectively with nonresponder parental B cells (Fig. 5).

However, the induction of a parental cell-induced allogeneic effect during priming of F_1 mice to GLT actually dictates the direction of cooperating preference that will be displayed by such F_1 helper cells for B cells of one parental type or the other. Thus, as shown in Fig. 5, F_1 T cells primed to GLT under the influence of an allogeneic effect induced by parental BALB/c cells developed into effective helpers for nonresponder A/J B cells but failed to develop effective helpers for responder BALB/c B cells, and vice versa. In contrast, F_1 T cells primed to GLT under the influence of an allogeneic effect induced by either parental type displayed significantly enhanced levels of helper activity for B cells derived from F_1 donors (Katz et al., 1979b).

These results were interpreted to reflect the existence of two interdependent events provoked by the allogeneic effect. One event augments the differentiation of GLT-specific helper T cells belonging to the subset corresponding to the opposite parental type; this would explain the development of increased helper activity provided to partner B cells of opposite parental type (as well as of F_1 origin). The second event, we postulated, involves the production of responses

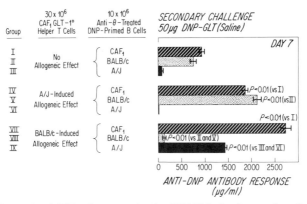

Fig. 5. Conventional CAF_1 mice were primed to GLT (1) in the absence of an allogeneic effect, with 50 μg of GLT in CFA followed 10 days later by a second injection of 50 μg in saline (groups I–III); or (2) under the influence of an allogeneic effect induced by i.v. injection (on day 10 after initial immunization with 50 μg of GLT in CFA) of 25×10^6 spleen cells from either parental A/J (groups IV–VI) or BALB/c (groups VII–IX) donors just prior to the second injection of 50 μg of GLT in saline. All GLT-primed spleen cells were recovered 7 days after the second injection to be used as helper cells. These cells were cotransferred with T cell-depleted DNP–Ascaris-primed B cells from conventional CAF_1, BALB/c, or A/J donors into 650-rad irradiated CAF_1 recipients. All adoptive recipients were secondarily challenged with 50 μg of DNP-GLT in saline shortly after cell transfer. The data are presented as geometric mean levels of serum anti-DNP antibodies of individual mice in groups of 5 mice each bled on day 7 after cell transfer and secondary challenge. Horizontal lines represent the range of standard errors, and relevant statistically significant differences versus corresponding control groups are indicated adjacent to the horizontal bars. The cells can thus be converted in their cooperating phenotype by priming under the influence of a parental cell-induced allogeneic effect. (Adapted from Katz et al., 1979b.)

against the receptors which normally self-recognize native CI determinants; this form of anti-αCI response is restricted against self-recognizing receptors of the same parental type used for induction of the allogeneic effect, hence explaining diminished helper activity of such F_1 cells for partner B lymphocytes of corresponding parental type. The existence of haplotype-specific anti-αCI receptor responses was postulated to explain the permissiveness of the development of one subpopulation of self-recognizing cells (corresponding to one of the parental haplotypes) in the face of nonpermissiveness of the development of the subpopulation of self-recognizing cells corresponding to the second haplotype involved. Moreover, it is not difficult to envisage that the existence of such a mechanism might explain mechanisms underlying environmental restraint as already described.

   The ability to orchestrate the cooperating phenotype of (responder × nonresponder) F_1 GLT-specific helper T cells, prompted us to investigate whether the success of such manipulations was unique to responses controlled by *H-2*-linked *Ir* genes, or whether priming F_1 lymphocytes to *any* antigen under the influence of a transient allogeneic effect would result in a similar deviation in cooperating preferences for partner cells of one or the other parent type. Additionally, it became important to ascertain whether F_1 B lymphocytes could be similarly directed in their cooperating partner cell preferences when primed under the influence of a parental cell-induced allogeneic effect. We have found that this ability to orchestrate the cooperating preferences of F_1 lymphocytes is not unique to antigen systems under *H-2*-linked *Ir* gene control, and is a property demonstrable in B lymphocytes as well as T lymphocytes (Katz *et al.*, 1980a).

   Thus, as shown in Fig. 6, the normally indiscriminate pattern of cooperation exhibited by F_1 lymphocytes can be deviated toward a pattern of cooperation manifesting preference for partner cells of one or the other parental type. This was accomplished by priming F_1 lymphocytes to conventional antigens under the influence of a transient allogeneic effect induced by the transfer of parental cells of one type or the other. The deviation in partner-cell preference resulting from such parental cell-induced allogeneic effects during antigen sensitization was displayed by both helper T lymphocytes and hapten-specific B lymphocytes, was shown to be exquisitely haplotype-specific, and did not reflect the activities of contaminating parental cells carried over into the final transfer assay system.

   This ability to orchestrate the cooperating phenotypes of F_1 lymphocytes derived from conventional, nonchimeric mice is compatible with the postulated concept of environmental restraint. The development of anti-αCI receptor responses was again considered as a likely mechanism involved in the process of environmental restraint and perhaps participating in the processes of adaptive differentiation that dictate the normally displayed self-receptor repertoire of interacting cells within the immune system.

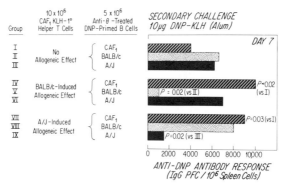

Fig. 6. CAF_1 mice were primed to KLH (1) in the absence of an allogeneic effect by immunization with 10 μg of KLH in alum followed 10 days later by 10 μg of KLH in saline (groups I–III); or (2) under the influence of an allogeneic effect induced by i.v. injection of 25×10^6 parental BALB/c (groups IV–VI) or A/J (groups VII–IX) donors 10 days after initial immunization and immediately prior to the secondary boost with 10 μg of KLH in saline. Spleen cells were obtained from such KLH-primed mice 7 days after the cell transfer and secondary boost and were cotransferred with T cell-depleted DNP–*Ascaris*-primed B cells from conventional CAF_1, BALB/c, or A/J donor mice into 650-rad irradiated CAF_1 recipients. All recipients were secondarily challenged with 10 μg of DNP-KLH in alum shortly after cell transfer. The data are presented as geometric mean levels of individual IgG DNP-specific plaque-forming cells/10^6 spleen cells of groups of 5 mice assayed on day 7 after cell transfer and secondary challenge. Statistically significant differences between experimental and control groups are indicated adjacent to the horizontal bars. The lymphocytes are specifically diminished in helper capacity for partner cells of the same parental type. (Adapted from Katz *et al.*, 1980a.)

Additional support for the idea pertaining to anti-αCI receptor responses came from experiments demonstrating that F_1–parent T–B-cell cooperation *in vivo* is significantly diminished by the presence of lymphoid cells of opposite parental type (Katz *et al.*, 1981a). This inhibition phenomenon is not a straightforward allosuppression mechanism, as it can be induced by parental lymphoid cells depleted of T cells, it does not operate on cooperative interactions between homologous T and B cells of opposite parental type, and it absolutely requires the presence of F_1 cells as participants in the reactions generated. Since the presence of parental lymphoid cells only affected cooperative interactions between F_1 T cells and B lymphocytes of opposite parental type, but had no inhibitory effect on cooperative interactions between homologous F_1 T and B cells, this strongly argues for the existence of one or more subsets of F_1-interacting partner cells that are uniquely specific for F_1, as distinct to either parental type CI determinants. Moreover, it again appears that the most likely mechanism underlying such parental cell-induced inhibitory effects on F_1–parent partner cell interactions is the development of anti-self CI receptor responses by F_1 cells against the relevant self receptors of the parental partner cells involved.

V. PARALLELISMS BETWEEN THE CELL INTERACTION AND IMMUNE-RESPONSE PHENOTYPES

Since the discovery of immune response (*Ir*) genes by Benacerraf and McDevitt and their colleagues (Benacerraf and McDevitt, 1972), much effort has been directed toward delineating the nature of these genes and the mechanism by which they determine the ability of an individual to develop an immune response to a specific antigen. The discovery of MHC-linked genetic control of interactions between T cells and B cells (Katz *et al.*, 1973a,b; Kindred and Shreffler, 1972) and between T cells and macrophages (Rosenthal and Shevach, 1973; Shevach and Rosenthal, 1973) added additional complexities to these questions, particularly when the *CI* genes were mapped to the *I* region of the murine *H-2* complex (Katz *et al.*, 1973c; Katz and Benacerraf, 1975; Katz *et al.*, 1975).

Even before the final mapping of *CI* genes to the *I* region had been accomplished, experimental evidence was obtained strongly indicating a crucial functional linkage between *CI* and *Ir* genes. The first such evidence was the observation, described above, that T cells from (responder × nonresponder) F_1 hybrids primed to the synthetic terpolymer GLT, to which responses are governed by *Ir-GLT* genes, were restricted in providing GLT-specific help for DNP-primed B cells only from phenotypic responder parental and F_1 donors in response to DNP-GLT; the same F_1 T-cell population was incapable of helping B cells obtained from nonresponder parental donors (Katz *et al.*, 1973c) (see Fig. 5, groups I–III). Since F_1 T cells can indiscriminately interact effectively with partner B cells from either parent when the carrier antigen employed is not one to which responses are governed by a known *Ir* gene (Katz *et al.*, 1973a,b), this restricted cooperating phenotype in the DNP-GLT experiment clearly signalled a role of *Ir* genes in determining the partner cell preferences in such cooperative interactions. For this and other reasons we have concluded that *Ir* and *CI* genes are one and the same. If this is true, then one would predict that the immune response phenotype should exhibit comparable plasticity to that already demonstrated for *CI* gene-determined cooperative phenotypes based on the environment in which stem cells differentiate. Indeed, several reports have appeared which indicate that this is so in bone marrow chimeras (Billings *et al.*, 1978; Hedrick and Watson, 1979; Kappler and Marrack, 1978; Longo and Schwartz, 1980; Miller, 1978; von Boehmer *et al.*, 1978).

The hypothesis that we have put to experimental testing can thus be stated as follows. Returning to the model schematically illustrated in Fig. 4, which illustrates at least three minimal subsets of self-specific interacting partner cell subsets in a conventional heterozygous (A × B)F_1 individual, it seems clear that when such an individual is immunized with an antigen to which there are no restrictions imposed on responses by *Ir* genes, all three subsets of interacting cells will be functionally expressed. Hence the cooperating phenotype of the lymphocyte population from this F_1 individual will appear totally unrestricted in terms of cooperating with partner cells of both parents as well as of F_1 origin.

In contrast, when an F_1 individual is exposed to an antigen to which response in one of the parental haplotypes is restricted by a specific *Ir* gene, the development of functionally interacting subsets follows a different course. Thus, as depicted schematically in Fig. 7, exposure of an (A × B) F_1 hybrid to GLT, to which parent A is a nonresponder, results in development of functional expression of only the B and A/B responder subsets of interacting cells; the parental A subset is functionally silent, as evidenced by the restricted phenotype of F_1 cells described in the original DNP-GLT studies (Katz *et al.*, 1973c). The question that we have specifically addressed is whether the functional silence of the parent A subset under these circumstances might be a manifestation of the development of an anti-αCI_A response provoked by exposure of the lymphoid system to GLT.

In order to explore the influence of the extrathymic corporeal environment on

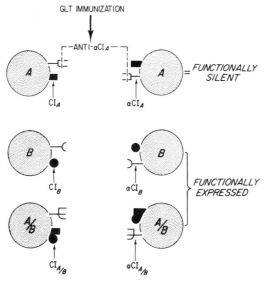

Fig. 7. As in Fig. 4, depicted are the minimal subsets of potential self-specific interacting partner cells in heterozygous (A × B) F_1 individuals, with the corresponding CI and αCI molecules displayed on the surfaces of such cells. In response to a conventional antigen, such as KLH, all three subsets of interacting cells would presumably be activated thus explaining the indiscriminate cooperative phenotype of F_1 T cells with partner cells of either parental type or of F_1 type. In contrast, in response to GLT (to which the parent A strain is a nonresponder) the model proposes that there develops a rather immediate anti-αCI$_A$ response which renders that particular subset functionally silent; the remaining B and A/B subsets are functionally expressed hence leading to the phenotype of effective cooperation by GLT-specific F_1 cells for partner cells of parent B and (A × B) F_1 type, but no cooperative activity for partner cells of parent A type.

David H. Katz

the immune response phenotype, thymic chimeras were constructed by reconstituting lethally irradiated, thymectomized (1) CAF_1, BALB/c, or A/J recipients with CAF_1 bone marrow cells and CAF_1 thymus grafts, or (2) CAF_1 recipients with both bone marrow cells and thymus grafts obtained from nonresponder parental A/J donors. These chimeras were then immunized with unconjugated GLT and analyzed for their capacities to develop GLT-specific antibody responses (Katz et al., 1981b).

As summarized in Fig. 8, $CAF_1 \rightarrow CAF_1$ and $CAF_1 \rightarrow$ BALB/c chimeras, both possessing CAF_1 thymus grafts, developed comparable GLT-specific antibody responses. In striking contrast, chimeras of $CAF_1 \rightarrow$ A/J type failed to produce detectable levels of anti-GLT antibody responses despite the fact that such chimeras possessed thymus grafts of CAF_1 origin. This failure to develop GLT-specific antibody responses did not reflect ineffective thymic reconstitution, since such mice were able to develop comparable KLH-specific antibody responses such as those displayed by corresponding $CAF_1 \rightarrow CAF_1$ controls when such animals were immunized with unconjugated KLH subsequent to the GLT

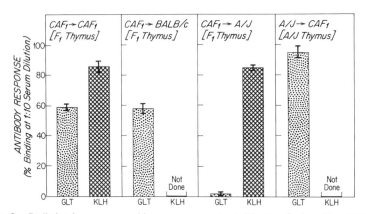

Fig. 8. Radiation bone marrow chimeras were constructed by transferring either CAF_1 bone marrow cells into thymectomized, lethally irradiated (950 rad) CAF_1, BALB/c, or A/J recipients (panels 1–3) or A/J bone marrow into thymectomized, lethally irradiated CAF_1 recipients (panel 4). Recipients were transplanted 2 weeks later with thymuses of the donor type indicated, under the kidney capsules. All mice were typed for H-2 3–4 months after reconstitution and were rested until 9 months after reconstitution before immunization with GLT and/or KLH. All mice were immunized i.p. with 50 μg of GLT in CFA on day 0 and boosted with 50 μg of GLT in saline on day 14. The data presented are mean percent binding of ^{125}I-labeled GLT of 1:10 dilution of individual serum samples from bleedings of groups of 4 mice each on day 24 (10 days after boosting). Standard errors are indicated by the vertical line on each bar. Mice immunized with KLH (panels 1 and 3) were immunized i.p. on day 30 (after initiation of GLT immunizations) with 20 μg of KLH in CFA and boosted on day 40 with 10 μg of KLH in saline. The data presented are mean percent binding of ^{125}I-labeled KLH of 1:10 dilutions of individual serum samples from bleedings on day 47. Thus, the corporeal environment dictates the *Ir* gene phenotype of thymic bone marrow chimeras in responses to GLT. (Adapted from Katz, 1980.)

immunization regimen. On the other hand, chimeras constructed with lymphoid stem cells and thymus grafts of nonresponder A/J parental origin, which had differentiated in the environment of CAF_1 hosts, developed excellent GLT-specific antibody responses.

This experiment may offer significant insight on the mechanism(s) by which Ir genes determine the immune response phenotype. The most pertinent findings, displayed by the $CAF_1 \rightarrow A/J$ and $A/J \rightarrow CAF_1$, indicate quite clearly that elements in the corporeal environment may determine the Ir phenotype of a given individual. This conclusion follows from the finding that stem cells from phenotypic responder F_1 donors that mature in an environment containing homologous F_1 thymus display the nonresponder phenotype characteristic of the remainder of the corporeal environment provided by the nonresponder parental host. Reciprocally, stem cells from phenotypic nonresponder parental donors differentiate in a corporeal environment provided largely by phenotypic responder F_1 elements, with the exception of the nonresponder parental thymus graft, to display phenotypic responsiveness to GLT. In other words, the Ir phenotypes in these circumstances reflect the permissiveness of the phenotypic responder F_1 environment, on the one hand, and the nonpermissiveness of the phenotypic nonresponder parental environment, on the other.

In order to ascertain to what extent lymphoid cells interact with other lymphoid as well as nonlymphoid elements in a chimeric environment, mixed-parent chimeras were constructed by reconstituting lethally irradiated CAF_1 recipients with equivalent numbers of responder BALB/c and nonresponder A/J parental bone marrow cells. Six months after reconstitution, these double-parent chimeras were primed with GLT in order to generate GLT-specific helper T cells. Spleens were removed from such mice, treated with BALB/c anti-A/J antibodies plus C to remove any cells of parental A/J type or of recipient F_1 type; the remaining "Chim.BALB/c" splenic cells were then tested for cooperative helper activity when co-transferred with DNP-primed B cells of CAF_1, BALB/c, or A/J origin in response to secondary challenge with DNP-GLT. The cooperative phenotype of "Chim.BALB/c" helper T cells was compared with that of GLT-primed helper T cells taken from conventional CAF_1 donors cotransferred with portions of the same populations of DNP-primed B cells.

As shown in Fig. 9, GLT-primed conventional CAF_1 helper T cells displayed the normal cooperative phenotype of providing good helper activity for B cells of responder F_1 and parental BALB/c origins, but not for B cells of nonresponder A/J origin (groups I–III). The cooperative phenotype of GLT-primed "Chim.BALB/c" helper T cells (groups IV–VI) provides a striking contrast. Such cells displayed excellent helper activity for DNP-primed partner B cells of CAF_1 type but failed to engage in effective interactions with either responder BALB/c or nonresponder A/J partner B cells.

The failure of "Chim.BALB/c" T cells to provide GLT-specific helper ac-

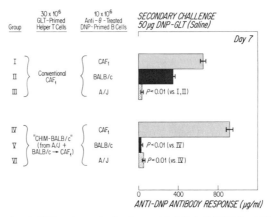

Fig. 9. Conventional CAF_1 mice and mixed parental A/J + BALB/C→CAF_1 chimeras were immunized with 50 μg of GLT in CFA followed by a single boost of GLT in saline 3 weeks thereafter. Spleen cells were obtained from such GLT-primed donor mice 3–4 weeks after the last saline boost to be used as helper cells. "Chim.BALB/c" spleen cells were obtained from such mixed parent→F_1 GLT-primed chimeras by treatment of the spleen cells *in vitro* with BALB/c anti-A/J antibodies + C. Then 30×10^6 GLT-primed conventional CAF_1 and "Chim.BALB/c" spleen cells were transferred together with T cell-depleted DNP–*Ascaris*-primed B cells from conventional CAF_1, BALB/c, or A/J donor mice into 650-rad irradiated CAF_1 recipients. All recipients were secondarily challenged with 50 μg of DNP-GLT in saline shortly after cell transfer. The data are presented as geometric mean levels of individual serum anti-DNP antibodies of groups of 5 mice each assayed on day 7 after cell transfer and secondary challenge. Horizontal lines represent standard errors, and relevant P values depicting statistically significant differences between experimental and control groups are indicated beside the corresponding horizontal bars. The cells thus provide GLT-specific helper activity to F_1, but not to either responder or nonresponder parental, B cells. (Adapted from Katz, 1980.)

tivity for either BALB/c or A/J partner B cells in response to DNP-GLT is not a reflection of some general abnormality existing in such mixed-parental chimeras. Nor are these data a reflection of some unusual properties of the partner B cells employed in this experiment with respect to their ability to interact with mixed-parental chimera T cells. Thus, "Chim.BALB/c" helper T cells obtained from the same group of mixed-parental→F_1 chimeras, but primed to KLH rather than to GLT, provided adequate helper T-cell activity for aliquots of the same B cells as those used in Fig. 9 in secondary adoptive responses to DNP-KLH, and such helper activity was comparable with the two parental-type partner B cells as well as with F_1 B cells.

The preceding experiments clearly demonstrate three critical points about the GLT system. First, the immune response phenotype of a given individual is not dictated by the nature of the thymic microenvironment; second, one or more elements in the extrathymic corporeal environment determine the permissiveness of immune response capability; and third, such elements are not derived primarily from the lymphoid stem cell pool, although lymphoid cells may interact with such corporeal elements in the determination of the immune response phenotype. It should be noted that no conclusion can be reached about the cellular locus at which the mechanism(s) determining *Ir* phenotype operates. For example, unresponsiveness to GLT displayed by $CAF_1 \rightarrow A/J$ chimeras could reflect a defect at the level of T cells, B cells, or macrophages, or any combination thereof, or at one or more of the requisite interactions between such cells. From the data in Fig. 9, it seems clear that nonpermissiveness can at least operate at the level of generation of a relevant subset of GLT-specific helper T cells, but again this could reflect a defect solely at the T-cell level or at the level of T-cell–macrophage and/or T-cell–T-cell interactions.

How, then, can such results be interpreted, and do they conform in any way to the hypothesis being tested—namely, that responses against CI molecules can determine the observed plasticity of the immune response phenotype? We believe that such anti-αCI responses could readily explain the mechanism by which *Ir* genes function to determine the immune response phenotype; it is only necessary to assume that *Ir* genes encode CI molecules. If one considers that CI molecules are distinct entities from antigen-specific receptors, then the manner in which *Ir* genes exert such exquisite specificity for antigen in responses over which they display control depends on whether *Ir* genes encode molecules serving as (a) αCI receptors alone (at least in part); (b) target CI molecules themselves; or (c) both αCI receptors and target CI molecules.

The above experiments are consistent with this notion. Thus, it is clear from these findings, as well as from our earlier studies in the *Ir-GLT* system (Katz *et al.*, 1973c) and from the work of others (Billings *et al.*, 1978; Kappler and Marrack, 1978; Longo and Schwartz, 1980), that expression of *Ir* phenotype is not a reflection of whether or not a given *I* region gene or genes is absent from the genome of an individual. Nor, for that matter, is there any structural evidence to indicate whether *Ir* phenotype is associated with the expression, or not, of the relevant *Ir* gene product(s). Data pertinent to this point arise from the results obtained with GLT-primed responder BALB/c T cells which had differentiated in the same environment with nonresponder A/J parental cells (Fig. 9). Such cells displayed a cooperating phenotype restricted for DNP-primed partner cells derived from conventional F_1 donors. The fact that differentiation and priming to GLT occurred in an environment where nonresponder parental lymphoid cells were also present obviously determined this unusual cooperating phenotype. We can think of no mechanism by which the presence of the cohabiting nonresponder A/J cells could have regulated expression of the relevant *I* region gene product by responder BALB/c cells that could account for functional deletion of the BALB/c-specific GLT helper T cell subset.

On the other hand, one *can* explain this observation by a mechanism involving responses against self-specific CI molecules. As shown schematically in Fig. 7, if a nonresponder individual displays that phenotype because, for whatever reason, exposure to GLT evokes a very strong (and early) anti-αCI response, this would, in turn, blunt any possible response to GLT from developing. Exposure of an individual of a responder phenotype to GLT, conversely, would not elicit this type of anti-αCI response under normal circumstances, and hence the environment of such an individual would be permissive for responses to GLT.

Why, then, does cohabitation of responder cells with nonresponder cells result in non-permissiveness for the population of responder self-specific cells? This could be explained by the fact that a state of mutual immunological tolerance exists between the cohabitating parental lymphoid cell populations in such chimeras (von Boehmer et al., 1975). A consequence of such mutual tolerance is the emergence within each parental lymphoid cell population of interacting subsets specific (in terms of CI molecules expressed and recognized) for the CI phenotype of the other parental population (Katz, 1980). Indeed, this point has been experimentally verified (Sprent and von Boehmer, 1979). It follows from this, therefore, that, for whatever reason GLT evokes a self-specific anti-αCI response in the nonresponder individual, the state of mutual tolerance in the mixed-parental chimeric environment would allow GLT to evoke a comparable response against the CI molecules displayed by the corresponding responder-specific subset reactive to GLT that originates from the responder stem-cell pool.

The fact that GLT-specific responder helper T cells capable of interacting with B cells of conventional F_1 donor origin were induced in such chimeras implies the existence of (1) an F_1-specific subset of T cells originating from the responder parental lymphoid population and, likewise, (2) a subset of F_1-specific partner B cells (distinct from the subsets corresponding in cooperating specificity to each of the two parental CI types) within the conventional CAF_1 partner B-cell population. Moreover, the presence of F_1-specific subsets of T and B cells within the mixed-parental chimera explains why such chimeras produced circulating anti-GLT antibodies in situ (not shown) despite the absence of detectable GLT-specific helper T cells of BALB/c-specific cooperating potential. The existence of F_1-specific cooperating helper T cells has recently been found in studies performed by Sproveira et al. (1980) and from our own laboratory (Katz et al., 1981).

Inherent in our thinking about CI molecules and their relationship to the immune response phenotype is the notion that in individual A there is heterogeneity among CI_A that we can denote $CI_{A1, A2, A3} \cdots {}_{An}$; for each CI_A specificity, there will be corresponding αCI_A receptors, i.e., αCI_{A1}, αCI_{A2}, αCI_{A3}, and so on (Katz, 1980). Moreover, one can further assume that within each CI_A subset are represented a given number of antigenic specificities in terms of distinct antigen-specific receptors. For example, let us assume that in a GLT nonresponder individual A, GLT-specific receptors may be affiliated on the same cells that belong to subset CI_{A1}. Anything that prevents the reaction

$\alpha CI_{A1} \rightarrow CI_{A1}$ could be manifested as specific unresponsiveness to GLT; for example, something analogous to an anti-αCI_{A1} reaction, as suggested earlier.

If this is the case, then one might anticipate that following immunization of a nonresponder individual with GLT, which might provoke such an anti-αCI_{A1} response, competence of that individual to mount responses against other antigenic determinants for which specific receptors are also affiliated with subset CI_{A1} might be, at least transiently, compromised. This speculation is very difficult to test experimentally at the moment, since the ability to detect such compromised responsiveness is hampered by the fact that a complex antigen, such as KLH, might display many major distinct antigenic determinants, receptors for each one of which could be affiliated with distinct CI_A subsets. Thus, temporary functional silence of subset CI_{A1} as a result of GLT immunization could indeed compromise the response to one of the major determinants displayed by KLH, but since responses against the other major determinants would not be similarly compromised, one would hardly detect any defect in the response to KLH under these circumstances. The collective results presented in Figs. 8 and 9 are compatible with this interpretation, since where responses to GLT were absent, there was no compromise noticeable in the ability of such animals to respond to KLH.

VI. CONCLUSIONS

The firmness of our grasp in understanding genetic control of lymphocyte recognition and differentiation processes has increased substantially in the 1980s. Thus, concepts which were hardly imagined in the mid-1970s concerning the role of the MHC in controlling cell–cell communication and certain aspects of recognition in the immune system have enabled us to view normal cell differentiation and its control with a quite different perspective. From these new perspectives have also developed new ideas in terms of the mechanisms by which immunocompetent cells transact their necessary and usually unmistakable communication processes which, we now know, determine the overall response pattern developed by the individual in both health and disease. It is probable that future studies will broaden our understanding of the genetic basis of self-recognition and cell–cell interactions which depend upon such self-recognition processes. Moreover, we should develop a clearer picture of the mechanisms underlying adaptive differentiation and the boundaries of the plasticity of phenotypic self-recognition. Finally, isolation and characterization of the CI molecules involved in such processes should clarify many of the existing ambiguities and questions that exist with respect to the general issue of MHC restrictions. In the broad sense, we might also expect that information obtained in studies such as these will be pertinent to furthering our basic knowledge of cell differentiation, receptor expression, self-recognition, and other developmental processes involved in multicellular organisms.

ACKNOWLEDGMENTS

Lucy Gunnill and Beverly Burgess provided the skilled assistance that made the preparation of this manuscript possible. This is publication number 59 from the Department of Immunology, Medical Biology Institute, La Jolla, California. The author's work was supported by National Institutes of Health grants AI 13781 and CA 25803 and March of Dimes–Birth Defects Foundation grant 1-540.

REFERENCES

Benacerraf, B. (1978). *J. Immunol.* **120,** 1809–1812.
Benacerraf, B., and Katz, D. H. (1975). *In* "Immunogenetics and Immunodeficiency," pp. 117–135. Medical and Technical Publ., London.
Benacerraf, B., and McDevitt, H. O. (1972). *Science* **175,** 273–279.
Bevan, M. J. (1975). *J. Exp. Med.* **142,** 1349–1364.
Bevan, M. J. (1977). *Nature (London)* **269,** 417–418.
Billings, P., Burakoff, S. J., Dorf, M. E., and Benacerraf, B. (1978). *J. Exp. Med.* **148,** 352–359.
Blanden, R. V., Doherty, P. C., Dunlop, M. B. C., Gardner, I. D., and Zinkernagel, R. M. (1975). *Nature (London)* **254,** 269–270.
Gordon, R. D., Simpson, E., and Samelson, L. E. (1975). *J. Exp. Med.* **142,** 1108–1120.
Hedrick, S. M., and Watson, J. (1979). *J. Exp. Med.* **150,** 646–652.
Hodes, R. J., Hathcock, K. S., and Singer, A. (1980). *J. Immunol.* **124,** 134–139.
Kappler, J. W., and Marrack, P. (1978). *J. Exp. Med.* **148,** 1510–1522.
Katz, D. H. (1972). *Transplant. Rev.* **12,** 141–179.
Katz, D. H. (1976). *Transplant. Proc.* **8,** 405–411.
Katz, D. H. (1977). *Cold Spring Harbor Symp. Quant. Biol.* **41,** 611–624.
Katz, D. H. (1980). *In* "Advances in Immunology," Vol. 29, pp. 137–150. Academic Press, New York.
Katz, D. H., and Benacerraf, B. (1974). *In* "The Immune System: Genes, Receptors, Signals," pp. 569–571. Academic Press, New York.
Katz, D. H., and Benacerraf, B. (1975). *Transplant. Rev.* **22,** 175–195.
Katz, D. H., and Benacerraf, B. (1976). *In* "The Role of Products of the Histocompatibility Gene Complex in Immune Responses," pp. 355–380. Academic Press, New York.
Katz, D. H., Hamaoka, T., and Benacerraf, B. (1973a). *J. Exp. Med.* **137,** 1405–1418.
Katz, D. H., Hamaoka, T., Dorf, M. E., and Benacerraf, B. (1973b). *Proc. Natl. Acad. Sci. U.S.A.* **70,** 2624–2628.
Katz, D. H., Hamaoka, T., Dorf, M. E., Maurer, P. H., and Benacerraf, B. (1973c). *J. Exp. Med.* **138,** 734–739.
Katz, D. H., Hamaoka, T., Dorf, M. E., and Benacerraf, B. (1974). *J. Immunol.* **112,** 855–857.
Katz, D. H., Graves, M., Dorf, M. E., DiMuzio, H., and Benacerraf, B. (1975). *J. Exp. Med.* **141,** 263–268.
Katz, D. H., Chiorazzi, N., McDonald, J., and Katz, L. R. (1976). *J. Immunol.* **117,** 1853–1859.

Katz, D. H., Skidmore, B. J., Katz, L. R., and Bogowitz, C. A. (1978). *J. Exp. Med.* **148,** 727–745.

Katz, D. H., Katz, L. R., Bogowitz, C. A., and Skidmore, B. J. (1979a). *J. Exp. Med.* **149,** 1360–1370.

Katz, D. H., Katz, L. R., Bogowitz, C. A., and Maurer, P. H. (1979b). *J. Exp. Med.* **150,** 20–30.

Katz, D. H., Katz, L. R., and Bogowitz, C. A. (1980a). *J. Immunol.* **125,** 1109–1116.

Katz, D. H., Katz, L. R., Bogowitz, C. A., and Bargatze, R. F. (1980b). *J. Immunol.* **124,** 1750–1757.

Katz, D. H., Katz, L. R., and Bogowitz, C. A. (1981a). *J. Exp. Med.* **153,** 407–422.

Katz, D. H., Katz, L. R., Bogowitz, C. A., and Maurer, P. H. (1981b). *J. Immunol.* **127,** 1103–1109.

Kindred, B., and Schreffler, D. C. (1972). *J. Immunol.* **109,** 940–943.

Koszinowski, U., and Ertl, H. (1974). *Nature (London)* **248,** 701–702.

Lengerova, A., Matousek, V., and Zeleny, V. (1973). *Transplant. Rev.* **15,** 89–122.

Longo, D. L., and Schwartz, R. H. (1980). *J. Exp. Med.* **151,** 1452–1467.

McDevitt, H. O., Deak, B. D., Schreffler, D. C., Klein, J., Stimpling, J. H., and Snell, G. D. (1972). *J. Exp. Med.* **135,** 1259–1278.

Miller, J. F. A. P. (1978). *Immunol. Rev.* **42,** 76–107.

Rosenthal, A. S. (1978). *Immunol. Rev.* **40,** 136–152.

Rosenthal, A. S., and Shevach, E. M. (1973). *J. Exp. Med.* **138,** 1194–1212.

Schmitt-Verhulst, A. M., and Shearer, G. M. (1975). *J. Exp. Med.* **142,** 914–927.

Sharkis, S. J., Cahill, R., Ahmed, A., Jedrzekczak, W. W., and Sell, K. W. (1979). *Transplant. Proc.* **II,** 511–514.

Shearer, G. M. (1974). *Eur. J. Immunol.* **4,** 527–533.

Shevach, E. M., and Rosenthal, A. S. (1973). *J. Exp. Med.* **138,** 1213–1229.

Singer, A., Hathcock, K. S., and Hodes, R. H. (1979). *J. Exp. Med.* **149,** 1208–1226.

Sprent, J. (1978a). *Immunol. Rev.* **42,** 108–137.

Sprent, J. (1978b). *J. Exp. Med.* **147,** 1838–1842.

Sprent, J., and von Boehmer, H. (1979). *J. Exp. Med.* **149,** 387–397.

Sproviero, J. F., Imperiale, M. J., and Zauderer, M. (1980). *J. Exp. Med.* **152,** 920–930.

von Boehmer, H., Sprent, J., and Nabholz, M. (1975). *J. Exp. Med.* **141,** 322–334.

von Boehmer, H., Haas, W., and Jerne, N. K. (1978). *Proc. Natl. Acad. Sci. U.S.A.* **75,** 2439–2443.

Waldmann, H. (1978). *Immunol. Rev.* **42,** 202–223.

Waldmann, H., Pope, H., Brent, L., and Bighouse, K. (1978). *Nature (London)* **274,** 166–168.

Zinkernagel, R. M., and Doherty, P. C. (1974a). *Nature (London)* **248,** 701–702.

Zinkernagel, R. M., and Doherty, P. C. (1974b). *Nature (London)* **251,** 547–548.

Zinkernagel, R. M., Callahan, G. N., Althage, A., Cooper, S., Klein, P., and Klein, J. (1978a). *J. Exp. Med.* **147,** 882–896.

Zinkernagel, R. M., Callahan, G. N., Althage, A., Cooper, S., Streilein, J. W., and Klein, J. (1978b). *J. Exp. Med.* **147,** 897–911.

II (CHAPTERS 5–8)

Receptors Involved in the Regulation of Development of Multicellular Organs and Organisms

REGINALD M. GORCZYNSKI

In the preceding chapters we followed an analysis of the development of a cellular recognition system in which cellular receptors recognize what is non-self and do not "see" what is self. In apparent contrast, the aggregation of similar cells to form tissues or organs seems to be critically dependent upon the mutual exclusion of dissimilar cells and the recognition of self. The process of mutual selection of similar cells results in intraorganismic selection in ontogeny, the basis of morphogenesis and organogenesis. There is now a growing body of evidence which argues that the linkage of the immune system with the polymorphic major histocompatibility complex (MHC) is not fortuitous, but is a reflection of the fact that the prime function of the MHC itself was initially to control cell–cell recognition within the immune system. In this function the MHC is postulated to play a role alongside other cell surface glycoproteins, the understanding of whose mode(s) of action is essential to develop any grasp of the phenomenon of correct "cell positioning" which underlies morphogenesis (Dausset and Cantu, 1980).

Steinberg (1970) early recognized the hierarchy in the sorting out of embryonic tissues in aggregates, such that it was possible to predict the sorting out

125

pattern, i.e., which cell type was external to the other. This in turn implied a quantitative property, shared by all cells but present in graded amounts. He suggested that this property was the interfacial cell tension existing between one cell type and another. Models of ligand–receptor interactions which seek to explain cell positioning have been forthcoming—e.g., Moscona (1973) and Burger (1978)—but there is little or no information available on *in vivo* tests of the putative aggregation factors involved in regulating cell–cell adhesion, or on the evidence for lectin-like receptors for these factors. Indeed, little attention has been paid to the need for such models to explain positioning as well as specific adhesion between aggregates of cells. Steinberg's differential adhesion hypothesis offers one possible mechanism for the dilemma, while the so-called interaction modulation hypothesis, which explains positioning as due to cellular responses (chemotactic; changes in adhesion, etc.) to chemical gradients of a diffusible signal molecule (or a nondiffusible signal fixed on the cell surface) is an alternative model (De Sousa, 1978). In an extension of this model to the interactions seen within the immune system in mammals, we can thus envisage recognition of non-self as a by-product of the inability of self-recognition (adhesion) due to the production of interaction modulation factors (IMFs) which diminish mutual adhesion. Lymphocyte IMFs are in turn presumed to be MHC products or MHC-linked in their action (Curtis, 1979).

As an aside, in any general discussion of organogenesis we can see that the mammalian immune system itself poses a particular problem in terms of cellular development in that it can be viewed as "a roving bag of cells without a fixed anatomy" (McConnell, 1983). Particular issues in terms of cell–cell recognition thus arise not merely to explain cellular interactions involved in producing immune reactions, as noted already in the discussion by Katz, but also to explain selective migration patterns of cells within the immune system itself. What is the nature of cellular interactions at the vascular endothelium? Is there specialization of different endothelial sites to allow for particular patterns of lymphocyte trafficking? Is there evidence for changes in the endothelium during disease processes (when lymphocyte trafficking is altered)? This fascinating field has received much attention recently with reports of specific acceptor and receptor sites on endothelial cells and lymphocytes respectively and with evidence that monoclonal antibodies can inhibit lymphocyte movement in a predictable and highly specific fashion (Chin *et al.*, 1982). These monoclonal reagents in turn recognize lymphocyte-derived glycoproteins which affect lymphocyte adhesion to endothelial surfaces (Gallatin, 1983).

Historically speaking, however, the critical insight into the important cellular events which occur during morphogenesis and organogenesis has come from nonimmunological fields, using model systems most familiar to the developmental biologist, e.g., embryogenesis and morphogenesis in nonmammalian systems. In the section that follows we shall be interested in discussing surface factors of importance in cell association, and in the processes whereby during cellular specialization and organogenesis "sorting out" of the different parts of the self structure occurs, as well as the differentiation and maturation of these same self-structures at the correct place and time in the developing organism.

The initial event in embryogenesis, fertilization, is regulated by a sperm-binding receptor at the oocyte cell surface, the bindin receptor (Gilbert, Chapter 5). This receptor binds a molecule on the sperm surface, the bindin protein, with molecular weight 30,000, which, though itself lacking in carbohydrate, apparently attaches to carbohydrates on the egg surface (Vacquier and May, 1971). A glycoprotein fraction with species-specific affinity for sperm-derived bindin has been obtained from the sea urchin egg surface and is apparently released after parthogenetic activation of oocytes [a finding which correlates with the observed natural loss of sperm receptors on the egg during normal fertilization (Glabe and Vacquier, 1978)].

An extremely interesting case of receptor controlled organogenesis is offered by the gonadal H-Y receptor system. The gonadal embryonic primordium is a unique germ tissue in that it can develop into either of two structures, an ovary or a testis, and in all mammals there is an "indifferent stage" where XX and XY gonads are histotypically indistinguishable. The normal hormonal environment of the embryo is female (and indeed embryos castrated at the indifferent gonad stage become phenotypic females). But if secretions from the testis (testosterone; anti-Mullerian duct factor) are so important in the development of maleness, what is it that causes the indifferent gonad to develop into a testis? Evidence from XX mice and male XX humans (Wachtel, 1979) show that the H-Y antigen is correlated with the presence of testes. *In vitro* experiments using dissociated gonadal tissue showed that if the H-Y antigen was removed from reaggregating testes cells by lysostripping with anti-H-Y antiserum, the cells form ovary-like structures [in the absence of this procedure, tubular aggregates with germ cells in the lumen were formed (Cullen *et al.*, 1978)]. This implies not merely that the H-Y antigen is involved in histotypic sorting out (as in embryogenesis) but that the gonads possess a receptor for H-Y which enables H-Y to perform this function—the experimental evidence to support this hypothesis is discussed further by Gilbert.

Analysis of cell surface interactions in mammalian embryogenesis is discussed in respect to recent work on the mouse T/t locus. Mutations at the T locus cause developmental, embryonic abnormalities in homozygous animals: the mutations are recessive lethals. There is a general sperm segregation–distortion occurring when a heterozygous male is used for fertilization, with some recessive t alleles transmitted to nearly 100% of the offspring (Hammerberg and Klein, 1975). Analysis of enzyme levels of wild-type and mutant sperm (which sperm can be distinguished by antisera, directed presumably to cell surface antigens) suggests an increased galactosyltransferase activity in the abnormal sperm (the zona pellucida is an acceptor for this enzyme), which may in itself be due to the interference by t alleles of the synthesis of diffusable galactosyltransferase inhibitors (Shur, 1984). Note once more the recurrent importance of carbohydrate recognition in developmentally regulated phenomena.

Finally Gilbert discusses what we now know of specific cell adhesion molecules from analysis of the neural cell adhesion system studied amongst others by Rutishauser and co-workers (Rutishauser, 1984). There is evidence for a role for such adhesion molecules in many aspects of neuron–target interactions and neural development (see also Riopelle and Dow, Chapter 8). In addition, it seems likely that the neural-cell adhesion molecule may be but one of a family of such molecules (glycoproteins) of functional relevance in cell–cell interaction in development (Rutishauser, 1984; Ogou et al., 1983). There is evidence that external glycosyltransferases may play a role in neural-cell adhesion (Balsamo and Lilien, 1980), as well as in the other cellular interactions occurring during embryological development discussed by Gilbert. Gilbert thus concludes by speculating on the possibility that we should envisage families of cell surface glycosyltransferases (or their evolutionary homologues) as being critically involved in cell–cell interaction/recognition per se.

Intercellular communication which regulates the differential gene activity contributing to the spatial organization essential for differentiation of tissues and organs, dictated somehow by positional information within morphogenetic fields, is discussed by Stocum (Chapter 6). The ability to reactivate the morphogenetic field of the limb of the urodele amphibian by limb amputation during the larval or adult life stage has made this an excellent system for experimental manipulation. By appropriate transplantation experiments it has been shown that the positional information constituting the autonomous regeneration field is inherited from parental limb cells by cells of the blastema. The information is neither primarily established nor secondarily reinforced upon those cells by a concentration gradient of signals emanating from the residual differentiated stump (Stocum, 1978). However, again by using a carefully designed transplantation protocol, it has also been shown that when a discontinuity is introduced between the blastema and adjacent stump tissue such that positional values which are not normally in opposition are made to confront one another, an interaction between the blastema and stump cells occurs. This results in cell proliferation and a subsequent alteration in positional values of progeny cells until the normal morphogenetic map is reestablished, which in turn can lead to the phenomenon of intercalary limb regeneration.

A three-dimensional model has been proposed [the outline model (Stocum, 1980)] which explains the regeneration field in terms of cellular responses to boundary positional values (wound epidermis, distal; blastema cells adjacent to dedifferentiating stump, proximal), newly juxtaposed by the introduction of a discontinuity (the effect of amputation). It has been suggested that cells change positional values in this field following an averaging rule (Maden, 1977), with a dividing cell continually assuming a position halfway between those of its neighbors until the discontinuity in the pattern is resolved. According to such a view, all blastema cells thus act as sources of boundary information during regeneration. Furthermore, an extension of this model at the molecular level suggests that a network of localized intercellular interactions between cell surfaces (rather than a diffusion gradient of morphogens) generates the molecular information system inherent to the morphogenetic field. A continual series of changes in the surface molecular array of adjacent blastema cells (ultimately transduced internally into differential gene activity) occurs until any discontinuity in the field is eliminated. Consistent with such a model is the evidence from studies with the tobacco hornworm *Manduca* (Mardi and Kafatos, 1976) which shows a gradient of cell adhesiveness in blocks of wing tissue isolated along a proximal–distal gradient. A similar conclusion can be drawn from the studies described by Nardi and Stocum (1983), as cited by Stocum in Chapter 6.

How can one study the molecular processes involved in changes in positional information in a developing system? It is known that the retinoids as a family of molecules are highly teratogenic to embryonic cells and tissue. Given at the stage of dedifferentiation and blastema formation in the regenerating amphibian limb, these compounds cause proximalization and posteriorization of positional values. Stocum discusses the possibility that those other retinoid-induced changes in the biochemistry of cell surface glycocalyx glycoconjugates which are known also to occur (Lotan, 1980; Levin *et al.*, 1983) are themselves causally related to the changes in positional information seen morphologically as proximalization/posteriorization.

An important feature of developing systems introduced by Stocum is this notion of plasticity, the mechanism by which some cells retain an ability to alter expression of cell-surface molecules which are used to communicate with neighboring cells in the local environment. We have already encountered one example of this phenomenon with the discussion by Katz of ''adaptive differentiation'' in the immune system. Another example which will be dealt with later (Raz and Fidler) centers around those changes in the cell surface or in noncollagen glycoproteins of the extracellular matrix (e.g., fibronectin, laminin) which may be causally related to the alterations in intercellular communication which are associated with neoplastic growth and the metastasis of tumors (Hynes, 1982).

One could argue that along with the immune system (which recognizes chemical differences between self and the environment) the next most plastic system in a multicellular organism is the nervous system (which in the case of higher vertebrates one might view as recognizing symbolic differences between self and environment). It is beyond the scope of this volume to discuss at any length the development of cell–cell communication within the nervous system and recent advances in our understanding of the biochemical/biophysical changes which occur in association with, for instance, memory and learning (for a review see Kandel and Schwartz, 1982). Suffice it to say at this stage that it seems that development of specific connections in neural development is a composite process of at least two different mechanisms. To a degree there is a preprograming to ensure connections are made to the appropriate partners. Further "fine-tuning" of this framework is also apparent as a result generally of electrical activity in the communicating cells, i.e., past functional experience plays a role (Purves and Lichtman, 1980). The contribution of these different mechanisms varies from one system to another. Thus some aspects of behavior will seem to be innate and depend very little on experience during development, while others will be critically dependent on such experience (e.g., development of visual acuity in the mammalian eye). Generation of specific neuromuscular junction connections seems to be one case where initially neuronal specificity is of key importance. After appropriate connections are made, large numbers of the motor-neuron synapses are redundant and eventually are lost. However, the muscle retains the capacity to release a factor (nerve growth factor in the case of smooth muscle) which can lead to reinnervation if some of the nerves innervating a muscle are destroyed (Brown *et al.*, 1981).

As in most cases within the nervous system, signal transmission at the neuromuscular junction between cells is highly localized, a function of the membrane distribution of the postsynaptic transmitter receptor (in this case the acetylcholine receptor). It is now apparent that there is a clear interaction between the local environment and cell receptor expression when the postsynaptic receptor at this junction is studied. Bloch and Steinbach (Chapter 7) discuss in detail the developmental regulation of acetylcholine receptors at the neuromuscular junction.

In vivo studies indicate that the normal adult muscle fiber shows a junctional aggregation of receptors whose appearance and further development is dependent on the presence of the presynaptic nerve, though not on nerve impulses. Extrajunctional receptors on innervated muscle occur as anticipated at higher density in a perijunctional gradient, though apparently in denervated tissue some receptor aggregates can also be seen. While the regulation of their appearance and the significance of these aggregates in denervated tissue *in vivo* is unclear, the production of similar, highly organized aggregates of acetylcholine receptors on cultured noninnervated myotubes in vitro suggests a specialized mechanism is involved in the production of these clusters which may reflect the normal *in vivo* developmental pathway (Anderson and Cohen, 1977). Studies on the control of production of receptor aggregates suggest a generalized role for an insoluble extracellular material (Jacob and Lentz, 1979), perhaps initially released in soluble form by muscle cells themselves, in the induction of receptor clustering. The signal(s) which determine cluster localization remain elusive. Extracellular material is believed to be important also in the formation of the cluster [as are cellular cytoskeletal proteins (Ben Ze'ev *et al.*, 1979)], but the intracellular molecules which may regulate localized receptor synthesis and/or lateral movement of receptors into clusters are undefined. However, Bloch and Steinbach offer some speculation concerning the molecular processes involved in the early events in receptors clustering, and suggest future approaches which will assist further in our elucidating the nature of the factors and/or membrane components involved perhaps both in neuromuscular and interneural synapse formation.

Perhaps one of the greatest challenges to biologists is to understand the intercellular signals which contribute to patterning within the nervous system, the mechanism by which highly specific axonal growth and synaptic mapping occurs during development. One theory of axonal guidance suggests the process is mediated by gradients of trophic materials (e.g., nerve growth factor, NGF) released from the target structure itself (Bonhoeffer and Gierer, 1984). A series of models views the development of neuronal specificity as the outcome of the relatively simple matching of recognition molecules on neuronal surfaces, a chemoaffinity hypothesis (Sperry, 1963). Rather more recently alternative models have been invoked, however, which take into account the likelihood that the final connectivity pattern established is the result of a multiplicity of factors which include intercellular competition and spatio-temporal opportunities (Edelman, 1984). Such models as the modulation or regulator hypothesis of Edelman predict the existence of a small number (family) of neural cell adhesion molecules (CAMs) which would perhaps even be shared across species. Even within a small family of glycoproteins, considerable structural micro-heterogeneity is possible (Berger *et al.*, 1982). Further modification of the extracellular milieu by, for instance, neural-derived glycosidases could add further to the diversity of interactions thought necessary to establish normal synaptic patterning. Riopelle and Dow discuss this and other aspects of Edelman's model, as well as an alternative hypothesis developed by Goodman which invokes selective fasciculation to explain neuronal patterning (Chapter 8).

REFERENCES

Anderson, M. J., and Cohen, M. W. (1977). *J. Physiol.* **268,** 757.
Balsamo, J., and Lilien, J. (1980). *Biochemistry* **19,** 2479.
Ben Ze'ev, Duerr, A., Solomon, F., and Renman, S. (1979). *Cell* **17,** 859.
Berger, E. G., Buddecke, E., Kamerling, J. P., Kobata, A., Paulson, J. C., and Vliegenthart, J. F. G. (1982). *Experientia* **38,** 1129.
Bonhoeffer, F., and Gierer, A. (1984). *Trends Neurosci.* **7,** 378.
Brown, M. C., Holland, R. L., and Hopkins, W. G. (1981). *Ann. Rev. Neurosci.* **4,** 17.

Burger, M. M. (1978). *Symp. Soc. Exp. Biol.* **32,** 1.
Chin, Y. H., Corey, G. D., and Woodruff, J. J. (1982). *J. Immunol.* **129,** 1911.
Cullen, S. E., Bamaco, D., Corbanora, A. O., Jacot-Guillaermond, H., Trinchieri, G., and Ceppellini, R. (1978). *Transplant. Proc.* **4,** 1835.
Curtis, A. S. G. (1979). *Dev. Comp. Immunol.* **3,** 379.
Dausset, J., and Cantu, L. (1980). *Hum. Immunol.* **1,** 1.
DeSousa, M. (1978). *In* "Immunopathology of Lymph Neoplasms" (R. A. Good and J. Twomey, eds.), p. 325. Plenum Press, New York.
Edelman, G. M. (1984). *Trends Neurosci.* **7,** 78.
Fuchs, E., and Green, H. (1981). *Cell* **25,** 617.
Gallatin, W. M., Weissman, I. L., and Butcher, E. C. (1983). *Nature (London)* **304,** 30.
Glabe, C. G., and Vaquier, V. D. (1978). *Proc. Natl. Acad. Sci. U.S.A.* **75,** 881.
Hammerberg, C., and Klein, J. (1975). *Nature (London)* **253,** 137.
Hynes, R. O. (1982). *In* "Cell Biology of Extracellular Matrix" (E. D. Hay, ed.), p. 295. Plenum Press, New York.
Jacob, M., and Lentz, T. L. (1979). *J. Cell Biol.* **82,** 115.
Kandel, E. R., and Schwartz, J. H. (1982). "Principles of Neural Science." Elsevier/North Holland, Amsterdam.
Levin, L. V., Clark, J. N., Quill, H. R., Newberne, P. M., and Wolf, G. (1983). *Cancer Res.* **43,** 1724.
Lotan, R. (1980). *Biochim. Biophys. Acta* **605,** 33.
Maden, M. (1977). *J. Theor. Biol.* **69,** 736.
Mardi, J., and Kafatos, F. J. (1976). *Embryol. Exp. Morphol.* **36,** 489.
McConnell, I. (1983). *Nature (London)* **304,** 17.
Moscona, A. A. (1973). *In* "Cell Biology in Medicine" (E. E. Bittor, ed.), p. 571. Wiley, New York.
Muller, U., Wolf, U., Siebars, J. W., and Gunther, E. (1979). *Cell* **17,** 331.
Nardi, J. B., and Stocum, D. L. (1983). *Differentiation* **27,** 13–28.
Ogou, S. I., Yoshida-Naro, C., and Takeichi, M. (1983). *J. Cell Biol.* **97,** 944.
Purves, D., and Lichtman, J. W. (1980). *Science* **210,** 153.
Rutishauser, U. (1984). *Nature (London)* **310,** 549.
Shur, B. D. (1984). *Mol. Cell Biochem.* **61,** 143.
Sperry, R. W. (1963). *Proc. Natl. Acad. Sci. U.S.A.* **50,** 703.
Steinberg, M. S. (1970). *J. Exp. Zool.* **173,** 395.
Stocum, D. L. (1978). *Science* **200,** 790.
Stocum, D. L. (1980). *Dev. Biol.* **79,** 276.
Thoms, S. D., and Stocum, D. L. (1984). *Dev. Biol.* **103,** 319.
Vacquier, V. D., and Moy, G. W. (1971). *Proc. Natl. Acad. Sci. U.S.A.* **74,** 2456.
Wachtel, S. S. (1979). *Cell* **16,** 691.

5

Cell–Cell Receptors in Embryogenesis

SCOTT F. GILBERT

I. INTRODUCTION

Embryonic cells continually change their ability to receive and respond to various stimuli. A cell can respond to the presence of another cell by forming adhesions to it, migrating over it, sliding under it, changing its rate of mitosis, dying, restricting its developmental potency, or initiating a phase of differentiation. Moreover, cells interact differently with each other at different times. This was first seen in 1931 by Just, who found that whereas the first four blastomeres

133

RECEPTORS IN CELLULAR RECOGNITION
AND DEVELOPMENTAL PROCESSES

of *Asterias* would normally adhere to each other when recombined, they would not adhere if the blastomeres were taken from different embryos fertilized only minutes apart. Townes and Holtfreter (1955) observed that cells derived from the prospective medullary plate region of amphibian embryos first sank into a mass of endodermal tissue and then became separated from it, thus mimicking normal gastrulation and neurulation. Both Just and Holtfreter postulated the existence of molecules on the cell surface which mediated these intercellular behaviors. It is only recently that the cell-surface receptors mediating such behavior have begun to be isolated and characterized.

Cell-surface embryonic receptors range from mythical constructs for explaining observed phenomena to relatively homogeneous (if large) isolated macromolecules. This review focuses on three such intercellular phenomena and the receptors mediating them: the bindin receptor on the egg cell surface thought to be responsible for mediating sperm–egg attachment, the H-Y receptor on the gonadal cell surface thought to be responsible for testis or ovary morphogenesis, and the neural retinal adhesion receptors thought responsible for the tissue organization of this relatively simple structure. The review will conclude with speculations on glycosyltransferases as molecules which may serve as receptors in all three cases and on cytoskeletal–receptor interactions. In order to limit the scope of material covered, I have excluded two sets of embryonic receptors from the discussion. The first set mediate the reception of hormonal stimuli and will be important in the adult animal as well as in the developing embryo. Receptors for testosterone and growth hormone obviously play important roles in development but will not be discussed here. The second set of receptors contained in the excluded volume are those mediating the attachment of cells to extracellular matrix proteins such as laminin and fibronectin. The strong affinity of epithelial cells for laminin and the relatively weak association between motile cells and fibronectin are certainly important for tissue formation and cell migration, respectively. This review, however, will focus on interactions between adjacent cell populations during embryogenesis. At the present time, none of these cell-surface receptors has been isolated as a single purified glycoprotein. There are several reasons contributing to this problem. As in all receptor studies, the solubilization procedure may be deficient, and as in most any study of embryonic biochemistry, the amount of material that one is able to obtain may not be sufficient. This latter restriction is not the case in the isolation of the bindin receptor from sea urchin eggs; however, here, the receptor appears to be isolated as a large complex aggregate which may reflect the need for a higher order structure to hold a moving sperm. In the case of the H-Y antigen receptor, the problems of receptor isolation are compounded by the lack of purity in existing preparations of the H-Y antigen itself. Finally, the attempts to purify the receptors (and soluble factors) mediating neural retina adhesion have been frustrated by several adhesion systems operating simultaneously between the same cells.

The dissection of these individual adhesion systems from each other should greatly aid the isolation of the component molecules. This review attempts to detail some of the progress made in the reification of these membrane receptors from hypotheses to characterized molecules.

II. THE SPERM BINDIN RECEPTOR OF THE OOCYTE

A. Introduction

Of all cell–cell interactions, perhaps the two best studied are those of fertilization and the development of lymphocyte immunocompetence. The similarities between these two systems were seen by developmental biologists seeking the mechanism of precise embryonic interactions. The first one to call attention to the similarity between fertilization and immunological phenomena was F. R. Lillie (1914) in his classical experiments on "fertilizin." This work was extended by Tyler, who provided evidence that egg and sperm hook on to each other, as it were, by the interlocking of surface substances of complementary configuration, acting precisely like antigen–antibody systems (Tyler, 1948, 1959). Indeed, the interaction of egg and sperm has long been seen as a model for embryonic cell interactions, as Harrison noted as early as 1910:

> There is nothing in the present work which throws any light upon the processes by which the final connection between the nerve fiber and its end organ is established. That it must be a sort of specific reaction between each kind of nerve fiber and the particular structure to be innervated seems clear from the fact that sensory and motor fibers, though running close together in the same bundle, nevertheless form proper peripheral connections, the one with the epidermis, the other with the muscle . . . The foregoing facts suggest that there may be a certain analogy here with the union of egg and sperm cell.

In Lillie's 1914 model of sperm–egg interaction, one can find such concepts as stereocomplementary configuration between cell-surface receptors, allosteric change in the receptor once the effector is bound, and competition for the cell-surface receptors. Lillie's hypothesis borrowed heavily from Ehrlich's sidechain theory of antibody production (Gilbert, 1984) and over the years has been severely modified (Tyler, 1948, 1956, 1959; Collins, 1961; Metz, 1978). The factors promoting sperm–egg interaction have been divided into those acting at a distance (chemotaxis, sperm agglutination, and the acrosome reaction) and those working at close range to bind the sperm to the egg (Metz, 1978). Although both types of reaction often show marked species specificity (implying receptors or agents on both the egg jelly and vitelline envelope), our present discussion shall concern only the latter class.

B. Sperm–Egg Attachment in Sea Urchins: Sperm Bindin

Lillie suggested that the moieties responsible for the species specificity of fertilization were gamete cell-surface proteins, and recent research has borne this out. When sea urchin sperm first contacts the ovum, it does so by adhering the spermatic acrosomal process to the outer surface of the vitelline envelope (VE), a glycoprotein layer closely apposed to the oocyte plasma membrane (Tegner and Epel, 1973; Summers *et al.*, 1975; Schatten and Mazia, 1976; Bellet, 1976). Vacquier and Moy (1977) have isolated a spermatic oocyte-binding protein by solubilizing the sperm membranes of *Strongylocentrotus purpuratus* in calcium-free sea water containing Triton X-100 and soybean trypsin inhibitor. The insoluble contents of the acrosome were released as granular spheres (approximately 0.2 μm diameter) which could be separated from sperm heads and whole sperm by glass-fiber filtration and centrifugation. Biochemical analysis revealed this material to be composed of a single major protein (30,500 daltons) having tyrosine as its sole N-terminus and no detectable carbohydrate or phospholipid (Vacquier and Moy, 1977; Vacquier, 1979). This particulate material was capable of agglutinating dejellied eggs and isolated vitelline envelopes. Although having no detectable carbohydrate of its own, this protein appeared to mediate attachment through carbohydrates on the egg surface. Glycoproteins digested from the VE inhibited this agglutination, and oxidation of the oocyte surface by sodium metaperiodate likewise rendered the eggs nonagglutinable by this particular acrosomal protein. This sperm protein was called bindin since it appeared to fulfill the requirements of being an acrosomal molecule capable of attaching to the surface of the VE (Vacquier and Moy, 1977).

Such sperm–egg attachment was found to be species-specific. Summers and Hylander (1975) showed that whereas echinoid sperm can undergo the acrosome reaction in the presence of a heterologous egg, the binding of the acrosome was specific for the homologous VE. Glabe and Vacquier (1977a) demonstrated that although homologous agglutination could occur at bindin concentrations as low as 3.5 ng/ml, there were no heterologous cross-agglutinations for immediately spawned oocytes of *S. purpuratus* and *S. franciscanus*, even at concentrations as high as 430 ng/ml. These two sea urchins are closely related (Whitely *et al.*, 1970), and the overall amino acid compositions of their respective bindins are similar. Tryptic peptide mapping, however, revealed 13 peptides unique to *S. purpuratus*, and 10 peptides unique to *S. franciscanus*, in addition to 24 shared in common (Bellet *et al.*, 1977). Glabe and Lennarz (1979) have quantitated the specificity of the bindin–bindin receptor interaction by measuring the number of egg–egg contacts mediated by particulate bindin spheres. In both *S. purpuratus* and *Arbacia punctulata*, the homologous eggs are aggregated to a marked degree compared with eggs from the heterologous species. When a limiting amount of *S. purpuratus* bindin was added to a suspension containing equal numbers of *S.*

purpuratus (*S.p.*) and *A. punctulata* (*A.p.*) eggs, 65 *S.p.–S.p.* attachments were formed to 4 *S.p.–A.p.* contacts and no *A.p.–A.p.* associations.

Evidence supporting the contention that bindin mediates sperm–egg attachment also came from studies using monospecific rabbit anti-bindin. Bindin (unlike any other protein believed to mediate specific intercellular adhesion) can be isolated in milligram quantities. When purified bindin was isolated and injected into rabbits in order to produce antisera, double immunodiffusion of this anti-bindin antiserum with bindin gave a single precipitin band which can be eliminated by the prior absorption of this antiserum to either purified bindin or sperm homogenate (Moy and Vacquier, 1979). After sea urchin eggs were inseminated and fixed, they were sequentially incubated in the rabbit anti-bindin antiserum and horseradish peroxidase-conjugated swine anti-rabbit immunoglobulin antiserum. Subsequent staining with 3,3-diaminobenzidine tetrachloride (DAB) and peroxide deposits a DAB precipitate wherever the peroxidase was localized (Moy and Vacquier, 1979). Electron microscopy showed that the acrosome was covered with DAB precipitate, and also showed that this reaction occurred for about 1–2 nm on the sperm plasma membrane adjacent to the acrosome. Thus, bindin is specifically located at the site of the natural sperm–egg interaction. Unfertilized eggs bound neither the anti-bindin nor the conjugated antiserum.

Bindins isolated from other marine organisms appear to have distinctly different biochemical characteristics. Aketa *et al.* (1978) have isolated a molecule from the sperm of the sea urchin *Hemicentrotus pulcherrimus* which causes species-specific agglutination of dejellied oocytes. This molecule, however, appears to be less than 5% protein and more than 95% carbohydrate. The insoluble bindin of *Crassostrea gigas* (oyster) acrosomes is a glycoprotein dimer containing about 15% carbohydrate (Brandiff *et al.*, 1978).

C. Sperm–Egg Attachment in Sea Urchins: The Bindin Receptor

Saturation binding of sperm to egg (Vacquier and Payne, 1973) indicated that while the sea urchin oocyte can bind about 1500 sperm, the entire oocyte surface was not completely covered. This suggested that sperm attachment cannot occur at any position on the oocyte, and that there is a limiting quantity of receptors for this function. In 1967, Aketa began a series of studies seeking to solubilize and isolate the sperm-binding substances from the VE of sea urchin oocytes. Initial investigations (Aketa, 1967) yielded a solution capable of binding sperm to air bubbles in a calcium-dependent manner (calcium ions being essential for the *in vivo* reaction). Immunodiffisuion revealed these 1 *M* urea eluates to be mixtures producing up to six precipitin bands (Onitake *et al.,* 1972; Aketa, 1973; reviewed in Glabe and Vacquier, 1978). More recently, Yoshida and Aketa (1978,

1979) have shown that species-specific sperm-binding factors in sea urchin oocytes can be localized by immunofluorescent probes and that univalent antibody against these sperm receptors destroyed the ability of those eggs to become fertilized.

Species-specific bindin receptor complexes have been partially purified by Glabe and Vacquier (1977b; 1978) and by Rossignol et al. (1981). Glabe and Vacquier (1978) isolated the VE of S. purpuratus in a hypotonic solution containing ethylene diamine tetraacetate (EDTA), Triton X-100, and a protease inhibitor. Using electron microscopy to distinguish the internal and external surfaces of the VE, they determined that the sperm could only bind to the external surface, even though both sides were accessible (Mazurkiewicz and Nakane, 1972). Thus, the sperm receptor (or its active site) resides on the external surface of the envelope. Surface labeling of the external side of the envelope by lactoperoxidase-catalyzed iodination (before the VE was isolated) yielded two high molecular weight bands separable by sodium dodecyl sulfate (SDS) polyacrylamide electrophoresis. Soon afterward, these authors reported the isolation of one of these, "a high-molecular-weight, trypsin-sensitive glycoprotein fraction from the sea urchin egg surface having species-specific affinity for bindin" (Glabe and Vacquier, 1977b). This glycoprotein could be iodinated on the surface of dejellied eggs by chloramine-T and was released after parthenogenetic activation of the egg. This correlates with the finding that during normal fertilization, the sperm receptors are destroyed, presumably by the proteases released from the cortical granules (Carroll and Epel, 1975; Vacquier et al., 1973).

This putative bindin receptor represents about 10% of the total iodinated surface molecules and passes into the void volume of Biogel A5m column, suggesting that the receptor is an aggregate. Isoelectric focusing separated the receptor into a protein fraction having a pI of 4.02 and containing galactose, mannose, and sulfate, but no sialic acid.

Kinsey et al. (1980) have shown that purified plasma membranes from S. purpuratus eggs will specifically bind homologous sperm which had undergone the acrosomal reaction. Neither reacted A. punctulata sperm nor S. purpuratus sperm with intact acrosomes would bind to these membrane fragments. Thus, a species-specific sperm receptor could be detected on the oocyte plasma membrane as well as on the vitelline envelope. Lennarz and his colleagues (Rossignol et al., 1981, 1984) have partially purified a 2×10^6 dalton glycoprotein from the S. purpuratus oocyte cell membrane. This material competes with the intact receptor for binding to homologous bindin, and antibodies against this glycoprotein coat the egg surface and prevent fertilization by homologous sperm (Rossignol and Lennarz, 1983).

The bindin receptors from S. purpuratus and A. punctulata are released from eggs only after solubilization in 4 M guanidine thiocyanate for 1–2 hr in the presence of dithiothreitol. After the protein portion of these receptors has been digested by pronase, extensive carbohydrate material still remains. The carbohydrate-rich receptor fragment from *S. purpuratus* is a sulfated, negatively charged polymer resembling a glycosaminoglycan having a very high molecular weight ($> 10^6$) and containing fucose, galactosamine, and iduronic acid. The resultant carbohydrate portion from the *A. punctulata* receptor, however, is a small molecular weight (6000 daltons) molecule having no net charge (Rossignol et al., 1984). While the isolated receptors were able to inhibit the binding of activated sperm to dejellied eggs in a species-specific manner, the carbohydrate portion of either receptor was able to block sperm binding to either egg. Thus, the carbohydrate moiety seems important for sperm–egg binding, but the species-specificity resides in the intact glyconjugate complex (or in the protein alone). Moreover, species-specificity may also reside in higher-ordered structures than bindin and its receptor, for while isolated receptors can recognize sperm in a species-specific manner to block fertilization in a bioassay, the receptors will both bind equally well to the bindin molecules isolated from the two urchin species.

These findings have led Rossignol and colleagues (Rossignol et al., 1984) to speculate that the hormone–receptor paradigm generally assumed for receptors involved in cell–cell interaction may not be applicable. Whereas the hormone receptor or drug receptor need only occupy a small area of the cell surface and have an enzyme-like binding constant, the receptors for actively wiggling sperm may have to be organized into a specific structure on the egg surface and may involve a series of kinetic intermediates. Thus, while studies of hormone receptors can use the isolated receptor and its ligand, the studies of cell–cell receptors may be much more complicated.

D. Sperm–Egg Attachment in Mammals and Other Species

Similar studies are being pursued in the case of mammalian sperm–egg interactions wherein the zona pellucida (ZP) appears to play a role analogous to that of the vitelline envelope. Binding of sperm to the ZP is relatively (but not absolutely) species-specific (Hanada and Chang, 1972; Hartmann, et al., 1972; Peterson, et al., 1979) and can be inhibited by proteases (Hartmann and Gwatkin, 1971; Barros and Yamagimachi, 1972), lectins (Oikawa et al., 1973), and antibodies can be made against the oocyte (Shivers et al., 1972). The presence of solubilized ZP materials is also seen to be able to inhibit the penetration of sperm into hamster eggs (Gwatkin and Williams, 1977), and Peterson et al. (1979) have shown that boar sperm plasma membranes will bind specifically to boar zona pellucida. The ZP appears to be modified upon fertilization such that sperm can no longer bind (Barros and Yamagimachi, 1972; Braden et al., 1954; Pikó, 1969; Inoue and Wolf, 1975), and it is thought that this involves the masking or the destruction of a ZP sperm receptor (Braden et al., 1954; Gwatkin, 1976; Gwatkin, 1977).

The role of the ZP sperm receptor may differ in various mammals. In guinea pigs, the egg appears to bind only those sperm which have already undergone the acrosomal reaction (Huang *et al.*, 1981) whereas in mice and hamsters the acrosome reaction takes place after the binding of sperm to the ZP receptor (Gwatkin, 1976; Saling and Storey, 1979; Saling *et al.*, 1979; Philips and Shalgi, 1980; Hall and Franklin, 1981). Florman and Storey (1982) have shown that when the mouse sperm acrosome reaction was inhibited by 3-quinuclidinyl benzilate, attachment to the ZP was still accomplished. It appears, then, that in some species, the ZP receptor(s) has a dual function: binding the sperm and then initiating the acrosomal reaction.

Bleil and Wassarman (1980a) have solubilized and characterized the ZP of mouse eggs and two-cell embryos. Of the three major glycoproteins isolated from the oocyte ZP, one of them, ZP3, is capable of inhibiting the *in vitro* binding of sperm to egg ZP. Moreover, when this 83,000-dalton protein is isolated from the two-cell embryo ZP, the protein no longer displays this property (Bleil and Wassarman, 1980b). Purified ZP3 from unfertilized eggs (but not from two-cell embryos) was found to bind to sperm and to initiate the acrosome reaction as well (Bleil and Wassarman, 1983). It appears, then, that the ZP3 glycoprotein is the sperm-binding receptor of the unfertilized mouse egg and that it becomes modified to lose its receptor ability after fertilization.

The mechanism of receptor function has focused on the carbohydrate portion of the ZP3 glycoprotein. The receptor activity is not affected by boiling, detergent treatment, or urea. Moreover, the 25-kDa (kilodalton) glycopeptides liberated by the digestion of ZP3 with CMC-linked pronase were just as effective as intact ZP3 in the fertilization-competition assay. The receptor activity was lost, however, when O-linked carbohydrate was removed from either the ZP or ZP3 by mild alkaline hydrolyses under conditions (5×10^{-3} N NaOH, 16 hr, 37°C) which did not destroy the protein (Florman and Wassarman, 1983). Removal of N-linked oligosaccharides from the ZP or purified ZP3 did not have any effect on sperm-receptor activity (Elder and Alexander, 1982). Recent work by Florman and co-workers (1984) shows that the carbohydrate portion of ZP3 is essential for sperm binding (a result paralleling the sea urchin studies), whereas both carbohydrate and polypeptide are needed for inducing the acrosome reaction of the sperm.

These data are consistent with those studies of Shur (reviewed in Shur, 1984) suggesting that carbohydrate-binding enzymes on the sperm cell surface are responsible for binding the mouse sperm to the egg. In these studies, three independent lines of evidence suggest that the zona pellucida sperm receptor is an anchor for repeating polymers ending in *N*-acetylhexosamine residues which are recognized by a sperm surface galactosyltransferase. The first evidence comes from mutations of the *T/t* complex in which a marked "segregation distortion" occurs. In such cases, heterozygous males preferentially transmit the mutant allele to their offspring, in some alleles with a frequency of more than

90% (Hammerberg, 1982). While eight other enzymatic activities are indistinguishable between wild-type and t-sperm, there is a fourfold increase in sperm-surface galactosyltransferase activity (Shur and Bennett, 1979; Shur, 1981). Those t and T alleles not showing segregation distortion did not show the elevated galactosyltransferase activity, suggesting that the observed segregation distortion may arise from increased galactosyltransferase activity at the time of sperm–egg binding.

The second line of evidence comes from studies of sperm capacitation. Freshly ejaculated mouse sperm are unable to bind to the ZP or undergo the acrosome reaction before they are capacitated. This phenomenon involves several changes in the structure of the sperm cell surface, including changes in membrane fluidity, ion permeability, and cell-surface composition. One of the first events of capacitation is the release of a glycoconjugate from the sperm surface. These glycoconjugate "coating" or "decapacitation" factors are secreted onto the sperm in the epididymis and can inhibit binding to the ZP when they are added to capacitated sperm in *in vitro* fertilization assays (Shur and Hall, 1982a). Shur and Hall isolated and characterized this glycoconjugate, demonstrating it to be a novel high-molecular-weight glycopeptide that could be galactosylated by the sperm cell-surface galactosyltransferase. Digestion of the pronase-treated glycopeptide with endo-β-galactosidase and precipitation of the galactosylated glycopeptide with antisera against F9 embryonal carcinoma cells that recognize a class of poly-N-acetyllactosamine glycoconjugates demonstrated that the epididymal decapacitation factor is a large glycopeptide characterized by long stretches of N-acetyllactosamine. This glycoconjugate was found to be an effective competitor for exogenous N-acetylglucosamine substrates of the sperm surface galactosyltransferase. Furthermore, the release of this glycoconjugate correlated with the ability of sperm to bind to the ZP. Thus, Shur and Hall present a model where the galactosyltransferase on the capacitated mouse sperm attaches to the egg ZP by binding to exposed carbohydrate residues terminating in N-acetylglucosamine. Before capacitation, however, the sperm galactosyltransferase is blocked by a glycopeptide which has N-acetyllactosamine residues (also ending in an N-acetylhexosamine) that compete for the active site of this enzyme.

The third line of evidence suggesting that sperm–egg attachment is due to interactions between the sperm galactosyltransferase and its ZP substrate comes from studies where the galactosyltransferase activity is specifically perturbed (Shur and Hall, 1982b). The milk protein α-lactalbumin specifically changes the substrate specificity of galactosyltransferases from N-acetylglucosamine to glucose (Ebner and Magee, 1975). The addition of α-lactalbumin to capacitated sperm simultaneously modified sperm surface galactosyltransferase activity from N-acetylglucosamine substrates to glucose substrates and inhibited sperm binding to mouse ZP in a dose-dependent fashion. Other substances such as bovine serum albumin did not interfere with transferase activity or sperm–ZP binding.

The galactosyltransferase activity of the sperm surface can also be inhibited by the competitive inhibitor UDPdialdehyde (Powell and Brew, 1976). This compound inhibited sperm–ZP binding and sperm surface galactosyltransferase activity to identical extents. Moreover, the covalent linking of UDPdialdehyde to sperm galactosyltransferases markedly inhibited sperm–egg binding, but eggs so treated were still able to bind untreated sperm. However, the intact ZP is a good substrate for the sperm galactosyltransferase, and pretreating the eggs with N-acetylglucosaminidase inhibited sperm–egg binding by 86%.

Lopez and co-workers (1985) have recently provided more evidence for this model. First, they found that the addition of UDPgalactose allowed the glycosyltransferase reaction to go to completion, thereby breaking the sperm-ZP attachment. The reaction was concentration-dependent and specific for UDP-galactose. Second, this group also found that monospecific antibody against the sperm surface galactosyltransferase inhibited sperm-ZP attachment and sperm surface galactosyltransferase activity to the same extent. These studies are all consistent with the hypothesis that the mouse ZP sperm receptor is a substrate for the sperm surface galactosyltransferase which is expressed during capacitation. At present, it is not known whether the 83-kDa ZP glycoprotein isolated by Bleil and Wassarman is, in fact, the receptor for the galactosyltransferase or whether there are more than one system of binding for the sperm and egg.

As pointed out by Shur (1984), sperm protein–egg carbohydrate interactions may mediate matings throughout the living kingdoms. De Santis et al. (1980) and Rosati and De Santis (1980) have also shown that species-specific attachment of sperm occurs on the chorion of tunicates. Fertilization was found to be inhibited by the addition of fucose to the sea water, and a fucosyl glycoprotein has been isolated from Ciona chorions, which acts to both bind the sperm and activate the acrosomal reaction (De Santis et al., 1983). Surface glycosyltransferases have also been seen to mediate the mating reaction of + and − gametes of Chlamydomonas (Bosmann and McLean, 1975). Here, the glycosyltransferases of each gamete are thought to bind to a carbohydrate receptor on the other cell. It appears likely, then, that protein–carbohydrate interaction may be a general mechanism for binding gametes throughout the living kingdoms.

III. THE GONADAL H-Y ANTIGEN RECEPTOR

A. Gonadal Development

There is widespread evidence for the existence of an antigen (or set of antigens) specific for the heterogametic (XY, ZW) sex in vertebrates. Since 1975, this substance, the H-Y antigen, has been proposed as the agent of sexual differentiation, inducing the indifferent gonad to form the structures characteristic of the heterogametic sex. Shortly thereafter, a more refined model was proposed which accounted for the ubiquitous expression of the H-Y antigen on all male mammalian somatic cells by postulating that the H-Y antigen receptor resided solely on the embryonic gonadal cells of both sexes. The conjunction of H-Y antigen and its receptor would occur only in the male gonad such that ''the ubiquitous expression of H-Y is compensated for by the organ-specific expression of its receptor, and the presence of the receptor in male and female is compensated for by the sex-specific expression of the inducer molecule itself'' (Wachtel *et al.*, 1980b). We will presently review gonadal development and the evidence that testis development in mammals is directed by the H-Y antigen. Thereafter, our focus will be directed on the data for the existence and developmental function of the gonad-specific receptor.

As a rule, each embryonic primordium has but one destiny. The hepatic diverticulum from the embryonic gut has no choice but to become a liver and not a thymus. The gonadal primordium, however, is an exception to this generality, as it can develop into either a testis or an ovary. In each mammal, there is a stage where XX and XY gonads are histotypically indistinguishable. Each consists of a surface epithelium connected to a central mesenchymal blastema by primitive sex cords. In male development, the sex cords proliferate internally as the central blastema becomes organized into seminiferous tubules (the presumptive Sertoli cells) and a sparce interstitium (presumptive Leydig cells). Male development then proceeds with the differentiation of these Sertoli and Leydig cells and with the formation of the tunica albuginea, the outer basement membrane surrounding the testis. In the female embryo, the indifferent stage persists longer. The internal sex cords eventually degenerate as thecal and follicular cells appear at the cortex (see Jost *et al.*, 1973; Hall and Wachtel, 1980).

The primordial germ cells, which reach the gonadal ridge coincident with the time at which the gonads are differentiating, probably play no role in the morphogenesis of the male gonad. When germ cells fail to reach the developing gonad, either naturally, as in the case of the *W* or *Sl* mutants of mice (Coulombre and Russell, 1954), or experimentally, when embryos are treated with busulfan (Marchant, 1975), no change is seen in the direction of gonadal differentiation. Evans *et al.* (1977) have also shown that an XY primordial germ cell can become an oocyte when organized in an XX gonad. (That germ cells may be essential in the *maintenance* of the gonad once it is formed has been proposed to explain certain abnormalities in ovarian dysgenesis.)

The normal hormonal environment of the mammalian fetus is female, and embryos castrated at the indifferent gonad stage will develop the female phenotype (Jost *et al.*, 1973). The developing testes alter this tendency by secreting anti-Müllerian duct factor (AMF) from the Sertoli cells and testosterone from the Leydig cells (Blanchard and Josso, 1974; Tran and Josso, 1977). The AMF causes the degeneration of the duct responsible for the formation of the female genitalia (uterus, Fallopian tubes, distal vagina) while testosterone (or its activated form, 5α-dihydrotestosterone) induces the differentiation of the Wolffian duct into the male genitalia (epididymis, vas deferens, and seminal vesicles) while causing the degeneration of the breast tissue primordia (Dürnberger and Kratochwil, 1980). It is only in the placental mammals that such a system is so defined. Other animals have more plastic mechanisms of sex determination. In birds, for instance, testicular development can be changed to ovarian development by the addition of female sex hormones (Müller *et al.*, 1979a; 1980), whereas no such change can occur in mammals.

B. The H-Y Antigen

Since the development toward maleness is dependent upon secretions from the testes, the problem then becomes: what induces the indifferent gonad to become a testis? The best candidate for this role is the H-Y antigen (for recent reviews, see Hall and Wachtel, 1980; Wachtel, 1983; Andrews, 1984). This "minor histocompatibility antigen" was first described when Eichwald and Silmser (1955) reported the unexpected finding that within an inbred strain of mice, females were consistently rejecting male skin grafts. They called this graft-defined antigen H-Y since it correlated with the presence of the Y chromosome. This antigen can be detected by various grafting and serological procedures (reviewed in Wachtel, 1977; Hall and Wachtel, 1980; Andrews, 1984) and has been found on all male mammalian cells, including sperm and primoridal germ cells, starting at the eight-cell embryo (Krco and Goldberg, 1976).

Immunological studies demonstrated that the H-Y antigen was conserved throughout vertebrate evolution on the cells of the heterogametic sex. Such conservation suggested an essential function, an in 1975, two series of investigations (Wachtel *et al.*, 1975; Bennett *et al.*, 1975) proposed that the H-Y antigen was responsible for mammalian testis morphogenesis. The first type of evidence demonstrated the independence of H-Y expression from the hormonal environment, and demonstrated that even individuals with androgen insensitivity syndrome expressed this antigen. The second type of evidence correlated the presence of the H-Y antigen with the Y-chromosome in various situations where Y chromosomes were lost or retained. Shortly thereafter, a third line of evidence demonstrated that an even better correlation could be made between the presence of the H-Y antigen and the presence of testes. In XX mice with the sex-reversed mutation and in male XX humans, the H-Y antigen correlated with the presence of testes (reviewed in Wachtel, 1979); even in the strange sex patterns of Scandinavian wood lemmings (where there exist fertile XY females) and mole voles (where both males and females can be XO) the correlation of testes and H-Y antigen was still upheld (Nagai and Ohno, 1977; Ohno, 1979).

The existence of fertile female XY wood lemmings suggested that in the absence of H-Y antigens, XY gonadal cells could organize into an ovary. This was demonstrated *in vitro* by dissociating gonadal tissue from mice (Ohno *et al.*, 1978) or rats (Zenzes *et al.*, 1978a,b) and allowing them to aggregate into histotypic reassortments (Moscona, 1957). Testes cells re-formed tubular aggregates incorporating several large (presumably germ) cells in their lumen, whereas reaggregated ovarian cells produced a spherical "follicular" envelope surrounding a large central cell. When the H-Y antigen was removed from the reaggregating testes cells by lysostripping the antigen with anti-H-Y antiserum (Cullen *et al.*, 1973), those lysostripped cells formed ovary-like structures. This alteration of *in vitro* morphogenesis by anti-H-Y antiserum was specific for testicular cells, as lysostripping H-Y from male epidermal cells did not impede their rearrangement from epidermis-like aggregate *in vitro*. This has important implications as it strongly suggests that the H-Y antigen is responsible for the histotypic sorting out that occurs in testis reaggregates and that such an assay accurately assesses embryonic interactions. It also suggests the existence of a receptor found in gonadal cells which enables the H-Y antigen to function in development.

There are several important caveats to keep in mind when discussing these and other experiments on the H-Y antigen. The first is that the antigen(s) defined by different procedures may be different molecules. Silvers and colleagues (1982) have proposed that the term H-Y be reserved for the original transplantation-defined antigen, while the serologically defined substance be termed SDM (serologically defined male-specific antigen). Andrews (1984) has proposed a nomenclature where $H-Y_t$, $H-Y_s$, and $H-Y_c$ are those male-specific antigens defined respectively by transplantation, serological reagents, and cell-mediated cytoxicity. In recent years, certain exceptional cases have implied that the $H-Y_t$ and $H-Y_s$ are not recognizing the same substance: $H-Y_t$ and $H-Y_c$ appear to correlate best with the presence of the Y-chromosome, whereas there are instances when $H-Y_s$ does not. It has recently been shown (Eicher *et al.*, 1982) that when two Y chromosomes derived from wild mice are introduced into C57BL/16 inbred mice, the production of XY phenotypic females with ovaries results. These XY females have been shown to express $H-Y_t$ antigen (Simpson *et al.*, 1983). More recently, McLaren and co-workers (1984) have shown an absence of $H-Y_c$ on certain male XX mice carrying the sex-reversed mutation. In any case, though, the expression of $H-Y_t$ appears to be associated with the presence of a Y chromosome rather than with the gonadal or phenotypic sex. Since the Y chromosome from C57BL/16 mice produces males, it appears that male development results from the interaction of the Y-linked gene with other non-Y-linked loci. The mere presence of $H-Y_t$ is not sufficient to organize male development. In other cases (Wolf *et al.*, 1980a,b; Engel *et al.*, 1981; Wachtel *et al.*, 1980a; Haseltine *et al.*, 1982; Koo *et al.*, 1981), the $H-Y_s$ antigen is seen to be expressed in certain females having complete or partial absence of the X chromosome. Thus it has been proposed (Wolf *et al.*, 1980b; Wachtel *et al.*, 1980a) that the structural gene for H-Y_S is in the X chromosome but is regulated negatively by a gene in the short arm of the X chromosome and positively by a gene in the short arm of the Y chromosome.

The second caveat is that the antisera used to identify the serologically defined H-Y antigen are extremely weak, having an affective titer of 1:4 to 1:8. The isolation of a monoclonal antibody against H-Y (Koo *et al.*, 1981) should soon get around these problems caused by the scarcity of antisera and its lack of constancy. The third caveat is related to the second, as only the polyclonal antisera have been used to try to isolate the H-Y antigen. Presently, the H-Y antigen has remained unpurified. Hall and Wachtel (1980) immunoprecipitated H-Y_S from ^{125}I-surface-labeled Sertoli cells, obtaining two labeled bands with molecular weights of 18,000 and 31,000. Another approach is to isolate H-Y antigen from the supernatant of the β_2-microglobulin-deficient Daudi cell line. Fellous *et al.* (1978) and Beutler (Beutler *et al.*, 1978) demonstrated that the Daudi human lymphoid cell line lacks the ability to produce β_2-microglobulin and sheds its HLA and H-Y antigens into supernatant. This H-Y antigen appears to be a protein with a molecular weight between 15,000 and 18,000 (Hall and Wachtel, 1980; Ohno, 1979; Nagai *et al.*, 1979; Hall *et al.*, 1981). Finally, H-Y antigen may be secreted by male gonadal cells themselves. Müller *et al.* (1978a,b) found that whereas supernatants from most H-Y-positive cells did not absorb out the cytotoxicity of anti-H-Y serum (against male BALB/c epidermal cells), the supernatant from cultured testis cells could. Furthermore, intact testis were seen to secrete H-Y antigen into the epididymal fluid. However, the secreted H-Y from testicular cultures was not able to be immunoprecipitated with the anti-H-Y antisera (Hall and Wachtel, 1980; Gore-Langton *et al.*, 1983).

The fourth caveat is that the proteins so defined may not be nearly as important as the carbohydrate groups attached to them. Shapiro and Erickson (1981) have demonstrated that H-Y_S can be eliminated from the surface of male cells by treating the cells with periodate, β-galactosidase, and galactose oxidase. Moreover, H-Y_S was resistant to various proteolytic digestions, suggesting that the antigenic determinant of H-Y_S is not polypeptide but carbohydrate. Such carbohydrate, moreover, need not be conjugated only to proteins, but could exist on glycolipid as well. The carbohydrate nature of H-Y_S would explain not only its evolutionary conservation but also why it has been so difficult to raise antibodies against. Aldofini and colleagues (1982) have attempted to harmonize the genetics of sex determination by speculating that H-Y_S is a complex oligosaccharide serially constructed by glycosyltransferases whose genes are located on the sex chromosomes and autosomes. They postulate that a core oligosaccharide is made under the control of autosomal genes. The Y chromosome would then encode a glycosyltransferase, which could convert the precursor into a substance found only in males.

Thus, in attempting to define the receptor for H-Y, one has three major problems. First, there is no agreement as to which H-Y the receptor may be binding; second, the nature of the H-Y molecule is not determined and may be glycoprotein or glycolipid; third, until recently there has been no preparation of H-Y to use to isolate the receptor. Be that as it may, these impure and uncharacterized reagents have been used to ascertain the existance of a gonad-specific H-Y receptor.

C. The Gonadal H-Y Antigen Receptor

Although the H-Y antigen is thought to be found on all male somatic tissues, Müller and colleagues (1978a) have evidence for the existence of a gonad-specific receptor. They find that only the gonadal cells are able to accept exogenously supplied H-Y. In these experiments, female cells from various organs were not seen to absorb out any significant amount of anti-H-Y antibody. Furthermore, these female tissues still do not absorb out any anti-H-Y antibodies after they have first been exposed to exogenous H-Y antigen. Thus, female cells do not accept exogenous H-Y. (Nor did exogenous H-Y increase the ability of male cells to absorb out the antiserum.) However, when ovarian cells were exposed to the H-Y antigen, they were capable of absorbing a significant amount of anti-H-Y antiserum (Table I). A similar phenomenon was observed in newborn, but not adult testes.

This still left unanswered whether the gonadal receptor was different from the β_2-microglobulin-MHC complex seen by Beutler and Fellous. Müller and co-workers determined that gonadal cells could still bind exogenous H-Y in the absence of the microglobulin (Müller *et al.,* 1979b). When the β_2-microglobulin was removed from the surface of newborn rabbit testes cells (as measured by both cytotoxicity and immunofluorescence), all the endogenous H-Y antigen was removed. When the lysostripped cells were then incubated with exogenous H-Y antigen (from either Daudi or epididymal sources), they still were able to accept the H-Y antigen.

The exclusivity of gonadal cells in binding a secreted product suggested that the H-Y antigen was acting like a hormone, albeit one that does not usually circulate through the blood (however, see later discussion). The hormone analogy is also suggested by other *in vivo* and *in vitro* evidence. Since 1968, it has been known that when mammalian and avian embryonic testes are cultured adjacently, the avian testis forms ovary-like structures (Akram and Weniger, 1968). Similarly, in XX/XY mosaics, the resultant gonads are usually testes, and even the enclosed XX cells are H-Y positive (Ohno, 1977; Ohno *et al.,* 1978).

The ability of H-Y to act as a hormone was demonstrated by adding exogenous H-Y antigen to dissociated indifferent gonad or embryonic ovarian cells and

TABLE I

Demonstration of Gonad-Specific H-Y Antigen Receptor by the Absorption of Anti-H-Y Antisera Cytotoxicity[a]

	Percentage cells dead when treated with anti-H-Y antiserum	
Age (days) of ovary	After absorption of serum by ovary cells	After absorption of serum by H-Y-pretreated ovary cells
1	58 + 6	23 + 3
11	47 + 4	20 + 5
50	62 + 4	28 + 3

Cytotoxicity of anti-H-Y antiserum before exposure to cells: 60 + 5%
Cytotoxicity of complement control (no antiserum): 20 + 2%

	Percentage cells dead when treated with anti-H-Y antiserum	
Age (days) organ	After absorption of serum by tissues	After absorption of serum by H-Y-pretreated tissues
Kidney	50 + 3	44 + 3
Liver	50 + 4	48 + 3
Brain	50 + 3	50 + 4
Epithelial cells	51 + 5	53 + 3

Cytotoxicity of antiserum before exposure to cells: 55 + 4%.
Cytotoxicity of complement control: 22 + 3%.

[a]From Müller *et al.* (1978a).

reaggregating them as previously described. Zenzes *et al.* (1978b) showed that whereas control ovarian tissues reaggregated to from the typical follicle type of arrangement, those ovarian cells incubated in H-Y antigen produced tubular structures. Moreover, this change in morphology could be prevented by the addition of anti-H-Y antiserum. Müller *et al.* (1978c) assayed a functional parameter, the appearance of the luteinizing hormone/chorionic gonadotropin receptor, rather than morphology alone. This receptor is expressed in newborn rat testes but not until later in the ovary. They found that whereas no receptor activity was observable in reaggregated ovarian cells, those ovary cells "converted" with H-Y began expressing gonadotropin-binding activity similar to that of reaggregated testes cells. No hormone binding was seen on spleen cells preincubated with H-Y antigen. However, Benhaim and co-workers (1982) were unable to find the induction of other markers of male gonad differentiation (testosterone and Müllerian duct inhibiting hormone) when they added Daudi supernatants to fetal rat or calf ovaries, even though they did observe some of the morphological changes reported earlier.

Nagai *et al.* (1979) have also seen the specific binding of H-Y antigen to fetal ovary cells (Table II) and have used this as a means of partially purifying the Daudi cell antigen. Fetal bovine ovary cells (and not other fetal cells) will absorb H-Y antigen from the Daudi cell culture medium. This antigen can then be eluted by lysing the cells. One of the major peaks obtained by this elution contains an 18,000-dalton hydrophobic protein which readily aggregates. When indifferent-stage gonads are cultured in what is thought to be the dimeric and trimeric aggregates of this protein, the XX gonadal cells develop into seminiferous tubules, and a tunica albuginea forms about these gonads. No such testicular development occurs in the absence of such protein additions.

Wachtel and Hall (1979) have obtained data suggesting that the H-Y antigen competes with an ovary-forming factor for this gonadal receptor. Anti-H-Y antisera were either left alone or else treated with (a) adult dog ovary cells, (b) adult dog ovary cells which had been exposed to exogenous H-Y antigen or (C) adult dog ovary cells exposed first to ovary supernatants and then to H-Y. As expected, the pretreated ovary cells were able to absorb out significant amounts of anti-H-Y cytotoxicity, showing that they bound the exogenously supplied antigen. However, when these cells had first been exposed to fetal ovary supernatant before exposure to H-Y, much less H-Y antigen was absorbed. Preexposure of the ovarian cells to brain, kidney, or spleen supernatants did not affect the binding of exogenous H-Y. Adult ovary cells, however, are able to bind exogenous H-Y indicating that such competition may not be complete, or that competition occurs only during a certain specified time. Zenzes *et al.* (1980) have extended these results to show that the reaggregation of newborn rat testicular cells into tubular aggregates was inhibited in the presence of newborn (but not adult) rat ovarian supernatants.

Evidence for the gonadal H-Y receptor also comes from studies of functional aberrations. There exists in humans a syndrome of XY gonadal dysgenesis which resembles the Turner (XO) gonadal dysgenesis syndrome. Some of these individuals, however, have not only an intact Y chromosome, but also an immuno-

TABLE II

Demonstration of Gonad-Specific H-Y Antigen Receptor by the Binding of Partially Purified Tritiated H-Y Antigen[a]

Target	Number of cells	cpm Bound to target	Total precipitable counts (%)
Bovine fetal ovary	7.5×10^5	57,392	11.3
	1.75×10^7	77,182	14.6
Adult mouse spleen	7.5×10^5	14,980	2.8
	1.75×10^7	6,157	1.2

[a]From Nagai *et al.* (1979).

logically detectable H-Y antigen on the surface of their leukocytes. Wachtel *et al.* (1980b) infer from these findings that the H-Y antigen is disseminated during embryogenesis, but that in these individuals the gonadal receptor is either lacking or dysfunctional. Thus, the gonad would differentiate into an ovary; but (as in the case of XO gonadal dysgenesis), the ovary degenerates. The alternative explanation—that the H-Y antigens of these individuals showed immunological cross-reactivity with wild-type H-Y, but were nonfunctional—has not yet been ruled out. Such a cross-reactive H-Y antigen which has lost its "receptor binding activity" has been postulated in the supernatant of a variant Daudi cell line (Iwata *et al.,* 1979; Nagai *et al.,* 1980).

Wachtel and his colleagues (1980b) have also presented data suggesting that the H-Y antigen can enter the blood circulation and that ovarian cells *in vitro* become H-Y-positive when exposed to the sera from a fetal bull or freemartin. Fetal cow serum was not capable of rendering the ovarian cells H-Y positive. Thus, the H-Y antigen may act as a true hormone when, secreted by the testes of the bull embryo, it circulates through the synchorial placenta and masculinizes the female gonad of its twin, producing a freemartin. While the mechanisms by which the H-Y antigen would enter the fetal circulation in the male twin and be picked up by the gonadal receptors in the female embryo have not been studied, there remains this evidence that soluble H-Y antigen can act as a fetal hormone which binds to an organ-specific receptor. Keeping in mind the caveats mentioned earlier, there appears to be a gonadal H-Y receptor that is essential for testis differentiation. The characterization of this receptor is still extremely uncertain. Ohno (1979) has calculated that this receptor has a binding constant in the order of $6.6 \times 10^{-9}\ M$ H-Y antigen. However, this is at best a rough approximation, as neither the molecular weight of the active antigen nor its percentage in the Daudi supernatant or epididymal fluid is known. What is known, though, is that as little as 0.146 gm of Daudi H-Y antigen fraction will saturate the receptors of 4 million fetal bovine ovary cells. It also appears that this binding is specific for the embryonic gonadal tissue of both sexes, and that binding of exogenous H-Y antigen can be seen on both Leydig and Sertoli cells of the developing testis (Wachtel *et al.,* 1980b). It seems that there is a specific gonadal receptor for something. But the nature of that "something" and whether it is indeed the heterogametic sex-determining substance has yet to be resolved.

IV. RECEPTORS IN CELL ADHESION

A. Introduction

The ability of cells to distinguish between homotypic and heterotypic cell types and to interact with these cells in a highly specific manner is one of the fundamental embryonic processes (reviewed in Maslow, 1976; Lilien *et al.,*

1979; Jones, 1980). This is best seen in experiments by Holtfreter, Trinkaus, and others concerning the changes in the affinity of embryonic tissues with time (Townes and Holtfreter, 1955; Trinkaus, 1963). Developmentally related changes in cell affinities have also been seen in avian and mammalian embryos where aggregates of mixed cells mimic the patterns of normal development by "sorting out" into homotypic cellular arrangements (Steinberg, 1964) or by forming histotypic structures *in vitro* (Moscona, 1957). Again, such histotypic patterns can be formed only between cells at the appropriate developmental age (DeLong and Sidman, 1970). The phenomenon of "sorting out" does not necessitate the existence of cell-specific effector or receptor molecules (Steinberg, 1970), and the search for such was not pursued in earnest until Roth (1968) presented evidence for their existence independent of the sorting out phenomenon.

B. Calcium-Dependent and Calcium-Independent Adhesion Molecules

While several embryonic systems provide evidence for the existence of specific cell-adhesion molecules, we will concentrate on those studies of neuronal tissue where such molecules have been best characterized. Since 1980, research on specific intercellular adhesion has been organized by the recognition of two types of adhesion mechanisms. The first involves calcium-independent adhesion, while the second type of adhesion is dependent on calcium ions. Dissociating cells with trypsin in the presence of calcium enables the cells to reaggregate in the presence of calcium, but not in its absence. The calcium-independent system (CIDS) has been removed, but the calcium-dependent system (CDS) remains. However when cells are dissociated with trypin in the absence of calcium, the cells cannot aggregate even if calcium is present. The CDS requires both Ca^{2+} and physiological temperatures to act, while the CIDS requires neither of these conditions (Takeichi *et al.*, 1979, 1982; Grunwald *et al.*, 1980; Brackenbury *et al.*, 1981; Magnini *et al.*, 1981; Thomas *et al.*, 1981).

Recent studies (reviewed in Rutishauser, 1984; Cook *et al.*, 1984) suggest that the cell-surface components of the CIDS and CDS are two molecular families. The CIDS family is represented by neural cell-adhesion molecule (N-CAM). N-CAM is an integral membrane glycoprotein with a single polypeptide chain. It is thought that the molecule is anchored into the cell membrane near the carboxyl end of the molecule and binds through its animo terminus (Cunningham *et al.*, 1983). N-CAM was first detected by the antibody-induced inhibition of reaggregation (Brackenbury *et al.*, 1977), and specific antibodies are able to inhibit N-CAM–cell binding and neural-cell–neural-cell adhesion (Rutishauser *et al.*, 1982; Rutishauser *et al.*, 1978). This evidence that N-CAM might serve as the ligand in cell–cell bonding was strengthened when it was demonstrated (Rutishauser *et al.*, 1978) that artificial vesicles composed solely of N-CAM can bind to neural cell surfaces, and that these vesicles have the same binding specificity as N-CAM-bearing retinal cells. Moreover, the N-CAM-containing vesicles are able to aggregate themselves. It was therefore proposed that the binding mechanism may involve direct interaction between N-CAM molecules (Rutishauser et al., 1982).

N-CAM is involved in several aspects of neural development and neuron–target interactions (reviewed in Rutishauser, 1984), including the formation of plexiform neuronal layers in the retina, neurite fasciculation in the dorsal root ganglia, tracking of growth cones along nerve fibers, nerve-myotube recognition in skeletal muscle innervation, and neuron–neuron recognition between retinal ganglia neurons and the nerves of the optic tectum. In all cases, both of the interacting cells have N-CAM and the interaction can be inhibited with anti-N-CAM Fab fragments.

The amount of sialic acid on the N-CAM molecule may be critical for its binding functions, removal of this negatively-charged sugar causing a nearly 10-fold increase in the rate of N-CAM binding to cells (Cunningham et al., 1983). These changes are dramatically seen in different regions of the visual system. At 10 days gestation, N-CAM isolated from chick neural retina has many fewer sialic acid residues than N-CAM isolated from the neural regions, (Hoffman et al., 1982; Schlosshauer et al., 1984), and it has been suggested (Rutishauser, 1984) that the gradual decline in sialic acid content from optic nerve and brain N-CAM during the final stages of visual system development may be involved in fixing the tentative retinotectal adhesions.

Other molecules involved in CIDS adhesion include nerve–glial cell adhesion molecule (Schlosshauer et al., 1984) and the L1 neural antigen (Rathgen and Schachner, 1984). Although these molecules are the products of different genes, they share immunological cross-reactivities with N-CAM (Grumet et al., 1984; Schachner et al., 1983). It is possible, then, that there is a family of such adhesion molecules which share a common epitope (Rutishauser, 1984; Schachner et al., 1983). These molecules all probably reside on their respective cell surfaces and may act as receptors for identical molecules on the surfaces of adjacent cells.

The calcium-dependent cell-adhesion molecules may also constitute a family of cell-surface glycoproteins. Antibodies made against cells dissociated by tryp-sin in the presence of calcium have identified glycoproteins having a molecular weight of approximately 130,000. These glycoproteins have been isolated from fibroblasts (Takeichi, 1977; Cook et al., 1984), embryonal carcinoma (Yoshida and Takeichi, 1982; Ogou et al., 1983) and neural retina cells (Cook et al., 1984). As the tryptic map of a 90-kDa protein released by neural retina cells into their media is similar to that of the 130-kDa calcium-binding glycoprotein de-rived from iodinated retinas, it is possible that other 90-kDa calcium-binding glycoproteins are also in this family. This would include uvomorulin, responsible for the calcium-dependent compaction of eight-cell mouse embryos (Hyafil *et al.*, 1980), and the calcium-dependent adhesion molecule of embryonic chick liver cells (Ogou *et al.*, 1983; Gallin *et al.*, 1983). In each case, Fab fragments directed against these molecules block calcium dependent adhesion.

It is possible that these receptors are modified by the action of glycosyltransferases (Lilien *et al.*, 1978). In 1970 Roseman proposed that cell-surface glycosyltransferases and their receptors (carbohydrate moieties) might mediate neuronal recognition. While this hypothesis has not been proven, various studies have found the cell surface to contain abundant oligosaccharides and surface-active glycosyltransferases (reviewed in Pierce *et al.*, 1980). Furthermore, developmental changes in cell-surface glycosyltransferase specificities and activities have been observed in chick embryos (Shur, 1977a,b). Roth *et al.* (1971) and Porzig (1978) have observed the ability of intact chicken neural retina cells to transfer galactose from its UDP-sugar form to a variety of acceptors including *N*-acetylglucosamine and an endogenous acceptor on the neural retina cell.

McDonough, Rutz, and Lilien (1977) and McDonough and Lilien (1978) presented evidence that turnover of the tissue-specific "ligand," from neural retina, alluded to earlier, is mediated by a β-*N*-acetylgalactosaminyltransferase. Release of endogenous ligand-like activity from the cell surface into the surrounding medium following preincubation of cells was inhibited by UDP (and no other nucleotide) and EDTA and stimulated by Mn^{2+} (but not Ca^{2+} or Mg^{2+}). This *N*-acetylgalactosaminyltransferase has now been demonstrated biochemically on the cell surface, where it transfers *N*-acetylgalactosamine from its UDP conjugate to an endogenous macromolecular acceptor which is concomitantly released from the cell surface (Balsamo and Lilien, 1980). This implies that "ligand" and receptor are related as transferase and acceptor or that the transferase is closely associated with the "ligand." In this manner, the maintenance of stable adhesions may be regulated by the ability of the glycosyltransferase to control ligand turnover.

V. THEORETICAL CONSIDERATIONS

A. Membrane–Cytoskeleton Interaction

The numerous specific intercellular interactions occurring during embryogenesis still demand molecular explanations. The review of five such agent–receptor systems—bindin and its vitelline envelope receptor, mouse sperm binding protein and its zona pellucida receptor, H-Y antigen and its cellular receptor, and the calcium-dependent and calcium-independent neural adhesion molecules—demonstrates the paucity of our knowledge concerning the molecular mechanisms of embryonic cell interactions. There are several conceptual matters into which these data are being organized, and two of them will be presented herein: (1) that adjacent cells interact through cell surface glycosyltransferases and (2) that cells effect changes in neighboring cells by altering the neighbor's membrane-bound cytoskeleton.

The binding of receptor to ligand is often the signal for initiating a series of preset events. The paradigms for such actions have been hormone receptors where the reception of hormone stimulates the activity of an adjacent adenyl cyclase enzyme or activates a pathway of enzymes allowing the influx of calcium ions into the cell (Sutherland, 1972; Majerus et al., 1984). While it is certainly possible for embryonic cell receptors to act in this manner (see Wudl et al., 1977), there is another mode of operation which may be used by the developing embryo: namely, cross-linking the cell membrane.

The cross-linking of cell-surface molecules has long been thought to stimulate the division and differentiation of B cells into plasmacytes, and Edelman's laboratory (Edelman et al., 1974) has shown that monovalent antibodies and lectins will not stimulate these phenomena while divalent ligands will. The T-cell-independent antigens are able to directly induce plasmacyte formation, presumably by their cross-linking ability (Dintzis et al., 1976; Dintzis et al., 1982), and the cross-linking of IgD molecules on T-cell-dependent B lymphocytes appears to trigger the appearances of the TRF lymphokine receptor (Yaffe and Finkelman, 1983). Cross-linking can activate development in at least two ways. First, it can activate cell membrane-bound enzymes, as in the case of mast-cell histamine release upon the cross-linking of membrane-bound IgE by allergens (Ishizaka et al., 1980). Second, it can restructure the cytoskeleton which is attached to the cell membrane. Hay's laboratory has shown that cytoskeletal rearrangements of corneal epithelial cells are occasioned by placing the cells on nonsoluble collagen and that these changes appear to initiate the program of corneal cytodifferentiation (Sugrue and Hay, 1982; Tomasek et al., 1982). In B-cell activation, both mechanisms may be used. When the cell-membrane immunoglobulins are cross-linked, the actin microfilaments of the cell are reorganized (Flanagan and Koch, 1978; Woda and Woodin, 1984), and phosphotidylinositol is hydrolyzed to diacylglycerol (Coggeshall and Cambier, 1984).

Such programming of nuclear events by cell-surface cross-linking is difficult to study in developing embryos, but Fujinami and Oldstone (1984) observed this phenomenon in studies of persistant viral infection. Cells infected with measles virus synthesize a glycoprotein antigen onto the cell surface. When antibodies are made by the host organism against this glycoprotein, they cap it. The process of capping appears to change the pattern of cellular synthesis such that one of the intracellular measles proteins becomes phosphorylated while the synthesis of another measles protein ceases altogether. Other cellular functions, however, proceed normally. It is possible that induction involves the binding of a particular molecule to the cell surface of the responding cell in a manner to "freeze" or alter the cytoskeleton of that cell coordinate regulation of cytoskeleton and gene expression by the extracellular matrix has also been observed in chick limb mesenchymal cells (Solursh et al., 1984) and 3T3-adipocytes (Spiegelman and Ginty, 1983).

B. Cell-Surface Glycosyltransferases

Lillie's hypothesis that gametes adhere to each other by complementary cell surface molecules has been an organizing principle throughout developmental biology. Weiss (1945, 1947), Tyler (1946), and Sperry (1951) extended it to account for cell migration, nerve–target interaction, and other embryonic phenomena. Roseman's hypothesis (1970) further refined this idea, predicting that glycosyltransferases would be able to bind their glycosyl receptors in the absence of the sugar donor (thereby forming intercellular adhesions) and catalyzing the addition of sugars in the presence of the sugar donor, thereby dissociating the cells. These hypotheses have been the subject of detailed reviews (Pierce et al., 1980; Shur, 1982a, 1984) during recent years.

Section II of this chapter presented evidence that cell-surface glycosyltransferases may be involved in the attachment of gametes in echinoderms, mammals, unicellular alga, and tunicates. In Section IV, external glycosyltransferases were seen to play an important (but as yet, undefined) role in the calcium-dependent neural adhesion system, and the amount of carbohydrate residues may also be important in the function of calcium-independent neural adhesion molecules and the H-Y antigen. Cell-surface glycosyltransferases have also been implicated as functioning in the cellular interactions occurring during gastrulation in the chick (Shur, 1977a,b, 1982b) where each of four cell-surface glycosyltransferases had a specific spatial and temporal pattern of activity. These activities were greatest on migrating cell types (neural crest cells, primordial germ cells, and primary mesenchyme) and those cells on interface of inductive interactions (notochord/somite, optic cup/head ectoderm, and medullary plate/presomitic mesoderm borders). The functioning of glycosyltransferases in cell migration has been observed in vitro by Turley and Roth (1979), who noted that migrating SV40-transformed fibroblasts covalently linked sugar groups onto the hyaluronidate matrix over which they travelled. In these studies, the rate of migration was inversely correlated with glycosylation of the substrate, suggesting that migration can only occur when there are free (unglycosylated) residues to function as receptors for the glycosyltransferase. The correlation of glycosyltransferase activity and induction has been studied with respect to limb bud chondrogenesis (Shur et al., 1982).

The activity of one of these enzymes, galactosyltransferase, increases as the subapical mesenchyme cells condense to form chondrocyte regions. Once the cells begin secreting cartilage matrix, the enzyme activity falls back to below its precondensation levels.

The T/t complex appears to be responsible for several intercellular interactions during early mouse embryogenesis. In addition to the earlier mentioned sperm segregation effect in heterozygotes (Hammerberg, 1982; Shur and Bennett, 1979; Shur, 1981), recessive mutants of the T/t complex, when homozygous, often abort development at specific times, depending on the allele. Although the possibility exists that some T/t-complex mutations are defects in intermediary metabolism (Wudl et al., 1977), Bennett and co-workers have assembled histological (Spiegelman and Bennett, 1974), cytological (Bennett, 1958; Yanagisawa and Fujimoto, 1977), and immunological (Bennett et al., 1972; Chang and Bennett, 1980) evidence that these alleles specify cell-surface antigens crucial for normal morphogenetic interactions. Somites isolated from homozygous T/T embryos fail to respond to normal inductive signals from wild-type neural tubes (Bennett, 1958) and the notochordal cells of such mice become interspersed with cells of both gut and neural tube (Dunn and Bennett, 1964). T/T and wild-type mesenchymal cells show strikingly different areas of intercellular contact (Spiegelman and Bennett, 1974) and aggregate differently in rotary cultures (Yanagisawa and Fujimoto, 1977; Yanagisawa and Fujimoto, 1978). Day 10 limb buds from T/T embryos have greater than six times the cell-surface galactosyltransferase activity as wild-type limb buds of similar size (Shur, 1982a).

Glycosyltransferases have also been implicated in several pattern-forming activities of early embryos. A gradient of galactosyltransferase activity has been found across the neural retina and optic tectum, and experiments using proteases and glycosidases suggest that retinotectal adhesion may be specified by protein in receptor-carbohydrate ligand interactions. Marchase (1977) demonstrated that the carbohydrate ligand of the tectum requires terminal N-acetylgalactosamine residues, and synaptonemal preparation of embryonic chick brain cells shows high glycosyltransferase activities (Den and Kaufman, 1968; Bosmann, 1973). This suggests a model where the migration of the growing axon tip is directed, in part, by the interactions of axon-tip glycosyltransferases with tectal carbohydrate ligands.

It has recently been proposed (Roth, 1985) that cell-surface glycosyltransferases are the evolutionary antecedents of immunoglobulins. Roth shows that glycosyltransferases fit five criteria that should be expected for such an evolutionary ancestor: (1) the ability to specifically bind to a wide range of ligands, (2) the existence of soluble and membrane-bound forms, (3) a tendency to be polymorphic, (4) involvement in species recognition in unicellular animals, and (5) the ability to modulate expression when a ligand is introduced. In addition, glycosyltransferases may still be active in cell–cell recognition in the mammalian immune system. Parish and colleagues (McKenzie *et al.*, 1977) have shown that the binding of erythrocytes to thymocytes is mediated by an erythrocyte carbohydrate and a thymocyte protein that recognizes it. The expression of both the carbohydrate and protein is controlled by the murine H-2 complex, suggesting (Parish *et al.*, 1981) that some of the MHC genes code for glycosyltransferases. This would not be unexpected if the MHC and the *T/t* complex (which appears to regulate galactosyltransferase activity) were derived from a common evolutionary ancestor (Bennett *et al.*, 1971). It has been suggested (Higgins and Parish, 1980) that the Ia genes of the MHC code for glycosyltransferases which create the Ia epitopes by specific glycosylation and Furukawa *et al.* (1985) have evidence for *N*-acetylgalactosaminyltransferase activity in an affinity-purified I-A antigen of the H-2^d haplotype. Moreover, the cell-surface galactosyltransferase of cytolytic T cells correlates with their activation (Baker *et al.*, 1980; Kurt *et al.*, 1985), although the role of this enzyme in lymphocyte function has not been determined.

Sequence homologies have been found between MHC proteins and immunoglobulins (Orr *et al.*, 1979), and it is plausible that homologies exist between glycosyltransferases and immunoglobulins, as well. Until the sequencing of cell-surface glycosyltransferases, there is no way of knowing for certain, but circumstantial evidence suggests that this is the case. First, polyclonal antibodies prepared against affinity-purified human serum galactosyltransferase cross-react strongly with immunoglobulins (Wilson *et al.*, 1982), while certain monoclonal antibodies to that transferase enzyme (Podolsky and Isselbacher, 1984) react strongly with human IgG. The authors of the latter study suggest that immunoglobulins and glycosyltransferases might share at least one common epitope.

The glycosyltransferase hypothesis is particularly pleasing in that it unites evolutionary and developmental perspectives in cell–cell recognition. A mechanism originally designed for species recognition during fertilization evolves into a mechanism for cell–cell interactions in metazoans (while keeping its gamete-recognition function). Later, as the immune system evolves, this cell–cell recognition function is modified into the cellular immune responses (controlled by the MHC) and the humoral antibodies (controlled by immunoglobulin genes). The chapters of this volume all concern themselves with various interactions of cell-surface receptors in phylogeny, development, and the immune response. It would seem appropriate if we were all studying the different mountain peaks of the same submerged island.

ACKNOWLEDGMENTS

Funds to produce this chapter were provided by National Institutes of Health grant HD 15032. I wish to thank Drs. K. Aketa, P. Andrews, R. Auerbach, J. Lilien, S. Ohno, S. Roth, B. Shur, V. Vacquier, and S. Wachtel for providing me with preprints of their unpublished studies.

158

Scott F. Gilbert

REFERENCES

Adinolfi, M., Polani, M., and Zenthon, J. (1982). *Hum. Genet.* **61**, 1–2.

Aketa, A. (1967). *Embryologia* **9**, 238–245.

Aketa, A. (1973). *Exp. Cell Res.* **80**, 439–441.

Aketa, K., Miyazaki, S., Yoshida, M., and Tsuzuki, H. (1978). *Biochem. Biophys. Res. Commun.* **80**, 917–922.

Akram, H., and Weniger, J. P. (1968). *Arch. Anat. Microsc. Microphol. Exp.* **57**, 369–378.

Andrews, P. (1984). *In* "Genetic Analysis of the Cell Surface" Receptors and Recognition, Series B, Vol. 16 (P. Goodfellow, ed.), pp. 159–190. Chapman and Hall, London.

Baker, A. P., Smith, W. J., and Holder, D. A. (1980). *Cell. Immunol.* **51**, 186–191.

Balsamo, J., and Lilien, J. (1980). *Biochemistry* **19**, 2479–2484.

Barros, C., and Yamagimachi, R. (1972). *J. Exp. Zool.* **180**, 251–265.

Bellet, N. F. (1976). *J. Cell Sci.* **22**, 547–562.

Bellet, W. F., Vacquier, J. P., and Vacquier, V. D. (1977). *Biochem. Biophys. Res. Commun.* **79**, 159–165.

Benhaim, A., Gangnerau, M. N., Bettane-Casanova, M., Felluws, M., and Picon, R. (1982). *Differentiation* **22**, 53–58.

Bennett, D. (1958). *Nature (London)* **181**, 1286.

Bennett, D., Boyse, E. A., and Old, L. J. (1971). *In* "Cell Interactions: Third Lepetit Coloquium" (L. G. Silvestri, ed.), pp. 248–262. North Holland/American Elsevier, New York.

Bennett, D., Goldberg, E., Dunn, L. C., and Boyse, E. A. (1972). *Proc. Natl. Acad. Sci. U.S.A.* **69**, 2076–2080.

Bennett, D., Boyse, E. A., Lyon, M. F., Mathieson, B. J., Scheid, M., and Yanagisawa, K. (1975). *Nature (London)* **257**, 236–238.

Beutler, B., Nagni, Y., Ohno, S., Klein, G., and Shapiro, E. (1978). *Cell* **13**, 509–513.

Blanchard, M. G., and Josso, N. (1974). *Pediatr. Res.* **8**, 968–971.

Bleil, J. D., and Wassarman, P. M. (1980a). *Dev. Biol.* **76**, 185–202.

Bleil, J. D., and Wassarman, P. M. (1980b). *Cell* **20**, 873–882.

Bleil, J. D., and Wassarman, P. M. (1983). *Dev. Biol.* **95**, 317–324.

Bosmann, H. B. (1973). *J. Neurochem.* **20**, 1037–1049.

Bosmann, H. B., and McLean, R. J. (1975). *Biochem. Biophys. Res. Commun.* **63**, 323–327.

Brackenbury, R., Thiery, J. P., Rutishauser, U., and Edelman, G. M. (1977). *J. Biol. Chem.* **252**, 6835–6840.

Brackenbury, R., Rutishauser, U., and Edelman, G. M. (1981). *Proc. Natl. Acad. Sci. U.S.A.* **78**, 387–391.

Braden, A. H. W., Austin, C. R., and David, H. A. (1954). *Aust. J. Biol. Sci.* **7**, 391–409.

Brandriff, B., Moy, G. W., and Vacquier, V. D. (1978). *Gamete Res.* **1**, 89–99.

Carroll, E. J., Jr., and Epel, D. (1975). *Dev. Biol.* **44**, 22–32.

Chang, C., and Bennett, D. (1980). *Cell* **19**, 537–543.

Coggeshall, K. M., and Cambier, J. C. (1984). *J. Immunol.* **133**, 3382–3386.

Collins, F. (1961). *Dev. Biol.* **49**, 381–394.

Cook, J. H., Pratt, R. S., and Lilien, J. (1984). *Biochemistry* **23**, 899–904.

Coulombre, J. C., and Russell, E. S. (1954). *J. Exp. Zool.* **126**, 277–295.

Cullen, S. E., Bernaco, D., Carbarnara, A. O., Jacot-Guillaermond, H., Trinchieri, G., and Ceppellini, R. (1973). *Transplant. Proc.* **5**, 1835–47.

Cunningham, B. A., Hoffman, S., Rutlshauser, U., Hemperly, J. J., and Edelman, G. (1983). *Proc. Natl. Acad. Sci. U.S.A.* **80**, 3116–3120.

DeLong, G. R., and Sidman, R. L. (1970). *Dev. Biol.* **22**, 584–600.

Den, H., and Kaufman, B. (1968). *Fed. Proc.* **27**, 346.

DeSantis, R., Janunno, G., and Rosati, F. (1980). *Dev. Biol.* **74**, 490–499.

DeSantis, R., Pinto, M. R., Cotelli, F., Rosati, F., Monroy, A., and D'Allessio, G. (1983). *Exp. Cell Res.* **148**, 508–513.

Dintzis, H. M., Dintzis, R. A., and Vogelstein, B. (1976). *Proc. Natl. Acad. Sci. U.S.A.* **73**, 3671–3675.

Dintzis, R. Z., Vogelstein, B., and Dintzis, H. M. (1982). *Proc. Natl. Acad. Sci. U.S.A.* **79**, 884–888.

Dunn, L. C., and Bennett, D. (1964). *Science* **144**, 260–267.

Durnberger, H., and Kratochwil, K. (1980). *Cell* **19**, 465–471.

Ebner, K. E., and Magee, S. C. (1975). *In* "Subunit Enzymes: Biochemistry and Function" (K. Ebner, ed.), pp. 137–139. Marcel-Dekker, New York.

Edelman, G. M., Spear, P. G., Rutishauser, U., and Yahara, I. (1974). *In* "The Cell Surface in Development" (A. Moscona, ed.), pp. 141–164. Wiley, New York.

Eicher, E. M., Washburn, L. L., Whitney, J. B., and Morrow, K. B. (1982). *Science* **217**, 535–537.

Eichwald, E. J., and Silmser, C. R. (1955). *Transplant. Bull.* **2**, 148–149.

Elder, J., and Alexander, S. (1982). *Proc. Natl. Acad. Sci. U.S.A.* **79**, 4540–4544.

Engel, W., Klemme, B., and Ebrecht, A. (1981). *Hum. Genet.* **57**, 68–70.

Evans, E. P., Ford, C. E., and Lyon, M. F. (1977). *Nature (London)* **267**, 430–431.

Fellous, M., Günther, E., Kemler, R., Weils, J., Berger, R., Guenet, J. L., Jakob, H., and Jacob, F. (1978). *J. Exp. Med.* **148**, 58–70.

Flanagan, J., and Koch, T. (1978). *Nature (London)* **273**, 278–281.

Florman, H. M., and Storey, B. T. (1982). *Dev. Biol.* **91**, 121–130.

Florman, H. M., and Wassarman, P. M. (1985). *Cell* **41**, 313–324.

Florman, H. M., Bechtol, K. B., and Wassarman, P. M. (1984). *Dev. Biol.* **106**, 243–255.

Fujinami, R. S., and Oldstone, M. B. A. (1984). *In* "Concepts in Viral Pathogenesis" (A. L. Notkins and M. B. A. Oldstone, eds.), pp. 187–193. Springer-Verlag, New York.

Furukawa, K., Higgins, T., and Roth, S. (1985). ASCB abstracts.

Gallin, W. J., Edelman, G. M., and Cunningham, B. A. (1983). *Proc. Natl. Acad. Sci. U.S.A.* **80**, 1038–1042.

Gilbert, S. F. (1984). *Perspect. Biol. Med.* **28**, 18–34.

Glabe, C. G., and Lennarz, W. J. (1979). *J. Cell Biol.* **83**, 595–604.

Glabe, C. G., and Vacquier, V. D. (1977a). *Nature (London)* **267**, 836–838.

Glabe, C. G., and Vacquier, V. D. (1977b). *J. Cell Biol.* **75**, 410–421.

Glabe, C. G., and Vacquier, V. D. (1978). *Proc. Natl. Acad. Sci. U.S.A.* **75**, 881–885.

Gore-Langton, R. E., Tung, P. S., and Fritz, I. B. (1983). *Cell* **32**, 289–301.

Grumet, M., Hoffman, S., and Edelman, G. M. (1984). *Proc. Natl. Acad. Sci. U.S.A.* **81**, 267–271.

Grunwald, G. B., Geller, R. L., and Lilien, J. (1980). *J. Cell Biol.* **85**, 766–776.

Gwatkin, R. B. L. (1976). *In* "The Cell Surface in Embryogenesis and Development" (G. Poste and G. L. Nicholson, eds.), p. 1. Elsevier-North Holland, New York.

Gwatkin, R. B. L., and Williams, D. T. (1977). *J. Reprod. Fertil.* **49**, 55–59.

Hall, M. V., and Franklin, L. E. (1981). *J. Androl.* **2**, 14.

Hall, J. L., and Wachtel, S. S. (1980). *Mol. Cell. Biochem.* **33**, 49–66.

Hall, J. L., Bushkin, Y., and Wachtel, S. S. (1981). *Hum. Genet.* **58**, 34–36.

Hammerberg, C. (1982). *Genet. Res.* **39**, 319–226.

Hanada, A., and Chang, M. C. (1972). *Biol. Reprod.* **6**, 300–309.

Harrison, R. G. (1910). *J. Exp. Zool.* **9**, 787–848.

Hartmann, J. F., and Gwatkin, R. B. L. (1971). *Nature (London)* **234**, 479–481.

Hartmann, J. F., Gwatkin, R. B. L., and Hutchinson, C. F. (1972). *Proc. Natl. Acad. Sci. U.S.A.* **69**, 2767–2769.

Haseltine, F. P., Vandyke, D. L., Breg, W. R., and Frake, U. (1982). *Am. J. Med. Genet.* **13**, 115–123.

Higgins, T. J., and Parish, C. R. (1980). *Mol. Immunol.* **17**, 1065–1073.

Hoffman, S., Sorkin, B., White, P., Brackenbury, R., Mailhammer, R., Rutishauser, U., Cunningham, B., and Edelman, G. M. (1982). *J. Biol. Chem.* **257**, 7720–7729.

Huang, T. T. F., Fleming, A. D., and Yanagimachi, R. (1981). *J. Exp. Zool.* **217**, 287–290.

Hyafil, F., Morello, D., Babinet, C., and Jacob, F. (1980). *Cell* **21**, 927–934.

Inoue, M., and Wolf, D. P. (1975). *Biol. Reprod.* **13**, 546–551.

Ishizaka, T., Hirata, F., Ishizaka, K., and Axelrod, J. (1980). *Proc. Natl. Acad. Sci. U.S.A.* **77**, 1903–1906.

Iwata, H., Nagai, Y., Stapleton, D. D., Smith, R. C., and Ohno, S. (1979). *Arthritis Rheum.* **22**, 1211–1216.

Jones, B. M. (1980). *Biol. Rev.* **55**, 207–235.

Jost, A., Vigier, B., Prepin, J., and Perchellet, J.-P. (1973). *Rec. Prog. Horm. Res.* **29**, 1–41.

Just, E. E. (1931). *Naturwissen* **19**, 953–1001.

Kinsey, W. H., Deker, G. L., and Lennarz, W. J. (1980). *J. Cell Biol.* **87**, 248–254.

Koo, E. C., Tada, N., Chiganti, R., and Hammerling, U. (1981). *Hum. Genet.* **57**, 64–67.

Krco, C. J., and Goldberg, E. H. (1976). *Science* **193**, 1134–1135.

Kurt, E. A., Shur, B. D., and Linquist, R. R. (1985). Manuscript submitted.

Lilien, J., Hermolin, J., and Lipke, P. (1978). *In* "Specificity of Embryological Interactions" (D. R. Garrod, ed.), pp. 131–155. Chapman and Hall, London.

Lilien, J., Balsamo, J., McDonough, J., Hemolin, J., Cook, J., and Rutz, R. (1979). *In* "Surfaces of Normal and Malignant Cells" (R. O. Hymes, ed.), pp. 389–424. Wiley, New York.

Lillie, F. R. (1914). *J. Exp. Zool.* **16**, 523–590.

Lopez, L., Bayna, E., Litoff, D., Shaper, N. L., Shaper, J. H., And Shur, B. D. (1985). *J. Cell Biol.* **101**, 1501–1510.

McDonough, J., and Lilien, J. (1978). *J. Supramol. Struct.* **7**, 409–418.

McDonough, J., Rutz, R., and Lilien, J. (1977). *J. Cell Sci.* **27**, 245–254.

McKenzie, I. F. C., Clarke, A., and Parish, C. (1977). *J. Exp. Med.* **145**, 1039–1053.

McLaren, A., Simpson, E., Tomonari, K., Chandler, P., and Hogg, H. (1984). *Nature* **312**, 552–555.

Magnani, J. L., Thomas, W. A., and Steinberg, M. S. (1981). *Dev. Biol.* **81**, 96–105.

Majerus, P. W., Neufeld, E. J., and Wilson, D. B. (1984). *Cell* **37**, 701–703.

Marchant, M. (1975). *Dev. Biol.* **44**, 1–21.

Marchase, R. B. (1977). *J. Cell Biol.* **75**, 237–257.

Maslow, D. E. (1976). *In* "The Cell Surface in Animal Embryogenesis and Development" (G. Poster and G. L. Nicholson, eds.), p. 697. Elsevier North Holland, New York.

Mazurkiewicz, J. E., and Nakane, P. K. (1972). *J. Histochem. Cytochem.* **20**, 969–974.

Metz, C. B. (1978). *Curr. Topics Devel. Biol.* **12**, 107–147.

Moscona, A. (1957). *Proc. Natl. Acad. Sci. U.S.A.* **43**, 184–194.

Moy, G. W., and Vacquier, V. D. (1979). *Curr. Top. Dev. Biol.* **13**, 31–44.

Müller, U., Aschmoneit, I., Zenzes, M. T., and Wolf, U. (1978a). *Hum. Genet.* **43**, 151–157.

Müller, U., Siebers, J. W., Zenzes, M. T., and Wolf, U. (1978b). *Hum. Genet.* **45**, 209–213.

Müller, U., Zenzes, M. T., Bauknecht, T., Wolf, U., Siebers, J. W., and Engel, W. (1978c). *Hum. Genet.* **45**, 203–207.

Müller, U., Zenzes, T., Wolfe, V., Engel, W., and Weniger, J. P. (1979a). *Nature (London)* **280**, 142–144.

Müller, U., Wolf, U., Siebers, J. W., and Gunther, E. (1979b). *Cell* **17**, 331–335.

Müller, U., Guichard, A., Royss-Brion, M., and Scheib, D. (1980). *Differentiation* **16**, 129–133.

Nagai, Y., and Ohno, S. (1977). *Cell* **10**, 729–732.

Nagai, Y., Ciccarese, S., and Ohno, S. (1979). *Differentiation* **13**, 155–164.

Nagai, Y., Iwata, H., Stapleton, D. D., Smith, R. C., and Ohno, S. (1980). *In* "Testicular Development, Structure and Function" (A. Steinberger and A. Steinberger, eds.), pp. 1–47. Raven Press, New York.

Ogou, S. I., Yoshida-Noro, C., and Takeichi, M. (1983). *J. Cell Biol.* **97,** 944–948.

Ohno, S. (1977). *Immunol. Rev.* **33,** 59–69.

Ohno, S. (1979). "Major Sex Determining Genes." Springer-Verlag, Berlin and New York.

Ohno, S., Ciccarese, S., Nagai, Y., and Wachtel, S. S. (1978). *Arch. Androl.* **1,** 103–109.

Oikawa, T., Yamagimachi, R., and Nicholson, G. L. (1973). *Nature (London)* **241,** 256–259.

Onitake, K., Tsuzuki, H., and Aketa, K. (1972). *Dev. Growth Differ. (Nagoya)* **14,** 207–215.

Orr, H. T., Lancet, D., Robb, R., Lopez de Castro, J. A., and Strominger, J. L. (1979). *Nature (London)* **282,** 266.

Parish, C., O'Neill, H. C., and Higgins, T. J. (1981). *Immunol. Today* **2,** 99–101.

Peterson, R. N., Russell, L., Bundman, D., and Freund, M. (1979). *Science* **207,** 73–74.

Philips, D. M., and Shalgi, R. M. (1980). *J. Exp. Zool.* **213,** 1–8.

Pierce, M., Turley, E. A., and Roth, S. (1980). *Int. Rev. Cytol.* **65,** 1–48.

Pikó, L. (1969). "Fertilization" (C. A. Metz and A. Monroy, eds.), Vol. 2, pp. 325–403. Academic Press, New York.

Podolsky, D. K., and Isselbacher, K. J. (1984). *Proc. Nat. Acad. Soc. U.S.A.* **81,** 2529–2533.

Porzig, E. F. (1978). *Dev. Biol.* **67,** 114–136.

Powell, J. T., and Brew, K. (1976). *Biochemistry* **15,** 3499–3504.

Rathgen, F. G., and Schachner, M. (1984). *EMBO. J.* **3,** 1–20.

Rosati, F., and DeSantis, R. (1980). *Nature (London)* **283,** 762–764.

Roseman, S. (1970). *Chem. Phys. Lipids* **5,** 270–299.

Rossignol, D. P., Roschelle, A. J., and Lennarz, W. J. (1981). *J. Supramol. Struct. Cell Biol.* **15,** 347–358.

Rossignol, D. P., and Lennarz, W. J. (1983). "Molecular Biology of Egg Maturation," Ciba Symposia, Vol. 98, pp. 268–285. Ciba Foundation, London.

Rossignol, D. P., Earles, B. J., Decker, G. L., and Lennarz, W. J. (1984). *Dev. Biol.* **104,** 308–321.

Roth, S. (1968). *Dev. Biol.* **18,** 602–631.

Roth, S. (1985). *Quart. Rev. Biol.* **60,** 145–153.

Roth, S., McGuire, E. J., and Roseman, S. (1971). *J. Cell. Biol.* **51,** 525–535.

Rutishauser, U. (1984). *Nature (London)* **310,** 549–554.

Rutishauser, U., Thiery, J. P., Brackenbury, R., and Edelman, G. M. (1978). *J. Cell Biol.* **79,** 371–381.

Rutishauser, U., Hoffman, S., Edelman, G. M. (1982). *Proc. Natl. Acad. Sci. U.S.A.* **79,** 685–689.

Saling, P. M., and Storey, B. T. (1979). *J. Cell Biol.* **83,** 544–555.

Saling, P. M., Sowinski, J., and Storey, B. T. (1979). *J. Exp. Zool.* **209,** 229–238.

Schachner, M., Faissner, A., Kruse, J., Linder, J., Meir, D. H., Rathgen, F. G., and Wernecke, H. (1983). *Cold Spring Harbor Symp. Quant. Biol.* **48,** 557–568.

Schatten, G., and Mazia, D. (1976). *J. Supramol. Struct.* **5,** 343–369.

Schlosshauer, B., Schwartz, U., and Rutischauser, U. (1984). *Nature (London)* **310,** 141–143.

Shapiro, M., and Erickson, R. P. (1981). *Nature (London)* **290,** 503–505.

Shivers, C. A., Didkiewicz, A. B., Franklin, L. E., and Fussel, E. N. (1972). *Science* **178,** 1211–1213.

Shur, B. D. (1977a). *Dev. Biol.* **58,** 23–39.

Shur, B. D. (1977b). *Dev. Biol.* **58,** 40–55.

Shur, B. D. (1981). *Genet. Res.* **38,** 225–236.

Shur, B. D. (1982a). *In* "The Glycoconjugates" (M. I. Horowitz, ed.), Vol. III, p. 145. Academic Press, New York.

Shur, B. D. (1982b). *Dev. Biol.* **91,** 149–162.

Shur, B. D. (1984). *Mol. Cell Biochem.* **61,** 143–158.

Shur, B. D., and Bennett, D. (1979). *Dev. Biol.* **71,** 243–259.

Shur, B. D., and Hall, N. G. (1982a). *J. Cell Biol.* **95,** 567–573.

Shur, B. D., and Hall, N. G. (1982b). *J. Cell Biol.* **95,** 574–579.

Shur, B. D., Vogler, M., and Kosher, R. A. (1982). *Exp. Cell Res.* **137,** 229–237.

Silvers, W. K., Glasser, D. L., and Eicher, E. M. (1982). *Cell* **28,** 439–440.

Simpson, E., Chandler, P., Washburn, L., Bunker, H., Eicher, E. M. (1983). *Differentiation* **23,** S116–S120.

Solursh, M., Jensen, K. L., Zanetti, N. C., Lisenmayer, T. F., and Reiter, R. S. (1984). *Dev. Biol.* **105,** 451–457.

Sperry, R. W. (1951). *In* "Handbook of Experimental Psychology" (S. Stevens, ed.), pp. 236–280. Wiley, New York.

Spiegelman, M., and Bennett, D. (1974). *J. Embryol. Exp. Morphol.* **32,** 723–738.

Spiegelman, B. M., and Ginty, C. A. (1983). *Cell* **35,** 657–666.

Steinberg, M. S. (1964). *In* "Cellular Membranes in Development" (M. Locke, ed.), pp. 321–366. Academic Press, New York.

Steinberg, M. S. (1970). *J. Exp. Zool.* **172,** 395–434.

Sugrue, S. P., and Hay, E. D. (1982). *Dev. Biol.* **92,** 97–106.

Summers, R. G., and Hylander, B. L. (1975). *Exp. Cell Res.* **96,** 63–68.

Summers, R. G., Hylander, E. D., Colwin, C. H., and Colwin, A. L. (1975). *Am. Zool.* **15,** 523–551.

Sutherland, E. W. (1972). *Science* **177,** 401–408.

Takeichi, M. (1977). *J. Cell Biol.* **75,** 464–474.

Takeichi, M., Ozaki, H. S., Tokunaga, K., and Okada, T. S. (1979). *Dev. Biol.* **70,** 195–205.

Takeichi, M., Atsumi, T., Yoshida, C., and Ogou, S. I. (1982). *In* "Teratocarcinoma and Embryonic Cell Interactions" (T. Muramatsu, ed.), pp. 283–293. Japan Science Society Press, Tokyo.

Tegner, M. J., and Epel, D. (1973). *Science* **179,** 685–688.

Thomas, W. A., Thomson, J., Magnani, J. L., and Steinberg, M. S. (1981). *Dev. Biol.* **81,** 379–385.

Tomasek, J. J., Hay, E. D., and Fujiwara, K. (1982). *Dev. Biol.* **92,** 107–122.

Townes, P. L., and Holtfreter, J. (1955). *J. Exp. Biol.* **128,** 53–120.

Tran, D., and Josso, N. (1977). *Nature (London)* **269,** 411–412.

Trinkaus, J. P. (1963). *Dev. Biol.* **1,** 512–532.

Turley, E. A., and Roth, S. (1979). *Cell* **17,** 109–115.

Tyler, A. (1946). *Growth* **10**(suppl. 6), 7–19.

Tyler, A. (1948). *Physiol. Rev.* **28,** 180–219.

Tyler, A. (1956). *Exp. Cell Res.* **10,** 377–386.

Tyler, A. (1959). *Exp. Cell Res. (Suppl.)* **7,** 183–199.

Vacquier, V. D. (1979). *Am. Zool.* **19,** 839–849.

Vacquier, V. D., and Moy, G. W. (1977). *Proc. Natl. Acad. Sci. U.S.A.* **74,** 2456–2460.

Vacquier, V. D., and Payne, J. E. (1973). *Exp. Cell Res.* **82,** 227–235.

Vacquier, V. D., Tegner, M. J., and Epel, D. (1973). *Exp. Cell Res.* **80,** 111–119.

Wachtel, S. S. (1977). *Immunol. Rev.* **33,** 33–58.

Wachtel, S. S. (1979). *Cell* **16,** 691–695.

Wachtel, S. S. (1983). *Curr. Top. Dev. Biol.* **18,** 189–216.

Wachtel, S. S., and Hall, J. L. (1979). *Cell* **17,** 327–329.

Wachtel, S. S., Ohno, S., Koo, G. C., and Boyse, E. A. (1975). *Nature (London)* **257,** 235–236.

Wachtel, S. S., Koo, G. C., de la Chapelle, A., Kallio, H., Hayman, J. M., and Miller, O. J. (1980a). *Hum. Genet.* **54,** 25–30.

Wachtel, S. S., Koo, G. C., Breg, W. R., and Grenel, M. (1980b). *Hum. Genet.* **56,** 183–187.

Wachtel, S. S., Hall, J. L., Müller, U., and Chaganti, R. S. K. (1980c). *Cell* **21,** 917–926.

Weiss, P. (1945). *J. Exp. Zool.* **100,** 353–386.

Weiss, P. (1947). *Yale J. Biol. Med.* **19,** 235–278.

Whitely, H. R., McCarthy, B. J., and Whitely, A. H. (1970). *Dev. Biol.* **21,** 216–242.

Wilson, J. R., Weiser, M. M., Albini, B., Schenck, J. R., Rittenhouse, H. G., Mirata, A. A., and Berger, E. G. (1982). *Biochem. Biophys. Res. Commun.* **105,** 737–744.

Woda, B. A., and Woodin, M. B. (1984). *J. Immunol.* **133,** 2767–2772.

Wolf, U., Fraccaro, M., Mayerova, A., Hecht, T., Maraschio, P., and Hameister, H. (1980a). *Hum. Genet.* **54,** 149–154.

Wolf, U., Fraccaro, M., Mayerova, A., Hecht, T., Zuffardi, O., and Hameister, H. (1980b). *Hum. Genet.* **54,** 315–318.

Wudl, L. R., Sherman, M. I., and Hillman, N. (1977). *Nature (London)* **270,** 237–240.

Yaffe, L. J., and Finkelman, F. D. (1983). *Proc. Natl. Acad. Sci. U.S.A.* **80,** 293–297.

Yanagisawa, K. O., and Fujimoto, H. (1977). *J. Embryol. Exp. Morphol.* **40,** 277–283.

Yanagisawa, K. O., and Fujimoto, H. (1978). *Exp. Cell Res.* **115,** 431–435.

Yoshida, M., and Aketa, K. (1978). *Acta Embryol. Exp.* **3,** 269–278.

Yoshida, M., and Aketa, K. (1979). *Dev. Growth Differ. (Nagoya)* **21,** 431–436.

Yoshida, C., and Takeichi, M. (1982). *Cell* **28,** 217–224.

Zenzes, M. T., Wolf, U., and Engel, W. (1978a). *Hum. Genet.* **44,** 333–338.

Zenzes, M. T., Wolf, U., Günther, E., and Engle, W. (1978b). *Cytogen. Cell. Genet.* **20,** 365–372.

Zenzes, M. T., Urban, E., and Wolf, U. (1980). *Differentiation* **16,** 193–198.

6

Retinoids: Probes for Understanding Pattern Regulation in Regenerating Amphibian Limbs

DAVID L. STOCUM

I. INTRODUCTION

One of the most fascinating and complex problems in developmental biology is to comprehend how the patterns of cell differentiation we recognize as tissues and organs are specified during ontogeny. The results of numerous experiments have indicated that embryonic organ-forming districts are formed by spatially and temporally patterned inductive interactions between heterogeneous cell populations. Once induced, organ-forming districts are able to self-organize their specific tissue patterns and shapes when isolated from the embryo or moved to an abnormal location within the embryo. The self-organizational mechanism is homeostatic, in that a normal pattern can be generated after removal of part of a district, adding two districts together, or interchanging parts of a district. This phenomenon, called *pattern regulation*, indicates that the regional pattern of the district is specified by boundary elements that outline the pattern as a whole, to

165

RECEPTORS IN CELLULAR RECOGNITION
AND DEVELOPMENTAL PROCESSES

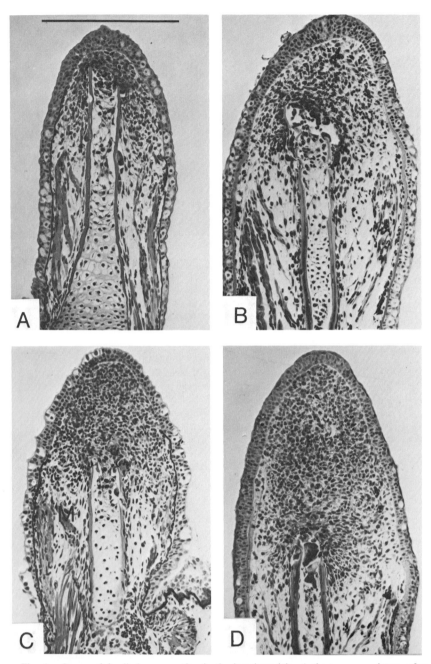

Fig. 1. Stages of forelimb regeneration in the larval urodele, *Ambystoma maculatum*, after amputation through the distal stylopodium, with (A–F) longitudinal sections stained with hematoxylin and light green, and (G and H) whole mounts stained with methylene blue to reveal cartilage matrix. Scale bar = 1 mm. (A) Five days postamputation. Cartilage, muscle and general connective tissue are dedifferentiating, and a small accumulation of blastema cells has appeared at the tip of the limb, under the wound epidermis. (B–D) The blastema increases in length and volume, by continued dedifferentiation and mitosis, to successively form the early bud, medium bud and late bud stages. M = stump muscle. (E–H) Stages of redifferentiation, in which the missing skeletal elements and

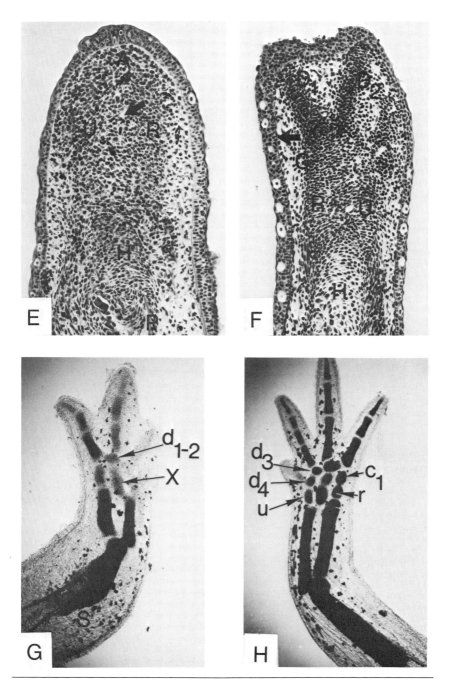

associated muscles are reformed in a proximal to distal, and anterior to posterior sequence. In (E), the thick arrow points to a blood vessel developing between the redifferentiating radius and ulna, and A indicates undifferentiated mesenchyme that will form the carpals and digits; in (F), the thick arrow points to the reforming basement membrane beneath the epidermis. H = humerus; R = radius; U = ulna; S = cartilage splint connecting the regenerated humerus to the stump humerus; X = block of carpal cartilage that will subsequently segregate into two carpals; u and r are basal carpals on the ulnar and radial sides, respectively, of the regenerate; c_1 and d_1–d_4 are remaining carpals.

which the cells of the district are subservient. Furthermore, after removal of some portion of these boundary elements, a whole set can be reconstituted from those remaining. Such a regulative pattern specification unit is called a *morphogenetic field* (Weiss, 1969). The pattern-regulating properties of most fields are transitory, and disappear as the pattern becomes determined.

Wolpert (1971) has defined the morphogenetic field in terms of *positional information*. According to this concept, each cell of a field senses its position with respect to adjacent cells and, ultimately, to boundary cells constituting the edges of the prospective pattern. The boundary cells are believed to impart information to the other cells of the field regarding their position with respect to the edges of the pattern, by acting as a source of or referent for informational signals. Cells differentiate to manifest the pattern according to the differential information content spread across the field. The existence of positional information is illustrated by the results of an experiment by Saunders *et al.* (1959) on embryonic chick limb buds. Prospective thigh mesoderm grafted from the leg bud to the digit-forming apical region of the wing bud differentiated in the pattern of *toes,* in accord with its new position within the wing bud. This result shows that, although leg cells were committed to form leg pattern at the time of transplantation, regional patterning of the leg had not yet occurred and was specified by the position of the grafted cells within the larger boundary of the whole.

The process of pattern specification imprints a memory on cells of their relative position within an organ district, termed *positional value.* The existence of positional memory has been demonstrated in the urodele amphibian limb, an example of a morphogenetic field that can be reactivated by amputation during larval or adult life so that the missing parts are regenerated. Regeneration occurs by the dedifferentiation of skeletal, muscle and connective tissue cells adjacent to the amputation plane, followed by their accumulation and division to form a *blastema,* which subsequently redifferentiates into just those parts that were amputated (Fig. 1).

The undifferentiated blastema can self-organize into the structures it normally forms *in situ,* if it is grafted to a nonregenerating territory, to a limb stump of different morphological character, or to a proximodistal (PD) level of the stump different than the PD level of origin of the blastema (Stocum, 1984). Self-organization proceeds in a proximal to distal sequence, and the apical wound epidermis of the blastema is required for the process (Stocum and Dearlove, 1972). Furthermore, if a blastema is grafted to the contralateral limb stump with either its anteroposterior (AP) or dorsoventral (DV) axes reversed, supernumerary limbs of stump handedness are formed where the poles of these axes meet the opposite poles of the corresponding stump axes (Carlson, 1975; Tank, 1978; Maden and Turner, 1978). These results suggest that (1) blastema cells inherit positional memories (values) of their points of origin, on all three axes of the stump, that specify the proximal and cross-sectional boundaries of the regenerate field at its PD level of origin (Stocum, 1983, 1984) and (2) the apical epidermis acts as the distal boundary of the field (Maden, 1977). The boundaries act as the sources of, or referents for, an intercellular communication network (positional signalling system) that results in re-imprinting of the positional values lost by amputation on the cells of the blastema as it grows (Stocum, 1983, 1984).

Because the regeneration blastema is a large-scale morphogenetic field, the components of which can be easily manipulated surgically and chemically, it is a good system with which to analyze the nature and organization of pattern regulating mechanisms. Although much is known about rules of pattern regulation on the tissue level, a comprehensive understanding of these rules requires knowledge of field structure and function on a cellular and molecular level. The purpose of this essay is to discuss the cellular and molecular nature of the positional memory (value) inherited by blastema cells, as revealed by studies of the effects of vitamin A and its derivatives on regenerating salamander limbs.

II. POSITIONAL MEMORY MAY BE ENCODED IN THE BLASTEMA CELL SURFACE

It is reasonable to assume that the molecular receivers of positional signals in blastema cells are the same as those that have been found to mediate physiological functions in other cells: i..e, glycoprotein or protein receptors located either in the cell membrane or cytosol, that transduce the signals into cell-specific patterns of gene transcription or enzyme activity (see Alberts *et al.,* 1983). For example, lipid-soluble signaling molecules, such as steroid hormones, exert their effects by increasing the binding affinity of cytosolic receptors to specific sites on chromatin. In contrast, the effects of some steroids and of water-soluble signaling molecules are mediated by second messengers such as cyclic AMP and calcium, the intracellular concentrations of which are increased or decreased after binding of the signalling molecule to plasma membrane receptors. Second messengers activate protein kinases, which increase or inhibit the activities of other cell enzymes.

The nature of the positional signals involved in regenerate patterning is unknown. They could be molecules diffusing over short ranges through the extracellular matrix or from cell to cell via gap junctions (see Alberts *et al.,* 1983), or molecules of the cell surface glycocalyx interacting with glycocalyx molecules of adjacent cell surfaces (Slack, 1980; Stocum, 1984). It is unlikely that they are extracellular matrix molecules themselves, since Tank (1981) has shown that only cells, not the extracellular matrix of the limb, are capable of evoking supernumerary limb regeneration when implanted in a position opposite their normal position on the AP axis. The available evidence suggests that the positional information system of the regeneration field uses local cell interactions to specify pattern (Bryant *et al.*, 1981).

Cell differentiation is known to be accompanied by the acquisition of phenotype-specific cell-surface antigens (Alberts *et al.*, 1983), and there is considerable indirect evidence that differences in the molecular composition and/or architecture of the cell surface are important in governing the spatial patterning of embryonic tissues (Townes and Holtfreter, 1955; Steinberg, 1970, 1978; Steinberg and Poole, 1982). The latter evidence has come from studies in which dissociated cells of different phenotypes were mixed *in vitro*, or different tissues juxtaposed either *in vitro* or *in vivo*. The general result of such experiments is that one cell type comes to surround the other. So far, the best scheme to account

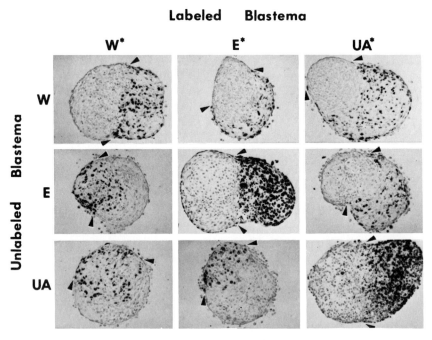

Fig. 2. Results of fusing two early or medium bud blastemas from the same or different limb levels in hanging drop cultures. Either the proximal or distal member of the pair (in the case of blastemas from different levels) or one member of the pair (in the case of blastemas from the same level) was labeled with tritiated thymidine (asterisk). Pairs were cultured for 3 days before fixation, sectioning and autoradiography. W = wrist blastema; E = elbow blastema; UA = upper arm blastema. Blastemas from the same level (diagonal row of photographs from upper left to lower right) fused in a straight line. In pairs of blastemas from different levels, the proximal blastema attempted to surround the distal one. The arrows indicate the maximum progress of the surrounding cells. Scale bar = 0.5 mm.

for this behavior is Steinberg's differential adhesion hypothesis, which postulates that cell or tissue mixtures adopt equilibrium patterns that minimize their interfacial free energy. These patterns are predictable on the basis of different adhesive strengths between the surfaces of isotypic and heterotypic cells (Steinberg, 1963; Trinkaus, 1984).

Likewise, there is evidence that the positional memory of origin of blastema cells is associated with cell-surface properties. When two axolotl or newt limb blastemas from the same level of origin are juxtaposed *in vitro,* they fuse in a straight line, but when blastemas from two different levels of origin are juxtaposed (Fig. 2), they adopt a pattern in which the proximal blastema surrounds the distal one (Nardi and Stocum, 1983). These results suggest that graded differences in the molecular composition and/or organization of the cell surface glycocalyx reflect the different positional memories possessed by blastema cells derived from different levels of the limb.

III. RETINOIDS ALTER POSITIONAL MEMORY IN REGENERATING LIMBS

Vitamin A is a polyisoprene molecule essential for vision, growth, reproduction, and maintenance of cell differentiation in vertebrates. In excess, vitamin A and its derivatives (retinoids) are teratogenic to embryonic cells and tissues. Retinoids have been shown to convert keratinizing epidermis to mucus epidermis, and scale epidermis to feather epidermis in chick embryos (Fell, 1957; Dhouailly *et al.,* 1980). In the mouse embryo, they convert vibrissae follicles to mucus glands (Hardy, 1983), and cause phocomelia and micromelia in limb buds about to initiate chondrogenesis (Kochhar, 1973, 1977). Retinoids also inhibit foot development in anuran tadpoles when given at a stage when the foot region has begun chondrogenesis (Jangir and Niazi, 1978; Niazi and Ratnasamy, 1984), and cause hypomorphism of anuran and urodele limb regenerates when administered at stages of redifferentiation (Jangir and Niazi, 1978; Niazi *et al.,* 1985).

However, retinoids have a different and unique effect on the stages of dedifferentiation and blastema formation in anuran and urodele limb regenerates. When given at these stages, the positional memory of the blastema cells is *proximalized,* and stump segments are duplicated in the proximodistal axis of the regenerate (Niazi and Saxena, 1978; Niazi and Alam, 1984; Maden, 1983; Thoms and Stocum, 1984).

We have recently measured the proximalizing effects of three different retinoids on axolotl limb regenerates (Kim and Stocum, 1986b). The retinoids were retinoic acid (RA), etretinate (ET), and arotinoid (AR) (Fig. 3). RA is the acid derivative of vitamin A, and ET and AR are synthetic retinoids with ethyl ester R

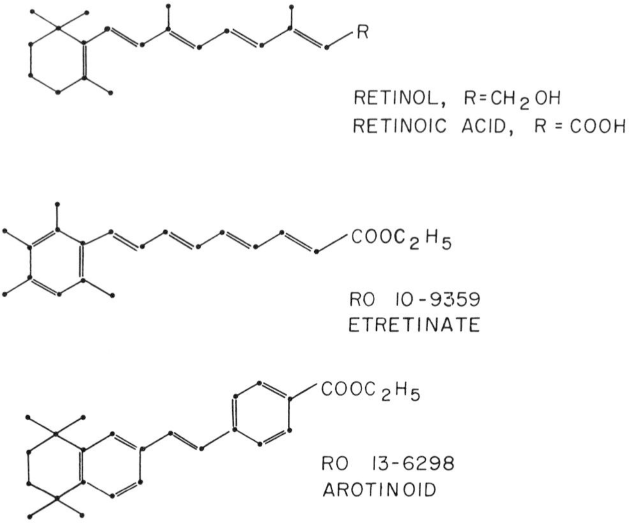

RETINOL, R=CH$_2$OH
RETINOIC ACID, R = COOH

RO IO-9359
ETRETINATE

RO I3-6298
AROTINOID

Fig. 3. Chemical structures of retinol, retinoic acid, etretinate (RO 10-9359, ethyl all-*trans*-9-(4-methyoxy-2,3,6-trimethylphenyl)-3,7-dimethyl-2,4,6,8-nonatetraenoate), and arotinoid [RO 13-6298, ethyl p-[(E)-2-(5,6,7,8-tetrahydro-5,5,8,8-tetramethyl-2-naphthyl)-1-propenyl] benzoate}. Etretinate and arotinoid are synthetic retinoids made by Hoffman-La Roche.

groups, with AR having two additional rings. The retinoids were dissolved in dimethyl sulfoxide (DMSO) and injected into the intraperitoneal cavity with a microliter syringe. Since they are water-insoluble, the retinoids precipitate after injection, where they slowly dissolve over a period of 7–10 days, probably by attaching to plasma retinoid-binding proteins which, in mammals, act to carry retinoids through the bloodstream to cells (see Lotan, 1980). During this time, regeneration is inhibited, and the inhibition is associated with depressed levels of blastema cell mitosis (Maden, 1983).

After amputation through either the distal zeugopodium (lower arm or leg) or distal stylopodium (upper arm or leg), the efficacy of the three retinoids in causing proximalization was in the order AR ≫ RA > ET. The level to which proximalization occurred was dose-dependent for RA and AR up to the dose producing maximal duplication (Figs. 4 and 5). Higher doses resulted in increasing inhibition and hypomorphism without duplication. The degree of proximalization was also dose-dependent for ET, but the dose producing regenerative inhibition was not determined, due to the very high dose required to achieve maximal duplication.

For a given dose, a greater magnitude of duplication was achieved when the level of amputation was through the distal zeugopodium. However, to strictly compare the proximalizing effects of retinoids on these two amputation levels, it is necessary to make the comparison at the point where zeugopodial and styl-opodial regenerates begin to duplicate the same segment—i.e., begin duplicating the stylopodium. To do this, the zero point of the stylopodial curve must be superimposed on the point of the zeugopodial curve where stylopodial duplication begins. When this is done, the two curves are nearly coincidental, indicating that a given dose of retinoid induces an equivalent amount of stylopodial duplication in both zeugopodial and stylopodial regenerates. Since more structures are duplicated from the zeugopodial level than from the stylopodial level, only a very small fraction of a given dose is required to duplicate the zeugopodium in a zeugopodial regenerate, with most of the dose going to duplicate the styl-opodium. Hence, we have suggested that either there is a difference in the sensitivity of zeugopodial- and stylopodial-derived cells to retinoids or that the

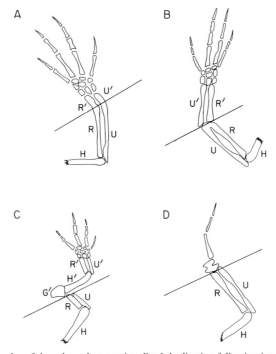

Fig. 4. Examples of dose-dependent proximodistal duplication following intraperitoneal injection of retinoids. The illustrations are tracings of axolotl (*Ambystoma mexicanum*) regenerate skeletons (after Thoms and Stocum, 1984). R, U, and H are stump radius, ulna, and humerus, respectively. R′, U′, H′ and G′ are duplicated radius, ulna, humerus, and girdle, respectively. The lines indicate the plane of amputation, which was through the wrist joint. (A) Duplication of the distal radius and ulna. (B) Duplication of a complete radius and ulna. (C) Duplication of a partial shoulder girdle, complete humerus, and complete radius and ulna. (D) Nonduplicated, hypomorphic regenerate formed after treatment with a dose of retinoid higher than the dose that causes maximum duplication.

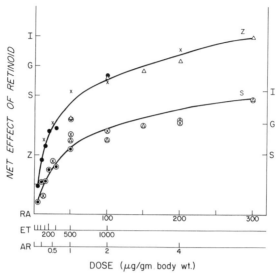

Fig. 5. Comparative efficacy of retinoic acid (RA), etretinate (ET), and arotinoid (AR) in inducing proximodistal duplication from the level of the zeugopodium (Z) or stylopodium (S). Retinoid dose is given on the abscissa, and the net effect of the retinoid is given on the ordinates in terms of the segments duplicated. On the ordinates, G = girdle, S = stylopodium, Z = zeugopodium, and I = inhibited regeneration. The left ordinate is for the zeugopodial curve, and the right ordinate is for the stylopodial curve. The circled symbols are for the stylopodial curve, and the symbols without circles are for the zeugopodial curve: x = etretinate, solid circles = arotinoid, and triangles = retinoic acid. For all three retinoids, a given dose produces more duplication from the zeugopodial level of amputation.

stylopodium is represented by more positional values than the zeugopodium (Kim and Stocum, 1986b).

Histological examination of retinoid-treated blastemas has revealed some interesting cellular features. Maden (1983) reported that blastema cells of duplicating regenerates form closely packed aggregates prior to redifferentiation. We have confirmed this observation, but in addition have noted that at doses where the limb would duplicate a girdle, the blastema shows a distinct subdivision into two regions differing in cell density (Kim and Stocum, 1986c). There is a region of low blastema cell density that always appears on the anterior side of the limb stump, and the girdle develops from this region. Blastema cells aggregate at high density under the apical epidermis, and this dense region forms the limb parts distal to the girdle. Furthermore, this part of the regenerate grows posteriorly, so that the duplicated limb is often oriented to the limb stump at an angle of 90°. What these observations mean with regard to specification of regenerate pattern is not yet clear.

Retinoids also modify positional value in the AP axis of the regenerating limb, as shown by the results of studies on half and double half limbs. Removal of the anterior or posterior half of the zeugopodium, followed by amputation through the distal end of the remaining half, results in regeneration of a half limb (Goss, 1957; Maden, 1979; Kim and Stocum, 1986a). Retinoid-treated anterior half limb regenerates not only duplicate structures proximal to the amputation plane, but also complete the AP pattern (Fig. 6). In contrast, retinoid-treated posterior half limbs fail to regenerate. The same results are obtained with half limbs *in situ*, or half limbs grafted to the orbit of the eye to ensure that regeneration cannot take place from a complete cross-section of limb (Kim and Stocum, 1986a).

Double anterior or posterior zeugopodia are made by grafting together the anterior or posterior halves of right and left limbs, so that the resulting limbs have two radii or two ulnae, and a symmetrical digit pattern. Both constructions regenerate a hypomorphic, symmetrical AP pattern of basipodial (carpal or tarsal) elements and digits, in which central structures are missing. Each half of a retinoid-treated double anterior regenerate duplicates structures proximal to the amputation plane, and also forms a normal AP pattern, resulting in the regeneration of right and left limbs from the amputation surface. However, retinoid-treated double posterior limbs fail to regenerate (Stocum and Thoms, 1984; Kim and Stocum, 1986a).

These results indicate that retinoids can *posteriorize* AP positional values when posterior or anterior values are removed from the AP axis, but cannot anteriorize positional values in this axis. Hence, retinoid-induced modification of positional value is unidirectional in both the AP and PD axes. The failure of retinoid-treated posterior half and double posterior limbs to regenerate, in contrast to their anterior counterparts, can be explained within the context of bound-

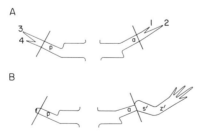

Fig. 6. Regeneration of (A) control and (B) retinoid-treated half zeugopodia amputated through their distal ends (level of amputation indicated by lines). A control anterior half regenerates a hand with digits 1 and 2, whereas a control posterior half regenerates a hand with digits 3 and 4. A retinoid-treated anterior half regenerates a complete limb with a normal anteroposterior pattern, whereas a retinoid-treated posterior half fails to regenerate. Symbols: a and p are anterior and posterior halves of the limb stump, respectively, s′ is duplicated stylopodium, and z′ is duplicated zeugopodium of the complete limb regenerated from a retinoid-treated anterior half.

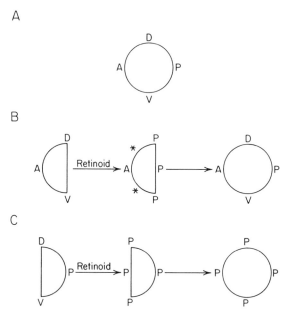

Fig. 7. Model to account for regeneration of normal anteroposterior pattern from a retinoid-treated anterior half zeugopodium and failure of regeneration from a retinoid-treated posterior half zeugopodium. The model assumes that (1) both poles of the anteroposterior axis must be present for regeneration to occur, (2) retinoid treatment posteriorizes midline, dorsal, and ventral cells, and (3) after posteriorization, the missing dorsal and ventral qualities (positional values) can be restored by intercalation between the anterior and posterior qualities. A = anterior; P = posterior; D = dorsal; V = ventral. The asterisk indicates the process of intercalation. (A) Normal cross-section of the zeugopodium with all four axial positional values present. (B) Amputated anterior half zeugopodium treated with retinoid. Midline, dorsal, and ventral cells acquire posterior positional values as a result of retinoid action. Dorsal and ventral positional values are restored by intercalation to give a complete cross-section, and regeneration of a complete anteroposterior pattern takes place. (C) Amputated posterior half zeugopodium treated with retinoid. Midline, dorsal, and ventral cells again acquire posterior positional values, but no regeneration takes place due to the lack of an anterior–posterior differential to stimulate intercalation.

ary models of short-range cell interactions (Bryant *et al.,* 1981; Lewis, 1981; Lheureux, 1977; Mittenthal, 1981; Meinhardt, 1982; Stocum, 1983). Posteriorization of positional value in anterior half and double half limbs restores the posterior boundary of the regenerate, allowing the intercalation of a full set of cross-sectional values, whereas in posterior half and double half limbs, the presence of an abnormal number of posterior positional values leads to inhibition of intercalation (Fig. 7).

We are currently analyzing the effects of retinoids on the DV axis, so that a complete three-dimensional picture of how these compounds modify positional value in regenerating limbs can be obtained.

IV. MODE OF ACTION OF RETINOIDS

No cellular or biochemical studies of retinoid-treated limb regenerates have been reported, but some of the more important cellular and biochemical effects of retinoids on mammalian cells are the modification of adhesive and growth properties (Lotan, 1980; Jetten, 1984), alterations in cytoskeletal organization (Lehtonen et al., 1983), and changes in proteoglycan, glycosaminoglycan and glycoprotein synthesis (Kochhar, 1977; Lewis et al., 1978; Lotan, 1980; Elias et al., 1983; Levin et al., 1983; Robinson et al., 1984). The mechanism by which retinoids induce these effects is unclear, but the majority of the evidence indicates they act by modifying gene expression. Fuchs and Green (1981) showed that retinoid-induced conversion of cultured keratinocytes to mucus epidermis is accompanied by a switch from the production of transcripts coding for a 67-kDa keratin to transcripts coding for 40- and 52-kDa keratins. In cultured fibroblasts, DNA synthesis is inhibited by retinoids, and so is transcription of message coding for ornithine decarboxylase (Russel and Haddox, 1981). Blalock and Gifford (1977) have shown that transcription of interferon message is induced in cultured mouse L-929 cells by retinoid treatment. The degradation of cultured rat limb bud cartilage by retinoic acid is negated by inhibitors of RNA and protein synthesis, indicating that this effect requires transcription induced by the retinoid (Gallandre and Kistler, 1980).

Since retinoids are lipid-soluble molecules with a molecular weight within the range of steroid hormones, it is possible they may act in the fashion of steroids, by binding to cytosolic receptors and increasing their binding affinity for nuclear DNA. Cytosolic retinoid binding proteins (CRBPs) are present in a wide variety of embryonic cells (Chytil and Ong, 1983), and have been shown in adult rat liver and testis to bind to chromatin following entry of the retinoid–CRBP complexes into the nucleus (Liau et al., 1981; Cope et al., 1984).

Other studies, however, suggest that retinoids may act through a calcium-activated, phospholipid-dependent protein kinase. In undifferentiated embryonal carcinoma cells, the activity of this enzyme is low but increases significantly when they are induced to differentiate to endoderm cells by retinoic acid (Kraft and Anderson, 1983). Finally, it is possible that retinoids could act directly on the glycosylation of proteins destined for the plasma membrane, as indicated by evidence that phosphorylated retinol can act as a carrier and donor of monosaccharides to growing oligosaccharides of glycoproteins (DeLuca, 1977).

V. USE OF RETINOIDS TO ANALYZE THE CELLULAR AND MOLECULAR BASIS OF POSITIONAL MEMORY IN LIMB REGENERATION

Our data (Nardi and Stocum, 1983) indicate that positional memory of the PD level of origin is associated with differences in blastema cell adhesive properties along the PD axis. Furthermore, Trisler *et al.* (1981) have demonstrated, using a monoclonal antibody, a graded distribution of a cell surface antigen along the ventroanterior–dorsoposterior axis in the synaptic layers of the developing avian retina. Such studies suggest, but do not prove, that cell surface antigens may be the basis of positional memory in morphogenetic fields.

Demonstrating a causal relationship between positional memory and cell surface properties in regenerating limbs requires use of an agent that modifies positional memory in a morphologically visible way that can be correlated with specific modifications in cell surface adhesivity, molecular composition, and/or architecture. Retinoids are agents that appear to fulfill these requirements, and are therefore potentially useful in analyzing the cellular and molecular basis of blastema cell positional value.

We have initiated such an analysis using an *in vivo* assay for position-related differences in blastema cell surface recognition and affinity (K. Crawford and D. L. Stocum, unpublished data) (Fig. 8). Axolotl forelimbs were amputated through the wrist, elbow, or mid-upper arm of the forelimb, and the animals were injected during the dedifferentiation stage of regeneration with 150 μg RA/g body weight, a dose known to cause proximalization of positional value to the shoulder girdle or proximal stylopodial level. After blastema formation, blastemas from each of these levels of the retinoid-treated animals were autografted to the blastema-stump junction of the ipsilateral hindlimb, which had been amputated through the mid-thigh several days prior to grafting. Controls consisted of similar grafts of DMSO-treated forelimb blastemas from the same three levels.

The results were quite striking (Fig. 8). Control blastemas derived from the wrist and elbow were displaced distally and formed regenerates that articulated with their corresponding level of origin in the host hindlimb regenerate (wrist with tarsus, and elbow with knee). Upper arm blastemas were not displaced, and formed regenerates that articulated with the host thigh. In contrast, blastemas from all three levels of the retinoid-treated forelimbs remained at the graft site, and formed regenerates in which the most proximal element was the humerus or shoulder girdle. These results suggest that blastema cells have surface properties defining their level of origin that promote their association with cells from corresponding levels of origin, and that these recognition properties are proximalized by retinoids.

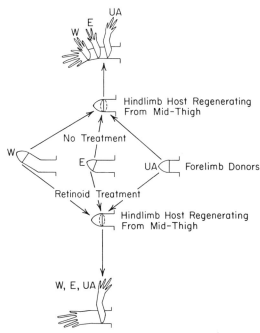

Fig. 8. Use of an *in vivo* "affinophoresis" system to demonstrate a correlation between modification of proximodistal positional value and modification of blastemal cell affinity properties. Blastemas derived from the wrist (W), elbow (E) or mid-upper arm (UA) levels of untreated or retinoid-treated animals are grafted to the blastema-stump junction of the hindlimb of untreated animals regenerating from the mid-thigh. The longitudinal axis of the grafted blastmas is perpendicular to the longitudinal axis of the host hindlimb, and the anteroposterior axes of graft and host are aligned. The untreated forelimb blastemas are displaced distally on the host hindlimb regenerate for a distance that brings them to their corresponding level of origin on the host. In contrast, retinoid-treated blastemas duplicate the upper arm and remain at the graft site, regardless of their level of origin.

Further work is required to characterize the molecular basis of these surface properties and determine the nature of the intracellular mechanism of action of retinoids in modifying the molecules involved. Our working hypothesis is that retinoids modify positional value through a steroid-like mechanism that controls patterns of gene expression involved in the production of cell-surface integral proteins, or the enzymes involved in their glycosylation. One reason for adopting this hypothesis is that we have recently detected apo (unoccupied) cytosolic retinoid binding proteins in blastema cells of the regenerating limb but not in unamputated limb tissue (A. M. McCormick and D. L. Stocum, unpublished data). However, other mechanisms of retinoid action, such as second messenger-dependent activation of protein kinases, cannot be ruled out.

VI. SUMMARY AND CONCLUSIONS

Limb regeneration is a pattern regulation process in which a regeneration blastema inherits, from parent limb cells, a set of positional values outlining the pattern to be regenerated. These values act as referents for a network of intercellular communication that restores the positional values lost by amputation. Positional value appears to be an affinity property of the cell surface that is graded along the PD and AP axes of the limb. Retinoids modify positional values in predictable ways and, in other systems, are known to alter the synthesis of glycocalyx glycoconjugates implicated in cell recognition and adhesion. Hence, retinoids are potentially promising agents with which to analyze the relationship between molecular composition and architecture of the cell surface, and morphological pattern. Such analyses will offer insights into understanding the nature and prevention of congenital and drug-induced embryonic defects, and perhaps into ways of inducing regeneration in non-regenerating or poorly regenerating tissues and organs.

REFERENCES

Alberts, B. A., Bray, D., Lewis, J., Raff, M., Roberts, K., and Watson, J. D. (1983). "Molecular Biology of the Cell." Garland, New York.
Blalock, J. E., and Gifford, G. E. (1977). *Proc. Natl. Acad. Sci. U.S.A.* **74**, 5382–5386.
Bryant, S. V., French, V., and Bryant, P. J. (1981). *Science* **212**, 993–1002.
Carlson, B. M. (1975). *Dev. Biol.* **47**, 269–291.
Chytil, F., and Ong, D. E. (1983). *Adv. Nutr. Res.* **5**, 13–29.
Cope, F. O., Knox, K. L., and Hall, R. C. (1984). *Nutr. Res.* **4**, 289–304.
De Luca, L. M. (1977). *Vitam. Horm.* **35**, 1–57.
Dhouailly, D., Hardy, M. H., and Sengel, P. (1980). *J. Embryol. Exp. Morphol.* **58**, 63–78.
Elias, P. M., Chung, J-C., Orozco-Topete, R., and Nemanic, M. K. (1983). *J. Invest. Dermatol.* **81**, 81s–85s.
Fell, H. B. (1957). *Proc. R. Soc. Lond. Biol.* **146**, 242–256.
Fuchs, E., and Green, H. (1981). *Cell* **25**, 617–625.
Gallandre, F., and Kistler, A. (1980). *Roux's Arch. Dev. Biol.* **189**, 25–33.
Goss, R. J. (1957). *J. Morphol.* **100**, 547–564.
Hardy, M. H. (1983). *In* "Epithelial-Mesenchymal Interactions in Development" (J. F. Fallon and R. H. Sawyer, eds.), pp. 163–188. Praeger, New York.
Jangir, O. P., and Niazi, I. A. (1978). *Indian J. Exp. Biol.* **16**, 438–445.
Jetten, A. M. (1984). *Fed. Proc.* **43**, 134–139.
Kim, W.-S., and Stocum, D. L. (1986a). *Dev. Biol.* **113** (in press).
Kim, W.-S., and Stocum, D. L. (1986b). *J. Embryol. Exp. Morphol.*, Submitted.
Kim, W.-S., and Stocum, D. L. (1986c). *J. Embryol. Exp. Morphol.*, Submitted.
Kochhar, D. M. (1973). *Teratology* **7**, 289–298.
Kochhar, D. M. (1977). *Birth Defects* **13**, 111–154.
Kraft, A. S., and Anderson, W. B. (1983). *J. Biol. Chem.* **258**, 9178–9183.
Lehtonen, E., Lehto, V-P., Badley, R. A., and Virtanen, I. (1983). *Exp. Cell Res.* **144**, 191–197.

Levin, L. V., Clark, J. N., Quill, H. R., Newberne, P. M., and Wolf, G. (1983). *Cancer Res.* **43,** 1724–1732.

Lewis, J. (1981). *J. Theor. Biol.* **88,** 371–392.

Lewis, C. A., Pratt, R. M., Pennypacker, J. P., and Hassel, J. R. (1978). *Dev. Biol.* **64,** 31–47.

Lheureux, E. (1977). *J. Embryol. Exp. Morphol.* **38,** 151–173.

Liau, G., Ong, D. E., and Chytil, F. (1981). *J. Cell Biol.* **91,** 63–68.

Lotan, R. (1980). *Biochim. Biophys. Acta* **605,** 33–91.

Maden, M. (1977). *J. Theor. Biol.* **69,** 735–753.

Maden, M. (1979). *J. Embryol. Exp. Morphol.* **52,** 183–192.

Maden, M. (1983). *J. Embryol. Exp. Morphol.* **77,** 273–295.

Maden, M., and Turner, R. N. (1978). *Nature (London)* **273,** 232–235.

Meinhardt, H. (1982). "Models Of Biological Pattern Formation." Academic Press, New York.

Mittenthal, J. E. (1981). *Dev. Biol.* **88,** 15–26.

Nardi, J. B., and Stocum, D. L. (1983). *Differentiation* **27,** 13–28.

Niazi, I. A., and Alam, S. (1984). *Roux's Archiv. Dev. Biol.* **193,** 111–116.

Niazi, I. A., and Ratnasamy, C. S. (1984). *J. Exp. Zool.* **230,** 501–505.

Niazi, I. A., and Saxena, S. (1978). *Folia Biol. (Krakow)* **26,** 3–11.

Niazi, I. A., Pescitelli, M. J., Jr., and Stocum, D. L. (1985). *Roux's Archiv Dev. Biol.* **194,** 355–363.

Robinson, J., Freinkel, R. K., and Gotschalk, R. (1984). *Br. J. Dermatol.* **110,** 17–27.

Russel, D. H., and Haddox, M. K. (1981). *Ann. N.Y. Acad. Sci.* **359,** 281–297.

Saunders, J. W., Jr., Gasseling, M. T., and Cairns, J. M. (1959). *Dev. Biol.* **1,** 281–301.

Slack, J. M. W. (1980). *J. Theor. Biol.* **82,** 105–140.

Steinberg, M. S. (1963). *Science* **141,** 401–408.

Steinberg, M. S. (1970). *J. Exp. Zool.* **173,** 395–434.

Steinberg, M. S. (1978). *In* "Cell-Cell Recognition" (A. S. G. Curtis, ed.), pp. 25–49. Cambridge University Press, Cambridge.

Steinberg, M. S., and Poole, T. J. (1982). *In* "Developmental Order: Its Origin and Regulation" (S. Subtelny and P. B. Green, eds.), pp. 351–378. Alan R. Liss, New York.

Stocum, D. L. (1983). *In* "Nerve, Organ and Tissue Regeneration: Research Perspectives" (F. J. Seil, ed.), pp. 377–406. Academic Press, New York.

Stocum, D. L. (1984). *Differentiation* **27,** 13–28.

Stocum, D. L., and Dearlove, G. D. (1972). *J. Exp. Zool.* **181,** 49–61.

Stocum, D. L., and Thoms, S. D. (1984). *J. Exp. Zool.* **232,** 207–215.

Tank, P. W. (1978). *Dev. Biol.* **62,** 143–161.

Tank, P. W. (1981). *Am. J. Anat.* **162,** 315–326.

Thoms, S. D., and Stocum, D. L. (1984). *Dev. Biol.* **103,** 319–328.

Townes, P. L., and Holtfreter, J. (1955). *J. Exp. Zool.* **128,** 53–120.

Trinkaus, J. P. (1984). "Cells into Organs: The Forces That Shape the Embryo," 2nd ed. Prentice-Hall, New York.

Trisler, G. D., Schneider, M. D., and Nirenberg, M. (1981). *Proc. Natl. Acad. Sci. U.S.A.* **78,** 2145–2149.

Weiss, P. (1969). "Principles Of Development." Hafner, New York.

Wolpert, L. (1971). *Curr. Top. Dev. Biol.* **6,** 183–224.

7

The Distribution of Acetylcholine Receptors on Vertebrate Skeletal Muscle Cells

JOE HENRY STEINBACH AND ROBERT J. BLOCH

I. INTRODUCTION

Signal transmission at most chemically transmitting synapses in the nervous system is a localized process. The presynaptic cell comes into close physical

183

RECEPTORS IN CELLULAR RECOGNITION
AND DEVELOPMENTAL PROCESSES

contact with the postsynaptic cell and transmitter release occurs in the regions of close contact. To complement this localized signal, the postsynaptic cell concentrates neurotransmitter receptors in its membrane near the sites of neurotransmitter release. We have been studying the process of receptor localization in the belief that elucidation of its molecular mechanism will reveal important principles of organization of the nervous system. Here we review what is known about the plasma membrane distribution of one neurotransmitter receptor and consider how localization of receptors at synapses may occur.

The nicotinic acetylcholine (ACh) receptor of vertebrate skeletal muscle and fish electric organs is the most extensively studied of the neurotransmitter receptors. Its function is to convert the binding of acetylcholine, the neurotransmitter released from the motor nerve terminal, into a change in the membrane potential at the motor endplate. It does so by opening a cation-specific channel in the membrane through which sodium and potassium ions pass, causing depolarization of the muscle fiber and subsequent contraction. The function of the ACh receptor (AChR) has been investigated by a number of techniques and is an area of active research. For some reviews of receptor activation, see Steinbach (1980) and Adams (1981). New techniques are beginning to give further insight into receptor function (Colquhoun and Sakmann, 1983; Montal *et al.*, 1984; Sine and Steinbach, 1984; Nass *et al.*, 1978), but the complete picture is not yet clear.

To perform its function the AChR has sites to bind ACh and a region to serve as the cation pore. The receptor is an integral membrane protein with a molecular weight of 250,000, is composed of several polypeptide chains, and is glycosylated. Structural studies have shown that portions of the AChR extend past the boundaries of the membrane lipid bilayer, 1.5 nm into the cytoplasm and 5 nm into the extracellular space (for reviews see Karlin, 1980; Changeux, 1981; Fairclough *et al.*, 1983). The primary amino acid sequences for receptor subunits from several species have been determined by nucleic acid cloning and sequencing (Claudio *et al.*, 1983; Noda *et al.*, 1983a,b; Ballivet *et al.*, 1983; Merlie *et al.*, 1983a). Computer analyses have shown that the subunits are similar to each other, both in primary sequence and in the predicted general tertiary structure of the proteins (e.g., Fairclough *et al.*, 1983). The ACh receptor is, in sum, a large, multimeric protein that crosses the cell membrane and has regions exposed on both the intracellular and extracellular surfaces.

The synthesis and degradation of acetylcholine receptors have been investigated by a number of laboratories, resulting in a fairly clear understanding of some of the basic steps in these processes (Fambrough, 1979; Pumplin and Fambrough, 1982). Receptor processing in cells and synthesis in cell-free systems are being investigated (Merlie *et al.*, 1983b; Anderson and Blobel, 1983). Together with recent success in cloning both the messenger RNA and the genomic sequence for receptor subunits, these results hold out the promise that the molecular processes involved in receptor synthesis and assembly will be worked out within the next few years.

A. The Distribution of Acetylcholine Receptors

Two techniques have been used to map the distribution of ACh receptors on muscle cells. The first is by iontophoresis. This consists of placing a micropipette filled with ACh outside the cell near the muscle surface, then releasing the ACh by a pulse of current through the ACh-containing micropipette and simultaneously measuring the membrane potential using a micropipette placed intracellularly. By moving the ACh-containing electrode along the length of the fiber, areas of receptor enrichment can be located, registered as greater depolarization per unit of current applied. Using this method it has been shown that in mature skeletal muscle, nearly all the AChRs are localized at the region immediately subjacent to the cholinergic nerve terminal (Dreyer and Peper, 1974; Kuffler and Yoshikami, 1975). The resolution of ionophoresis is limited by the diffusion of the released ACh, to about 5 μm.

The use of snake α-neurotoxins has given higher resolution in mapping receptor distributions. Alpha-bungarotoxin (αBT), the most commonly used α-neurotoxin, is an 8000-dalton polypeptide isolated from the venom of the banded krait, *Bungarus multicinctus,* and binds specifically and essentially irreversibly to the nicotinic AChR of skeletal muscle and electric organ (Changeux *et al.,* 1970). Because it retains its specificity and biological activity after reaction with radio-iodine, electron-dense macromolecules, or fluorophores, it can be used as a probe for AChRs in both light and electron microscopic studies of AChR distribution. It has also been used to show that the large angular particles seen in freeze fracture images of the neuromuscular junction and of cultured muscle cells contain nicotinic AChRs (Cohen and Pumplin, 1979). Quantitative estimates of receptor number or density have only been obtained using α-neurotoxins. The number of toxin-binding sites per receptor is not certain, although it is likely to be two (Sine and Taylor, 1980).

B. General Framework

Acetylcholine receptors are not randomly distributed over the surface of muscle cells, but often occur in aggregates of high density. In innervated adult muscle cells most ACh receptors are aggregated at the neuromuscular junction. These observations prompt the question: How is such a nonuniform distribution produced and maintained? In particular, what stimulates the muscle to produce a receptor aggregate? How is a particular region (the neuromuscular junction) singled out? What mechanism is used by the muscle cell to produce the aggregate? This chapter will consider what is known about the answers to these questions. The next two sections will present experimental observations of receptor distributions on muscle cells *in vivo* and *in vitro.* The chapter will conclude with a consideration of possible mechanisms for receptor clustering.

The adult neuromuscular junction is a highly organized structure. In most skeletal muscles the shape of the junction is convoluted, suggesting that some of the biochemical specializations at the junction may be required for reasons other than receptor aggregation. The junction itself is composed of the nerve terminal, the muscle cell and an intimately associated glial cell, the Schwann cell. In addition, there are nearby connective tissue cells. These facts, the presence of several specializations and of several cell types at the junction, need to be kept in mind when we weigh the evidence about mechanisms for receptor aggregation. Finally, we will not discuss additional biochemical differentiation at the junction although it is present as well. In particular, acetylcholinesterase activity is concentrated at the junction and shows interesting changes during development and innervation (e.g., Lomo and Slater, 1980a; Weinberg and Hall, 1979a).

II. MUSCLE CELLS *IN VIVO*

The ACh receptors of skeletal muscle *in vivo* have generally been classified as junctional or extrajunctional. These terms refer to the location of the receptors either at the motor endplate or elsewhere on the muscle fiber surface, respectively. Receptors located in these different regions differ significantly in a number of properties, including function (Neher and Sakmann, 1976), metabolism (Berg and Hall, 1975b; Merlie *et al.*, 1979), antigenic determinants (Weinberg and Hall, 1979b; Hall *et al.*, 1983), and isoelectric point (Brockes and Hall, 1975b). It is now clear that there is no mandatory relationship between the location of a receptor and its other properties, for receptors at the neuromuscular junction show different properties at different stages of development (see below). The relationship, if any, between the mechanisms that control receptor distribution and those that cause changes in other receptor properties is not known. In this review we will concentrate on the distribution of receptors and discuss other receptor properties only in this context.

A. Adult Muscle

In normal adult skeletal muscle the majority of AChRs is junctional, that is, confined to the postsynaptic membrane. Quantitative measurements of the number of αBT-binding sites indicate that there are about 4×10^7 binding sites per mammalian junction (Fambrough and Hartzell, 1972; Courtney and Steinbach, 1981). At the higher resolution afforded by the electron microscope, these sites are found to be localized at the tops of the postsynaptic folds, where they are present at densities of about $2 \times 10^4/\mu m^2$ (10^4 receptors/μm^2) (Fertuck and Salpeter, 1976; Matthews-Bellinger and Salpeter, 1978, 1983). The receptors at the adult junction are very tightly bound and are essentially immobile, unable to diffuse in the plane of the membrane (Stya and Axelrod, 1983b). Freeze fracture studies of the neuromuscular junction have also shown that characteristic large, angular particles are concentrated at the tips of the postsynaptic folds, and are absent from the depths of the folds. However, the density of membrane particles is generally reported to be only $3-6 \times 10^3/\mu m^2$ of membrane surface (Peper et al., 1974; Heuser and Salpeter, 1979; Rosenbluth, 1975). This difference has led to the suggestion that each particle contains several ACh receptors (for discussion see Matthews-Bellinger and Salpeter, 1978; Heuser and Salpeter, 1979).

Most adult muscle fibers have only one neuromuscular junction and one receptor cluster. Muscle fibers can produce more than one cluster, however, since some types of fibers normally are multiply innervated (Bennett et al., 1973; Orkand et al., 1978; Betz et al., 1980). Also, during reinnervation of denervated muscle more than one junction often forms on a single fiber (Kuffler et al., 1977; Lomo and Slater, 1980b), and on inactive muscle fibers several junctions may form, each with its receptor aggregate (Eldridge et al., 1981). Presumably, the observation that most normal adult fibers have only one receptor aggregate results from the fact that they have only one neuromuscular junction. The reasons for the presence of only one junction are not yet understood (e.g., Cangiano et al., 1980).

Much lower levels of AChR are found elsewhere on the muscle fiber. The muscle membrane between the neuromuscular and myotendonous junctions (the extrajunctional membrane) has a very low density of AChRs, as judged by small or absent iontophoretic responses, at most 0.1% that of the postsynaptic region (Katz and Miledi, 1964; Albuquerque and McIsaac, 1970). The muscle membrane at the myotendonous insertion of some muscle fibers has a slightly higher ACh receptor density than the general extrajunctional membrane, but still only about 0.1–1% of that at the junction (Miledi and Zelena, 1966; Steinbach, 1981a). There is also a region of increased extrajunctional receptor density for several hundred micrometers from the edge of the junction (Dreyer and Peper, 1974; Hartzell and Fambrough, 1972; Steinbach, 1981a). The maximal density of receptors in this "perijunctional gradient" is, again, less than 1% of the receptor density at the junction. When all nonjunctional receptors are considered, 70% or more of the total receptor content of the innervated adult muscle cell is concentrated in the neuromuscular junction, which comprises only about 0.1% of the muscle surface area.

The nerve terminal can be physically removed from adult muscle by sectioning the nerve trunk to examine the role of the nerve in maintaining receptor aggregates. Junctional receptor aggregates persist for long periods after surgical denervation, showing a slow decrease in the number of receptors in the aggregate (Frank et al., 1975; Steinbach, 1981a). The decrease in number of aggregated receptors is slower than can be accounted for by the degradation of individual receptors in the aggregate (Levitt et al., 1980; Brett et al., 1982; Bevan and Steinbach, 1983). Therefore, the stability of the aggregate cannot depend solely on the continued presence of the receptors which formed it when the nerve was present. Material at the neuromuscular junction can also promote the formation of new receptor aggregates on muscle fibers *in vivo* (Burden *et al.*, 1979; Bader, 1981). The aggregating activity is associated with extracellular basal lamina at the junction (McMahan and Slater, 1984), and will be discussed further in the final section of this chapter.

The extrajunctional AChR density increases several hundredfold after denervation, to about 5% of that found at the junction itself (Hartzell and Fambrough, 1972). The appearance of extrajunctional receptors results from the synthesis of new receptor protein molecules (Brockes and Hall, 1975a; Devreotes and Fambrough, 1976). This can be blocked by electrical stimulation of the muscle (Lomo and Rosenthal, 1972; Linden and Fambrough, 1979). Conversely, extrajunctional receptor synthesis by innervated muscle can be elicited by blocking nerve-induced stimulation of innervated muscle (Lomo and Rosenthal, 1972; Berg and Hall, 1975a; Lavoie *et al.*, 1977; Eldridge *et al.*, 1981). Extrajunctional receptor density, therefore, is controlled to a large degree by changes in the rate of receptor synthesis, which in turn appears to be sensitive to muscle activity.

In contrast, nerve or muscle activity does not appear to play a major role in controlling the density of junctional receptors. Direct stimulation of denervated muscle does not alter the slow decrease in the number of receptors aggregated at previously junctional regions (Frank *et al.*, 1975). Furthermore, nerve terminals in which the nerve impulse activity has been virtually eliminated can produce and maintain junctional receptor aggregates (Eldridge *et al.*, 1981).

The extrajunctional receptors seen after denervation are present in a generally uniform distribution, but occasional aggregates of nonjunctional AChR can be found with densities nearly as high as those seen at the neuromuscular junction (Ko *et al.*, 1977; Steinbach, 1981a; McMahan and Slater, 1984). These ectopic receptor aggregates disappear upon reinnervation of the muscle or when denervated muscle is directly stimulated (Lomo and Slater, 1980b). It is not known whether their disappearance is due to specific suppression or merely follows because the total extrajunctional receptor density decreases (see above). Ectopic aggregates appear rather rapidly after denervation (Ko *et al.*, 1977), and have a fairly nondescript appearance resembling aggregates at developing junctions (see below; Ko *et al.*, 1977; Steinbach, 1981a). The receptors in ectopic clusters do not seem to be as tightly bound as do junctional receptors (Stya and Axelrod, 1983b).

B. Development of the Neuromuscular Junction

Studies of receptor distributions during development of the neuromuscular junction provide information on the nature and timing of the changes which occur. This information, in turn, allows some insight into possible mechanisms for the control of receptor distribution. This section will concentrate on the development of the mammalian neuromuscular junction. In general what is known about the development of avian and amphibian junctions is similar, although some clear differences exist between junctional development in different vertebrate families (for comparison see, for avian junctions, Burden, 1977a, b; Jacob and Lentz, 1979; Harvey and van Helden, 1981; Ishikawa and Shimada, 1982; Smith and Slater, 1983; for amphibian junctions, Kullberg *et al.*, 1977, 1981; Morrison-Graham, 1983; Cohen, 1980; Chow and Cohen, 1983). The localization of AChRs at the mammalian neuromuscular junction is the result of a process that begins shortly after the onset of myogenesis. Within a day or so after myotube formation begins, receptor aggregates are seen at developing junctions. These aggregates assume the properties of the adult junctional receptor aggregate over several weeks. The density of extrajunctional ACh receptors on the muscle fiber surface decreases during this period (Diamond and Miledi, 1962; Bevan and Steinbach, 1977), probably because of muscle activity (see above).

Fibers from muscles less than a day after fibers first appear show a uniform and rather high density of receptors, with no sign of junctional aggregates (Bevan and Steinbach, 1977; Harris, 1981; Ziskind-Conhaim and Dennis, 1981). On the next day the first aggregates are seen. The receptor density in the first aggregates is higher than the uniform density the day before, indicating that the first aggregates do not form simply as the result of stabilization of an initially very high receptor density (Bevan and Steinbach, 1977). Indeed, the formation of the initial loose cluster involves, at least in part, the aggregation of receptors that had been diffusely distributed on the muscle surface before any receptor-dense regions had formed (Ziskind-Conhaim *et al.*, 1984). The appearance of the junctional receptor aggregate changes greatly during development, as seen with fluorescent snake toxins. At the adult junction the region of high receptor density basically follows the nerve terminal branches, showing an irregular shape with crisp outlines. In contrast, the receptor aggregates at very young junctions are a relatively loose accumulation of small regions of high receptor density (about 0.5 μm in diameter, called "speckles" because of their appearance in the fluorescence microscope). The initially loose association of speckles changes to a more compact and clearly outlined "plaque" over a day or so. The plaque still shows speckles, but the speckles are more tightly clustered together. Eventually the plaque-like receptor aggregate is transformed to an adult junctional aggregate, by 2–3 weeks after birth (Steinbach, 1981b; Slater, 1982a).

In the loose aggregate and the plaque the receptor-dense membrane is near but not necessarily directly under nerve terminals (Steinbach, 1981b; Slater, 1982a; Matthews-Bellinger and Salpeter, 1983). In the loose aggregate some receptor-dense regions are not covered by either a nerve terminal or a Schwann cell, whereas in the plaque all are covered by either nerve or glia and in the adult almost all are covered by nerve terminal (Matthews-Bellinger and Salpeter, 1983). The density of receptors in the loose aggregate also is only about one-fourth that of the adult junction (Bevan and Steinbach, 1977; Matthews-Bellinger and Salpeter, 1983; Ziskind-Conhaim et al., 1984; Reiness and Weinberg, 1981).

The initial loose aggregate, the plaque, and the adult junctional receptor aggregate differ from each other in several respects. The initial loose aggregate morphologically resembles the receptor clusters seen on noninnervated muscle cells in culture (see below). The loose receptor aggregates also can be reversibly dispersed by treatments that disperse noninnervated receptor aggregates on cultured cells (Bloch and Steinbach, 1981). Finally, the receptors in the loose aggregate are rapidly metabolized (Reiness and Weinberg, 1981), in contrast to the low degradation rate of junctional receptors on adult muscle.

The plaque aggregate can be distinguished from both the initial loose aggregate and the adult aggregate. In addition to differences in appearance, the plaque cannot be disrupted by treatments that disperse loose junctional aggregates (Bloch and Steinbach, 1981). However, when muscles are denervated during the first week or so after birth the junctional receptor aggregates are not stable as in the adult but slowly disperse along the muscle fiber (Slater, 1982b). Finally, the receptors aggregated in a plaque are metabolically as stable as are the receptors at adult junctions (Steinbach et al., 1979; Michler and Sakmann, 1980; Reiness and Weinberg, 1981), but differ from adult junctional receptors in their functional properties (Michler and Sakmann, 1980; Fischbach and Schuetze, 1980) and an antigenic determinant (Hall et al., 1983).

Taken together, these observations show that during junctional development the properties of the receptor aggregate change greatly. The nature and timing of the changes suggest (but do not establish) that the different types of aggregate are produced by different mechanisms, possibly acting sequentially during development.

After either denervation or inactivity, ectopic receptor clusters appear on developing muscle (Harris, 1981; Ziskind-Conhaim and Bennett, 1982). If developing muscles are physically denervated after innervation but before aggregates form, no junctional receptor clusters develop, although they do appear if nerve impulse activity is blocked with intact physical innervation (Harris, 1981; Ziskind-Conhaim and Bennett, 1982). These observations suggest that the junctional cluster develops in response to the presence of the nerve, and that muscle activity suppresses the formation of ectopic clusters while junctional clusters are protected. However, it has been reported that when spinal motorneurons are destroyed with β-bungarotoxin before innervation is established, receptor clusters form in the appropriate position on the muscle fiber to have been "junctional" clusters (Braithwaite and Harris, 1979; Harris, 1981). The reason for this difference is not known.

The formation of receptor aggregates at new neuromuscular junctions on denervated adult muscle fibers has also been studied (Weinberg *et al.*, 1981a; Brenner and Sakmann, 1983). The sequence seen when new junctions are formed between adult nerve and muscle cells is very similar to that during junctional development. This finding suggests that the sequence reflects changes in the junction, rather than the general maturation of the nerve or muscle cells during development. As ectopic junctions grow, new receptors appear preferentially at the perimeter, consistent with the idea that extrajunctional AChR are recruited into the junctional aggregate (Weinberg *et al.*, 1981b).

There is some evidence for preferential synthesis and insertion of receptors near junctional regions. The concentration of receptor messenger RNA is higher in junctional regions (Sanes and Merlie, 1984). Also, Olek and Robbins (1983) have reported that there is an increase in the number of AChR in regions near the junction within 2 days after denervation, and suggest that this reflects preferential local synthesis and incorporation of receptors. However, ultrastructural studies of the location of newly incorporated junctional receptors have not revealed sites of preferential insertion (Salpeter and Harris, 1983). In sum, there is evidence for both accumulation and local insertion of receptors *in vivo* and *in vitro* (see below). The relative importance of the two processes is not clear as yet, nor is it known whether one or the other predominates at different developmental stages.

C. Summary of Acetylcholine Receptor Distribution on Muscle Cells *in Vivo*

The majority of the ACh receptors on a normal adult muscle fiber are clustered at high density at the neuromuscular junction. The adult junctional aggregate is relatively stable after the removal of the nerve terminal, persisting for a longer period of time than the ACh receptors that make up the aggregate. The development of the junctional receptor aggregate follows a fairly slow and complicated sequence, the nature of which suggests that muscle cells can produce several types of receptor aggregates differing in shape and stability and in the properties of the receptors in the aggregate. Some of the early junctional aggregates resemble the receptor clusters found on noninnervated muscle *in vivo* and *in vitro*. Nerve terminals do not seek out receptor accumulations on developing muscle fibers; instead, the formation of the junctional aggregate follows innervation. Both the initial appearance and the further development of the junctional aggregate apparently require the presence of the nerve, although neither production nor maintenance requires nerve impulses.

Extrajunctional receptors are not uniformly distributed on muscle cells. On innervated cells extrajunctional receptors occur at a higher density in a perijunctional gradient and at the myotendonous junctions of some muscle fibers. On denervated fibers ectopic receptor aggregates can be found, although most extra-junctional receptors are uniformly distributed. The control and the functional significance of these nonrandom distributions of extrajunctional receptors are not understood as yet.

III. CULTURED MUSCLE CELLS

Muscle cultures established from a number of mammalian, avian, and amphibian sources synthesize AChRs and insert them into their surface membranes. Studies of receptor distributions on cultured cells have provided a great deal of information about receptor clustering, because cells in culture may be experimentally manipulated more easily and in better defined ways than muscle cells in the animal. Most of the muscle cell cultures used in these studies have been primary cultures. In primary cultures, the cells are obtained from embryoes or young animals and allowed to differentiate *in vitro*. Some work has used clonal cell lines. A clonal cell line is a population of cells that comes from a single ancestral cell and that is maintained in cell cultures. The surface ACh receptors on cultured muscle cells can be diffusely distributed, or they can form clusters of varying size even in the absence of nerve cells.

A. Cultures without Nerve Cells

We will first discuss the receptor clusters that form on cultured muscle cells when no nerve cells are present in the culture. Large aggregates are obvious when observing a culture of muscle cells after labeling the cells with α-BT derivatives. Aggregates are defined as intensely stained regions showing well defined boundaries. The size of a "large" aggregate varies, and depends on the cells being studied, but they are 2 μm or more wide and up to 40 μm long (areas of 10–1000 μm^2). Such aggregates have been observed in primary cultures of rat, chick, and toad cells (Sytkowski *et al.*, 1973; Fischbach and Cohen, 1973; Axelrod *et al.*, 1976; Anderson and Cohen, 1977) and in cultures of some clonal muscle cell lines (Land *et al.*, 1977; Kidokoro and Patrick, 1978). Quantitative studies have found that the density of αBT-binding sites in the aggregates is 2–5 $\times$ 10^3/μm^2 (Axelrod *et al.*, 1976; Land *et al.*, 1977; Salpeter *et al.*, 1982), although it is possible to increase the density with a soluble factor (see below; Salpeter *et al.*, 1982). This is about 10–20% the density at the mature neuromuscular junction, suggesting that receptors are relatively loosely packed in these clusters. Indeed, large gaps of receptor-poor areas have been observed within AChR clusters in rat and toad cells (Axelrod *et al.*, 1976; Peng *et al.*,

1980; Bloch and Geiger, 1980). Gaps in chick myotubes seem less prevalent (D. Pumplin, personal communication). At most, only about 10% of the receptors on a cultured myotube are present in aggregates (Axelrod et al., 1976).

AChRs in aggregates seem to be highly organized: in rat myotubes they may appear as linear arrays of distinct smaller aggregates (Bloch and Geiger, 1980), referred to as speckles. The existence of such arrays cannot be accounted for by a random aggregation process of individual receptors. Axelrod et al. (1976) have shown that clustered AChRs are immobile: they are not free to diffuse in the membrane. [Other membrane components do, apparently, diffuse freely in receptor clusters (Axelrod et al., 1978a)]. Receptor immobilization is reversible, however. Exposure of myotubes to energy metabolism inhibitors, AChR agonists, or Ca^{2+} chelators causes clusters to disperse reversibly, apparently in part by movement of AChRs along the plane of the membrane (Bloch, 1979, 1983; Bursztajn et al., 1984). Electrical stimulation of cultured muscle cells also results in a loss of receptor clusters (Axelrod et al., 1978a).

Most of the AChRs in the surface membranes of cultured muscle cells are not present in clearly defined aggregates, and have been termed diffuse. Estimates of the overall density of diffuse receptors in myotubes are 30–500/μm^2 (Sytkowski et al., 1973; Axelrod et al., 1976; Land et al., 1977; Kidokoro and Patrick, 1978). The receptor distribution in the diffuse regions of L6 myotubes appears to be random, but that of primary rat myotubes appears not to follow a Poisson distribution (Land et al., 1977). Heterogeneity of diffuse AChRs has also been seen using fluorescence photobleaching, which reveals that up to 50% of the receptors on primary rat myotubes are unable to diffuse laterally within the plane of the membrane (Axelrod et al., 1976). The remaining diffuse receptors move through the membrane with an apparent diffusion constant of about $5 \times 10^{-11} cm^2/sec$. This value is not significantly altered by a variety of drugs or enzymatic treatments, suggesting that restraint on the diffusion of these receptors is minimal (Axelrod et al., 1978a,b). The fraction of immobile receptors does increase with time after myotube formation, however (Stya and Axelrod, 1983a), and can be increased by a soluble factor from neurons (Axelrod et al., 1981; see below).

The difference in mobility between diffuse and clustered receptors might indicate some fundamental differences between them. This seems not to be the case, however. Clustered and diffuse receptors are both rapidly metabolized by cultured myotubes (Scheutze et al., 1978; Axelrod, 1980; Salpeter et al., 1982). Also, clusters can form from diffusely distributed receptors (Bloch, 1979; Stya and Axelrod, 1983c), so clustered and diffuse receptors seem to be interconvertible.

Fluorescence observations of cultured muscle cells further reveal that small areas (about 1 μm^2) of higher receptor density occur in otherwise diffuse regions, which appear as speckles when stained with fluorescent derivatives of αBT. Because their small size makes them difficult to see, speckles are hard to study quantitatively, and no evidence is available on the relationship of speckles to diffuse receptors. It is possible that AChRs in speckles account for the immobile receptors found in the diffuse receptor regions. Some speckles are probably related to the larger receptor aggregates, since fluorescence and freeze fracture images show that the large receptor clusters are composed of groups of discrete aggregates less than 1 μm in diameter which resemble the speckles seen in diffuse receptor areas (see above). Also, sequential studies of the formation and dissolution of large aggregates on cultured cells suggest that a class of speckle may be a constituent unit of large aggregates (Olek *et al.*, 1983, 1984; Kuromi and Kidokoro, 1984a). Speckles of a different nature may represent structures involved in AChR removal from the cell membrane, as cross-linking of receptors by extracellular macromolecules is accompanied by a large increase in fluorescently labeled receptor speckles (Axelrod, 1980; Bursztajn *et al.*, 1983) and an increase in the number of aggregates of membrane particles (Pumplin and Drachman, 1983). Finally, some speckles may reflect the mechanism of receptor insertion into the membrane (Bursztajn and Fischbach, 1984; Bursztajn, 1984).

The receptors away from large aggregates on cultured chicken muscle cells are not randomly distributed. By analyzing the density of large angular particles in freeze fracture images, Cohen and Pumplin (1979) showed that neither the small aggregates, consisting of only a few particles, nor the larger aggregates, with many hundreds of particles, could have arisen by chance association. This observation implies that the receptor distribution is not random, even in an apparently diffusely distributed receptor population.

B. Innervated Muscle Cells in Culture

The discussion of AChR aggregation *in vitro* has so far concentrated on aggregation that occurs spontaneously in the absence of neuronal influences. However, many attempts have been made to duplicate *in vitro* the nerve-induced formation of the postsynaptic AChR aggregate observed *in vivo*. Attempts with primary cultures of myotubes and spinal-cord cells obtained from mammals have failed: the sensitivity of the muscle membrane under nerve processes to iontophoretically applied ACh is indistinguishable from control areas not contacted by nerve (Kidokoro, 1980). Using co-cultures of myotubes and spinal cord from chick, however, increased local ACh sensitivities have been recorded at some sites of contact between nerve and muscle cells. Study of single muscle cells over several days in culture has indicated that at least some of the regions of high ACh sensitivity develop after the nerve contacts the muscle (Frank and Fischbach, 1979).

The most fully characterized system for studying nerve–muscle interactions *in vitro* utilizes tissue from *Xenopus* tadpoles (Cohen, 1980). Upon co-culture of myocytes with spinal cord explants, the AChRs in the myocyte membrane accumulate to some regions at which the nerve contacts the muscle cell, while aggregates elsewhere disintegrate (Anderson and Cohen, 1977; Anderson et al., 1977; Kuromi and Kidokoro, 1984b). At least some of this receptor accumulation is due to lateral movement of AChRs already in the myocyte membrane to the postsynaptic region (Anderson et al., 1977). The neurite does not preferentially contact areas already enriched in receptors. These results clearly imply that the nerve directly influences the state of AChR aggregation in muscle membrane.

The nerve-induced receptor aggregates on *Xenopus* cells disintegrate when the nerve is destroyed, although they do seem to become somewhat more stable to denervation when the neurons have had a longer time in the culture (Kuromi and Kidokoro, 1984a). An interesting observation is that aggregation does not usually occur at all places where a nerve process lies over the muscle, and often the aggregate is found where the neurite is physically most tightly attached to the muscle surface (Harris et al., 1971a). Some types of nerve cells are not effective at inducing receptor aggregates, either in *Xenopus* primary cultures (Cohen and Weldon, 1980; Kidokoro et al., 1980) or using clonal mammalian cells (Busis et al., 1981).

Although there is clear evidence for the lateral motion and accumulation of receptors at nerve-induced aggregates, some evidence supports the idea that there is preferential use of newly synthesized receptors in forming nerve associated receptor aggregates on myotubes in culture (Bursztajn et al., 1982; Fischbach et al., 1984; Bursztajn, 1984). As is the case for junctions *in vivo*, the mechanisms for accumulation and selective insertion need to be worked out and their relative importance determined.

A final question is whether muscle cells in culture produce more than one type of receptor aggregate. Does the observation that noninnervated aggregates disappear as nerve-induced ones appear imply that the mechanisms of aggregation differ? Aggregates on myotubes can differ in appearance (e.g., Salpeter et al., 1982), and nerve-associated clusters change in appearance with time (Anderson et al., 1977; Kidokoro et al., 1980). The population of receptors recruited into high density regions is reported to differ between types of aggregate (Fischbach et al., 1984), and the density of receptors can be altered by a soluble factor (Salpeter et al., 1982). This question will be resolved only as a more detailed understanding of the molecular basis of clustering is reached.

C. Summary of Acetylcholine Receptor Distribution on Cells *in Vitro*

ACh receptors on cultured myotubes are not randomly distributed even in the absence of innervation. They are found in several distinct but interconvertible populations: those in large clusters (area of $10-1000$ μm^2), those in speckles (area about 1 μm²), and those that are diffusely distributed. The clusters present a highly organized appearance: in rat myotubes the aggregates are often composed of parallel rows of speckles. This distinctive arrangement suggests that a specialized mechanism is involved in immobilizing AChRs into clusters. Whether similar mechanisms participate in the formation of speckles seen in the diffuse receptor regions is not known. While several factors, described below, have recently been shown to be important for the formation and stability of large AChR aggregates, those involved in receptor speckling have not been systematically studied.

The addition of competent nerve cells induces receptor clusters at some regions of contact. There is evidence both for rearrangement of existing receptors and for preferential use of newly inserted receptors in forming the nerve-induced aggregate. In addition, noninnervated receptor patches disintegrate during formation of nerve-induced aggregates, suggesting the existence of intracellular mechanisms to control aggregate stability.

IV. THE CONTROL OF ACETYLCHOLINE RECEPTOR CLUSTERING

The density of AChRs in the plasma membrane of muscle cells is clearly nonrandom. A hierarchy of receptor distributions can be assigned, from a "diffuse" distribution to small aggregates ("speckles") to larger aggregates ("clusters"). The rest of this chapter is concerned with the control of receptor clustering. It is possible, based on the sequence of junctional development *in vivo,* that a muscle cell can produce more than one type of receptor cluster. The validity of this hypothesis, however, has not yet been demonstrated and requires experimental testing. Receptor clustering on cells in culture is similar to the initia! stages of junctional development, at least, and observations made on cultured muscle cells have provided much of our information about receptor clustering on muscle cells. Our discussion will emphasize the common themes emerging from receptor clustering both in the animal and in cell cultures.

We distinguish two processes in receptor clustering: a signal that stimulates clustering and a mechanism that actually produces the cluster. An example of the first could be a macromolecule released by a nerve terminal, whereas an example of the second could be an alteration of the cytoskeleton of the muscle cell with a resultant redistribution of associated receptors. It is possible that the two processes are the same: a hypothetical example is that the nerve terminal releases a very large multivalent molecule that binds to external moieties of many AChRs and immobilizes them directly in a cluster. However, it is useful to consider the signal and the mechanism as distinct processes. It is even possible that more than one signal exists: there could be one signal that stimulates the muscle to produce a cluster anywhere on its surface, and a second (or separate) signal that indicates the location for the cluster.

A. Is There a Signal Inducing Clustering?

What is the evidence that a signal exists that stimulates cluster formation, and what might the signal be? In the case of clusters under nerve terminals, the observation that clusters develop after nerve contact certainly suggests the presence of a signal. Clusters on noninnervated muscle cells in culture often occur at locations recognizable by criteria other than the presence of a cluster, and the frequency of clusters on cells can be changed (see below). However, the facts that denervated muscle fibers *in vivo* and noninnervated cells *in vitro* can produce clusters at apparently random locations on the cell surface implies either that muscle cells can "spontaneously" produce clusters or that a signal is generally present (perhaps produced by muscle cells).

The possibility that a signal exists and is chemical in nature has received considerable experimental support. Work in several laboratories has found that extracts prepared from the central nervous system (Podleski *et al.*, 1978; Jessell *et al.*, 1979; Olek *et al.*, 1983; Buc-Caron *et al.*, 1983) and extracts from *Torpedo* electric organ (Godfrey *et al.*, 1984) or medium conditioned by neuronal cells (Christian *et al.*, 1978; Bauer *et al.*, 1981; Schaffner and Daniels, 1982) are capable of promoting the formation of large receptor aggregates in cultured myotubes. Some of this increased aggregation is associated with an overall increase in the density of receptors in the membrane (Podleski *et al.*, 1978; Jessell *et al.*, 1979). However, at least one preparation is capable of promoting receptor aggregation without a significant increase in receptor number (Bauer *et al.*, 1981). Such factors appear to be found only in neuronal cells, or tissues with many synapses (Christian *et al.*, 1978; Schaffner and Daniels, 1982; Godfrey *et al.*, 1984). The active agents in these preparations are now being sought. There are several different candidates, some of high molecular weight (Podleski *et al.*, 1978; Bauer *et al.*, 1981; Schaffner and Daniels, 1982; Nitkin *et al.*, 1983), and another of low molecular weight which is nevertheless inactivated by heat and by proteases (Jessell *et al.*, 1979). Until the promoters of receptor aggregation are purified and characterized, their role in nerve-induced receptor clustering at the maturing neuromuscular junction remains problematic.

Some interesting observations have already been made using crude or partially purified preparations of these factors. Receptor aggregates appear with a lag after addition of a factor from neuronal cultures, and clusters appear even when cultures are washed during this period (Vogel *et al.*, 1983). This implies either that the factor binds tightly to the muscle cell or that processes are set in action which do not require the continued presence of the factor. A similar material produces as one of its earliest effects a decrease in the fraction of freely diffusing receptors in the membrane of cultured muscle cells, as though it promoted the attachment of receptors to an immobile structure (Axelrod et al., 1981). Brain extract has been reported to increase the density of ACh receptors in noninnervated receptor clusters (Salpeter et al., 1982). Finally, there are a number of provocative experiments that suggest that there is an interaction between the soluble factors which promote receptor clustering and other extracellular macromolecules which make up the basal lamina (see below).

Acetylcholine itself is not responsible for inducing AChR clustering, although cholinergic neurons are more efficient at inducing clusters in *Xenopus* cultures (Kidokoro et al., 1980). Nerve-induced receptor clustering *in vitro* can occur if noncholinergic neurons are used or if ACh synthesis is blocked (Steinbach et al., 1973). Similarly, the interaction of ACh with ACh receptors can be blocked with no effect on nerve-induced receptor clustering (Steinbach et al., 1973; Anderson et al., 1977).

The factors just discussed are soluble in physiological solutions, but there is evidence, provided by studies of regenerating muscle fibers, that insoluble extracellular material may be involved in inducing cluster formation as well. If adult muscle is denervated and the muscle fibers are killed, new muscle fibers will appear and grow inside the old basal lamina sheaths of the original muscle fibers. Clusters of AChR form at the old junctional region even when reinnervation is prevented and glial cells are killed (Burden et al., 1979; Bader, 1981; McMahan and Slater, 1984). The density of receptors in these clusters is lower than that at the adult junction (McMahan and Slater, 1984), but comparable to that at early junctions (Matthews-Bellinger and Salpeter, 1983). The basal lamina is the only structure seen to survive degeneration in these experiments. so it has been inferred that this material and associated extracellular macromolecules plays a role in inducing receptor clustering. Indeed, a preparation of insoluble material from *Torpedo* electric organ is able to promote receptor aggregation on cultured chicken myotubes (Godfrey et al., 1984). Less direct evidence is provided by the observation of a cap of amorphous extracellular material over receptor clusters on myotubes developing *in vivo*, even when the cluster is not covered by a nerve terminal (Jacob and Lentz, 1979; Matthews-Bellinger and Salpeter, 1983; Burrage and Lentz, 1981), and over receptor aggregates on cultured cells (Weldon et al., 1981; Burrage and Lentz, 1981; Salpeter et al., 1982). The extracellular material and the region of high receptor density are usually not perfectly coextensive, however, and at early stages of development the extracellular material is often sparse and patchy.

Recently a major effort has been made to identify the components of the junctional basal lamina and to study their interaction with the muscle membrane. This area has been reviewed by Sanes (1983), so we treat it here only briefly. Several extracellular macromolecules have been found to be concentrated in the

junctional region. In addition to acetylcholinesterase, a heparan sulfate proteoglycan (Anderson and Fambrough, 1983) is also enriched extracellularly at
the neuromuscular junction and at receptor clusters on cultured cells (Bayne *et
al.*, 1984). Other extracellular macromolecules have been recognized by immunological means but have not yet been identified (Sanes and Chiu, 1983). These
macromolecules are not the exclusive components of the junctional basal lamina,
however. They are embedded in a matrix of common extracellular constituents,
including collagen, fibronectin, and laminin (Sanes, 1982).

It is still not known how these molecules are organized within the extracellular
matrix to allow their effective interaction with the postsynaptic membrane. One
approach to this question has been to isolate from the basal lamina of Torpedo
electric organ constituents that induce AChR clustering in cultured myotubes.
One such constituent has been partially purified and characterized (Nitkin *et al.*,
1983; Godfrey *et al.*, 1984).

A second approach is that of Sanes and his colleagues (Weinberg *et al.*, 1981a;
Chiu and Sanes, 1984), who have examined the sequence of accumulation of
junctional basal lamina components in the junctional regions of rats. They have
found that junctional components are incorporated sequentially after innervation
and the onset of AChR clustering. During this period, nonjunctional basal lamina
antigens are eliminated from regions undergoing synaptogenesis (Chiu and
Sanes, 1984). It is now clear that components of the synaptic basal lamina are
made by aneural muscle cells *in vitro* (Silberstein *et al.*, 1982; Sanes and Lawrence, 1983). The distribution of these antigens is influenced by muscle activity
(Sanes and Lawrence, 1983) and receptor aggregating factors (Sanes *et al.*,
1984; Feldman *et al.*, 1984) in a fashion similar to ACh receptors.

The addition of ascorbic acid increases the production of mature extracellular
collagen in primary cultures of rat myotubes (which also contain fibroblasts) and
also increases the number of AChR aggregates (Kalcheim *et al.*, 1982). Ascorbic
acid also enhances the activity of one of the aggregating factors, and the factor,
conversely, increases synthesis of collagen by cultures (Kalcheim *et al.*, 1982).
Very recently it has been reported that a low-molecular-weight factor isolated
from brain tissue, which increases receptor synthesis by cultured myotubes, is
actually ascorbic acid (Knaack and Podleski, 1984). The higher molecular-
weight aggregation-stimulating factors and the protease-sensitive factors appear
to be different. Related experiments have examined the effects of increasing the
levels of known basal lamina components on AChR clustering in culture. Addition of laminin promotes clustering and enhances the action of a neuronal aggregation-promoting factor, (Vogel *et al.*, 1983), and laminin is found enriched
in regions over ACh receptor clusters (Daniels *et al.*, 1984; Bayne *et al.*, 1984;
but see Sanes, 1982; Chiu and Sanes, 1984).

A number of important questions are still unresolved. The extracellular molecules need to be identified, and their organization needs to be clarified. The
relationships and interactions between insoluble extracellular material and the

soluble factors need to be analyzed, and the roles of molecules in providing a scaffold as opposed to presenting a specific stimulus must be determined.

The possible role of insoluble extracellular materials in promoting clustering may be quite general, since the density of receptors increases in muscle fiber membrane underneath foreign objects placed on the surface of the muscle (Jones and Vrbova, 1974; Jones and Vyskocil, 1975). Cultured rat myotubes place nearly all their large AChR clusters in regions of the cell closely apposed to the tissue culture substrate (Axelrod *et al.*, 1976; Bloch and Geiger, 1980). Comparisons of the distribution of fluorescent-labeled AChRs with the sites of cell-substrate apposition (revealed by interference reflection microscopy) show that within these regions of cell-substrate adhesion clustered receptors and contact sites do not coincide but instead interdigitate (Bloch and Geiger, 1980). Similar results have been reported for *Xenopus* myocytes, in that receptor clusters on the bottom of cultured cells are associated preferentially with regions where the cell adheres strongly to the substrate (Moody-Corbett and Cohen, 1982). Receptor clusters also are rapidly induced on *Xenopus* myocytes by latex beads (Peng and Cheng, 1982). The nerve terminal, in addition to releasing soluble factors and carrying specific macromolecules, also comes into close physical contact with the muscle during early stages of synaptogenesis (James and Tresman, 1969; Kelly and Zacks, 1969; Jacob and Lentz, 1979; Matthews-Bellinger and Salpeter, 1983; Bursztajn and Fischbach, 1984).

Neither specific chemical factors nor non-specific physical contact alone seems able to account for all the data regarding signals stimulating AChR cluster formation. Some of the large regions of close myotube–glass contact seen in culture do not have associated receptor clusters (Bloch and Geiger, 1980), and clusters can appear on the upper surface of noninnervated cells in culture (Anderson and Cohen, 1977; Burrage and Lentz, 1981; Salpeter *et al.,* 1982). Similarly, receptor clusters are not found at all places where nerve processes contact muscle cells *in vitro* (Steinbach *et al.,* 1973; Anderson and Cohen, 1977; Frank and Fischbach, 1979). A model depending on physical contact alone is unable to explain this failure of receptor to cluster when contacts are made, or clustering in the absence of apparent contacts. On the other hand, a nerve-derived or systemic receptor aggregation factor cannot account for the localization of receptor clusters at regions where muscle cells contact the substrate or foreign objects. Something special about regions of contact must also exist. Although it is possible that one or the other set of observations is artifactual, it seems more likely that a combination of chemical and physical factors is involved in inducing receptor aggregation.

B. How Does a Muscle Produce a Cluster?

Muscle cells can respond to environmental signals by producing ACh receptor aggregates. How do the signals result in localized receptor aggregation? Nothing is known about the immediate response of the muscle to these signals, but some information is available about the mechanism used to produce receptor clusters.

To produce a region of increased receptor density, receptors must be accumulated at a greater rate than in surrounding regions or lost at a lower rate (or both). The evidence indicates that a lower rate of loss alone cannot account for the increased density, so receptors must be added to clusters selectively. There is some evidence to support a role for both a locally increased rate of receptor insertion and a gathering together of receptors from neighboring regions, but the relative importance of the two processes is not known. The mechanism by which receptors are gathered to aggregates is not understood, but it seems likely that an active cellular process is involved. Edwards and Frisch (1976) proposed that receptors could passively diffuse into a junction from extrajunctional regions and be "trapped" there. For the adult neuromuscular junction the predicted number of junctional receptors is too low by about 100-fold, when the diffusion constant of extrajunctional receptors is taken to be 2×10^{-11} cm^2/sec (Stya and Axelrod, 1983b). The passive entry model also seems unlikely even for clusters on cells in culture. One circumstantial point is that cluster assembly seems to occur using smaller units of receptor-dense membrane (Olek *et al.*, 1983, 1984). Another is that some drugs that affect cytoskeleton assembly also block cluster formation (see below). Finally, one of the factors that increase receptor aggregation actually produces a decrease in the fraction of mobile receptors as one of its early effects (Axelrod *et al.*, 1981), which might be expected to reduce the formation of large clusters by passive diffusion. In sum, passive entry of receptors into aggregates may occur (see Young and Poo, 1983), but it is unlikely to account for clustering as an exclusive mechanism.

Receptors within a cluster are organized and held immobile. Conceivably, direct receptor–receptor interactions could organize a cluster. This does not explain the fine structure of clusters composed of speckles, nor the distribution of intramembranous particles seen in freeze-fractured images of receptor aggregates. It also cannot explain why only one aggregate forms on singly innervated muscle fibers during development. Instead, there is now considerable circumstantial evidence to support the idea that clustering involves specific interactions of receptors with other macromolecules.

Clustered receptors could be bound by either extracellular or intracellular molecules. We have already mentioned the association of extracellular material with receptor clusters, when possible signals for clustering were discussed. These observations are also consistent with the idea that extracellular macromolecules play a role in the actual processes of cluster formation and maintenance. Indeed, if receptors were tightly bound by them it might be expected that the diffusion constant and degradation rate could be altered. There is reason to think that intracellular components are involved in clustering, however, and it seems likely to us that they will be more actively involved than external macromolecules.

The evidence for the presence of interactions between cytoplasmic macromolecules and receptors in clusters comes in three basic categories: pharmacological manipulations, structural studies, and immunological and biochemical studies.

The pharmacological work has been aimed at demonstrating a role for structural proteins, broadly called "cytoskeletal" proteins. Cytoskeletal proteins are involved in the movement and capping of surface proteins on other cells (Schreiner and Unanue, 1976), processes which have some similarities to receptor clustering (Bloch, 1979). These proteins also play roles in maintaining cell shape and in cell adhesion (e.g., Porter, 1984). Pharmacological manipulations have suggested a role for both microfilaments and microtubules in the formation or maintenance of AChR clusters. Drugs like colchicine depolymerize microtubules, whereas cytochalasins disrupt microfilaments. Neither class of drug disperses clusters on rat myotubes in culture during a 6-hr exposure to the agent (Bloch, 1979, 1983; Axelrod et al., 1978a). Colchicine blocks the reappearance of clusters that had been disrupted by poisoning energy metabolism or removing Ca^{2+} from the medium, whereas cytochalasins have no effect on reformation (Bloch, 1979, 1983). Colchicine also blocks the addition of acetylcholine receptors into clusters after the previously aggregated receptors had been labeled with fluorescent αBT, then the bound fluorophore bleached (Stya and Axelrod, 1983c). Exposure of cultured chick myotubes to cytochalasins for 24 hr disperses clusters, but colcemid has no effect (Connolly, 1984). Both types of drug block the increase in clustering on chick cells caused by brain extract (Connolly, 1984). Some of the differences in these reports may result from the fact that the cultured cells come from different species, while others clearly result from the different culture conditions and durations of exposure. All of these drugs have multiple effects in the cell, and some can change ACh receptor or energy metabolism. The observations do suggest, however, that microfilaments and/or microtubules are involved in cluster formation or maintenance.

Electron microscopic studies have provided evidence for intracellular specializations under regions of receptor-dense membrane. There is a patch of filamentous material under the postsynaptic receptor-dense membrane (Rosenbluth, 1974; Ellisman et al., 1976; Fertuck and Salpeter, 1976; Heuser and Salpeter, 1979). Similar structures are seen associated with ACh receptor clusters in cultured myotubes and at the developing neuromuscular junction (Jacob and Lentz, 1979; Matthews-Bellinger and Salpeter, 1983; Salpeter et al., 1982; Burrage and Lentz, 1981). Plastic beads can induce receptor clusters on cultured Xenopus myocytes; intracellular filaments accumulate at these sites before receptors do, suggesting a possible causal relationship (Peng and Phelan, 1984). Images of quick-frozen, freeze-etched frog neuromuscular junctions have shown bundles of thin filaments branching in close proximity to the postsynaptic membrane, with the smaller branches terminating at the membrane (Hirokawa and Heuser, 1982). The number of filament attachment points is much smaller than the number of receptors, however, so the anchorage of receptors must involve mechanisms in addition to filament attachment. Intermediate filaments are also present postsynaptically, but at a greater distance from the membrane (Hirokawa and Heuser, 1982; Burden, 1982). These ultrastructural studies are suggestive, but their interpretation is limited by the fact that often regions of receptor accumulation are specialized in other ways: the postjunctional membrane is a complicated structure, for instance, and presumably needs an intracellular scaffold simply to keep its shape. However, intracellular specializations are associated with receptor-dense membrane even when no other specializations are obvious, for example, at clusters on the upper surface of cultured muscle cells (Weldon *et al.*, 1981; Salpeter *et al.*, 1982; Burrage and Lentz, 1981).

Biochemical and immunological work has provided evidence that at least six proteins are enriched in the cytoplasm under receptor-rich membranes. Most of the studies have used specific antibodies to show that the protein (or a cross-reacting antigen) is concentrated in postsynaptic regions or at receptor aggregates on myotubes in tissue culture. A cytoplasmic form of actin (Hall *et al.*, 1981) and the actin-binding proteins vinculin, filamin, and α-actinin are found at both embryonic and adult rat neuromuscular junctions, and at adult mouse, chick, and *Xenopus* junctions (Bloch and Hall, 1983). A protein thought to be related to tonofilament (intermediate filament) protein is also concentrated at frog neuromuscular junctions, in an amorphous layer within 50 nm of the AChR-rich membrane (Burden, 1982).

The clearest biochemical evidence for a direct interaction between receptors and nonreceptor proteins concerns a 43,000-dalton protein found in electric organs (Barrantes, 1983). Immunological studies have found that this protein, or a protein with cross-reacting antigenic regions, is also concentrated at the neuromuscular junction in rat muscle fibers (Froehner *et al.*, 1981). The 43,000-dalton protein has a similar molecular weight to actin, but biochemical studies have shown that it is a distinct protein (Strader *et al.*, 1980; Gysin *et al.*, 1981; Porter and Froehner, 1983). It is present in nearly the same molar amount as receptor in membrane fragments from *Torpedo* electric organ (Sobel *et al.*, 1978) and codistributes with AChR along the postsynaptic membrane (Ngiem *et al.*, 1983; Sealock *et al.*, 1984). It apparently is not covalently linked to the receptor since it can be removed from membrane fragments at high pH; removal does not alter receptor function, so far as is known (Neubig *et al.*, 1979). However, removal of the 43,000-dalton protein does change the organization of receptors in the membrane (Barrantes, 1982; Cartaud *et al.*, 1981) and increases the rotational diffusion constant for receptor (Barrantes *et al.*, 1980; Lo *et al.*, 1980). Recent chemical cross-linking experiments indicate that this protein is within 1.2 nm of the beta subunit of the membrane-bound AChR (Burden *et al.*, 1983), and other cross-linking experiments suggest that it has the capacity to form a protein lattice resembling that seen in the red blood cell cytoskeleton (Cartaud *et al.*, 1982). The 43,000-dalton protein therefore is intimately associated with ACh receptors in receptor-dense membrane and is probably involved in maintaining receptor organization at mature synapses. It is important to determine how it participates in the early stages of cluster formation and to define its role more clearly in maintaining receptor aggregates.

Efforts are now under way in several laboratories to learn how these proteins interact with one another and with AChRs to promote clustering. In these experiments, as with the 43,000-dalton protein, progress has been facilitated by studying isolated membrane fragments containing AChR clusters. Such fragments are readily obtained from cultures of rat myotubes, since many of the AChR clusters in these cultures are found where the myotubes attach to the substrate. The bulk of the myotube can be removed by gentle shearing in the presence of a low concentration of formaldehyde, leaving behind the attachment sites and associated receptor clusters. The fragments obtained have their intracellular regions exposed to the medium and retain the two-dimensional array of AChR and cell–substrate contact domains found in the intact cell. Using this preparation, it has been found that vinculin, which is associated with the intracellular aspect of membrane–substrate contact regions in other cells (Geiger, 1979; Burridge and Feramisco, 1980), is also enriched over the contact domains of AChR clusters (Bloch and Geiger, 1980). Little or no vinculin is associated with the AChR-rich regions of clusters. It is therefore unlikely that vinculin associates directly with AChRs in clusters *in vitro*. A second method for preparing membrane fragments containing receptor clusters has recently been described, using the detergent saponin (Bloch, 1984). In fragments prepared with saponin no cytoskeletal proteins are found to be associated with the cell–substrate contact domain (probes to actin, vinculin, α-actinin, myosin, and filamin have been used). However, actin (Bloch and Resneck, 1983) and myosin (R. J. Bloch, in preparation) codistribute with AChR over AChR domains. Several treatments that remove actin from the isolated clusters result in extensive redistribution of AChRs within the plane of the membrane (Bloch and Resneck, 1983). This suggests that actin is directly or indirectly involved in maintaining the organization of AChRs in isolated clusters (see also Connolly, 1984). The relationship of the actin in isolated clusters to the cytoplasmic actin present in the postsynaptic region of the rat neuromuscular junction is not yet known. The way in which actin interacts with the membrane and with other cytoskeletal proteins near the AChR clusters in intact cells must also still be learned. It appears, however, that studies of the AChR-rich membrane fragments of the *Torpedo* electric organ and AChR clusters isolated from cultured myotubes will contribute greatly to our understanding of the organization of the cytoplasm underlying the postsynaptic membrane.

C. Summary of Receptor Clustering

Both chemical factors and physical contact are able to stimulate receptor clustering. At present, the factors have not been purified and the mechanisms by which they increase clustering are not known, nor is the mechanism by which receptors aggregate at contact sites. It is appealing to speculate that the two sets of observations are related: that the physiological stimulus for clustering involves the presentation of a specific signal on an extracellular scaffolding and that either the scaffold or the stimulus alone is able to promote some degree of aggregation.

The actual production and organization of a region of high receptor density in the membrane is very likely to involve a number of cytoskeletal proteins and a limited number of unique "linking" proteins. A large amount of circumstantial evidence argues in favor of this idea, but direct tests have not yet been possible. It is not yet clear which proteins are required for production (i.e., for increasing the rate of addition of receptors to the cluster) and which are utilized in organizing the aggregated receptors.

V. SUMMARY AND SPECULATION

We have broken the topic of acetylcholine receptor clustering into two broad portions: the stimulation of aggregation and the formation and organization of the cluster. We think that receptor clustering will share many features with other cellular specializations that require regional differentiation of the cell surface. One of the interesting points will be to determine which processes are unique to the control of neurotransmitter receptors. Indeed, at present it is not even known which specializations associated with receptor aggregates are required for clustering per se and which subserve other functions, such as adhesion.

A. Summary of Observations

1. The junctional receptor aggregate shows progressive changes during development. The changes suggest that there are several kinds of nerve-induced receptor aggregates, with different mechanisms for forming and maintaining the aggregate. Noninnervated muscle cells, both in the animal and in culture, can also form receptor clusters. The clusters are not identical to adult postjunctional regions, but show similarities to developing junctions at various stages.

2. A number of extracellular macromolecules are found at a high density at neuromuscular junctions or receptor clusters. Some are components of the basal lamina of other cells or regions, whereas others (not yet fully characterized) occur preferentially at receptor clusters.

3. The muscle cytoplasm shows ultrastructural specializations immediately under receptor clusters. At least five cytoskeletal proteins are concentrated in this region. One additional protein, the 43,000-dalton protein, may be uniquely associated with postjunctional membrane.

4. Several external factors can stimulate formation of receptor clusters. Insoluble material at adult junctions induces aggregation; soluble factors from several

neuronal sources do so, as well. In addition, physical stimuli such as plastics or glass result in formation of receptor aggregates at some regions of contact. Physical stimuli induce re-organization of cytoplasmic components before receptor clusters appear.

Not all nerve cells are competent to induce receptor clusters. The nerve and muscle cells come in close physical contact, especially early during their interaction. Muscle activity acts to reduce the number of noninnervated clusters, while the junctional clusters are protected. Both nerve cells and plastic beads cause a loss of non-contacted clusters on cultured cells.

External chemical stimuli can result in the redistribution of extracellular material, as well as receptor clustering. Activity also affects the amount and distribution of extracellular material on cultured muscle cells.

B. Speculation

Our speculation takes the form of a proposed sequence of events in receptor clustering, especially the early phases of clustering.

1. When the nerve first contacts the myotube, adhesion dominates the interaction and triggers receptor aggregation. This is consistent with the observation that clustering occurs at sites of muscle attachment to glass or plastics, and that the nerve and muscle are closely associated at early stages of synaptogenesis. At the earliest stages of junctional development relatively little basal lamina or extracellular material is present at the synapse. Some specificity in cell–cell adhesion is also necessary, since not all neurons induce aggregation. However, the fact that receptor clusters initially can be found without nerve terminals over them means, if this idea is correct, either that the nerve moves over muscle surface somewhat or that the response is not precisely localized.

2. Molecules on the surface of the muscle that are involved in adhesion accumulate at the region of nerve–muscle contact. The accumulation organizes the cytoskeleton by transmembrane effects, resulting in an association of microfilaments and other proteins with the internal face of the muscle membrane. This sequence is consistent with the segregation of membrane components involved in cell–cell adhesion in other systems (Ocklind et al., 1983; Singer, 1982) and the observation that extracellular proteins can be aligned with the cytoskeleton (Singer, 1979; Burridge and Feramisco, 1980). It includes the finding that vinculin, actin, and other actin-binding proteins are found at the developing junction. Vinculin may mediate the interaction between the membrane components and the cytoskeleton (Geiger, 1979; Singer, 1982).

3. The microfilaments and other proteins associated with the internal face of the muscle membrane bind and immobilize AChRs, possibly through a unique linking protein. This is consistent with the effects of the removal of membrane-bound actin on receptor organization in isolated AChR clusters. It includes the

observation that, when receptor clustering is stimulated by plastic beads, cytoplasmic organization precedes receptor aggregation.

4. The initial accumulation of receptors and other proteins forms a template
onto which other components are added and from which some constituents are
removed. This is consistent with the sequential changes in the aggregates and in
the basal lamina components seen at junctions during development. The idea of
an initial scaffold, on both faces of the membrane, also accounts for the interaction of aggregation-promoting factors with elements of the basal lamina and the
ability of soluble factors to increase the density of receptors in aggregates on
cultured myotubes.

VI. PROSPECTS

In the last few years, our knowledge of the factors controlling AChR distribution on skeletal muscle cells has progressed beyond phenomenology. We can
now discuss some of the molecular entities that affect AChR clustering and we
are beginning to test specific hypotheses regarding their modes of action. The
involvement of cytoskeletal proteins in AChR clustering should receive further
support as additional monospecific antibodies to various cytoskeletal proteins are
used to correlate the spatial distribution of receptors with structural entities
within the muscle cell. Monoclonal antibodies to extracellular macromolecules
have begun to reveal which antigens codistribute with receptors on the external
face of the cell. As the technology develops, these macromolecules will be
purified and characterized, and the way in which they influence membrane
structure will be studied. Purification of the factors which influence AChR
clustering will open two additional avenues. One is the use of antibodies to these
factors to determine their distribution at junctions *in vivo* and, potentially, to
perform antibody blocking experiments as were done for nerve growth factor.
The second is to analyse the mechanism of signal transduction *in vitro,* since
purified signal molecules will provide a temporally defined stimulus so that the
sequence of events leading to receptor clustering can be studied in detail. A
temporally defined stimulus will also assist greatly in sorting out which of the
intra- and extracellular molecules are directly involved in the production of
clusters. Finally, biochemical characterization of the cluster itself should provide
complementary information regarding receptor-associated macromolecules and
the membrane components mediating signal transduction in response to innervation. All these approaches are currently under intensive investigation in a number
of laboratories and should soon be fruitful.

Complications are sure to arise, however. One need only recall the several
stages through which AChR aggregates pass during neuromuscular junction maturation to realize that several mechanisms may control the process of AChR

aggregation and the nature of the aggregate. It seems to us that while we will soon know a great deal about the initial clustering of AChRs in response to innervation, we will need a considerably longer time to learn about the subsequent steps that generate the mature motor end plate.

Synapses between two neurons seem to be organized along the same general principles as the neuromuscular junction. For example, the receptor density is increased in postsynaptic regions at some synapses (Harris *et al.*, 1971b; Jacob *et al.*, 1984) and postsynaptic regions at central synapses are often enriched in large intramembranous particles (Gulley and Reese, 1981; Landis *et al.*, 1974; Landis and Reese, 1974). Synapses often have associated regions of increased cytoplasmic density and are enriched in cytoskeletal proteins (Gulley and Reese, 1981; Kelly and Cotman, 1978; Blomberg *et al.*, 1977). Synapse formation between two neurons has not been as well characterized as the development of the neuromuscular junction, but similarities are clearly apparent (Landmesser and Pilar, 1978).

How will an understanding of the details of AChR distribution on skeletal muscle assist in the study of the organization of neurotransmitter receptors in the central nervous system? Clearly, it is still too early to say. If nothing else, the results from muscle cells will be used to design experiments on neuronal cells. We think it more likely that there will be direct parallels between the control of acetylcholine receptor distribution and that of transmitter receptors on neurons. Certainly, the process of receptor clustering at developing junctions encompasses almost any degree of plasticity in neurotransmitter receptor distribution that could be desired.

ACKNOWLEDGMENTS

Research in the authors' laboratories was supported by grants from the National Institutes of Health and the Muscular Dystrophy Association. R. J. Bloch holds a Research Career Development Award.

REFERENCES

Adams, P. R. (1981). *J. Membrane Biol.* **58,** 161–174.
Albuquerque, E. X., and McIsaac, R. J. (1970). *Exp. Neurol.* **26,** 183–202.
Anderson, D. J., and Blobel, G. (1983). *Cold Spring Harbor Symp. Quant. Biol.* **48,** 125–134.
Anderson, M. J., and Cohen, M. (1977). *J. Physiol.* **268,** 757–773.
Anderson, M. J., and Fambrough, D. M. (1983). *J. Cell Biol.* **97,** 1396–1411.
Anderson, M. J., Cohen, M., and Zorychta, E. (1977). *J. Physiol.* **268,** 731–756.
Axelrod, D. (1980). *Proc. Natl. Acad. Sci. U.S.A.* **77,** 4823–4827.

Axelrod, D., Ravdin, P., Koppel, D. E., Schlessinger, J., Webb, W. W., Elson, E. L., and Podleski, T. R. (1976). *Proc. Natl. Acad. Sci. U.S.A.* **73**, 4594–4598.

Axelrod, D., Ravdin, P. M., and Podleski, T. R. (1978a). *Biochim. Biophys. Acta* **511**, 23–38.

Axelrod, D., Wight, A., Webb, W., and Horwitz, A. (1978b). *Biochemistry* **17**, 3604–3609.

Axelrod, D., Bauer, H. C., Stya, M., and Christian, C. N. (1981). *J. Cell Biol.* **88**, 459–462.

Bader, D. (1981). *J. Cell Biol.* **88**, 338–345.

Ballivet, M., Nef, P., Stalder, R., and Fulpius, B. (1983). *Cold Spring Harbor Symp. Quant. Biol.* **48**, 83–88.

Barrantes, F. J. (1982). *J. Cell Biol.* **92**, 60–68.

Barrantes, F. J. (1983). *Int. Rev. Neurobiol.* **24**, 259–341.

Barrantes, F. J., Neugebauer, D.-Ch., and Zingsheim, H. P. (1980). *FEBS Lett.* **112**, 73–78.

Bauer, H. C., Daniels, M. P., Pudimat, P. A., Jacques, L., Sugiyama, H., and Christian, C. N. (1981). *Brain Res.* **209**, 395–404.

Bayne, E. K., Anderson, M. J., and Fambrough, D. M. (1984). *J. Cell Biol.* **99**, 1486–1501.

Bennett, M. R., Pettigrew, A. G., and Taylor, R. S. (1973). *J. Physiol.* **230**, 331–357.

Berg, D. K., and Hall, Z. W. (1975a). *J. Physiol.* **244**, 659–676.

Berg, D. K., and Hall, Z. W. (1975b). *J. Physiol.* **252**, 771–789.

Betz, H., Bourgeois, J. P., and Changeux, J. P. (1980). *J. Physiol.* **302**, 197–218.

Bevan, S., and Steinbach, J. H. (1977). *J. Physiol.* **267**, 195–213.

Bevan, S., and Steinbach, J. H. (1983). *J. Physiol.* **336**, 159–177.

Bloch, R. J. (1979). *J. Cell Biol.* **82**, 626–643.

Bloch, R. J. (1983). *J. Neurosci.* **3**, 2670–2680.

Bloch, R. J. (1984). *J. Cell Biol.* **99**, 984–993.

Bloch, R. J., and Geiger, B. (1980). *Cell* **21**, 25–35.

Bloch, R. J., and Hall, Z. W. (1983). *J. Cell Biol.* **97**, 217–223.

Bloch, R. J., and Resneck, W. G. (1983). *Abstr. Soc. Neurosci.* **9**, 757.

Bloch, R. J., and Steinbach, J. H. (1981). *Dev. Biol.* **81**, 386–391.

Blomberg, F., Cohen, R., and Siekevitz, P. (1977). *J. Cell Biol.* **74**, 204–225.

Braithwaite, A. W., and Harris, A. J. (1979). *Nature (London)* **279**, 549–551.

Brenner, H. R., and Sakmann, B. (1983). *J. Physiol.* **337**, 159–171.

Brett, R. S., Younkin, S. G., Konieczkowski, M., and Slugg, R. M. (1982). *Brain Res.* **233**, 133–142.

Brockes, J. P., and Hall, Z. W. (1975a). *Proc. Natl. Acad. Sci. U.S.A.* **72**, 1368–1372.

Brockes, J. P., and Hall, Z. W. (1975b). *Biochemistry* **14**, 2100–2106.

Buc-Caron, M. H., Nystrom, P., and Fischbach, G. F. (1983). *Dev. Biol.* **95**, 378–386.

Burden, S. (1977a). *Dev. Biol.* **57**, 317–329.

Burden, S. (1977b). *Dev. Biol.* **61**, 79–85.

Burden, S. (1982). *J. Cell Biol.* **94**, 521–530.

Burden, S. J., Sargeant, P. B., and McMahan, U. J. (1979). *J. Cell Biol.* **82**, 412–425.

Burden, S. J., DePalma, R. L., and Gottesman, G. S. (1983). *Cell* **35**, 687–692.

Burrage, T. G., and Lentz, T. L. (1981). *Dev. Biol.* **85**, 267–286.

Burridge, K., and Feramisco, J. R. (1980). *Cell* **19**, 587–595.

Bursztajn, S. (1984). *J. Neurocytol.* **13**, 503–518.

Bursztajn, S., and Fischbach, G. (1984). *J. Cell Biol.* **98**, 498–506.

Bursztajn, S., Berman, S. A., McManaman, J., and Appel, S. H. (1982). *Abstr. Soc. Neurosci.* **9**, 1059.

Bursztajn, S., McManaman, J. L., Elias, S. B., and Appel, S. H. (1983). *Science* **219**, 195–197.

Bursztajn, S., McManaman, J. L., and Appel, S. H. (1984). *J. Cell Biol.* **98**, 507–517.

Busis, N. A., Daniels, M. P., Bauer, H. C., Sonderagger, P., Schaffner, A. E., and Nirenberg, M. (1981). *Soc. Neurosci. Abstr.* **7**, 766.

Cangiano, A., Lomo, T., Lutzemberger, L., and Sveen, O. (1980). *Acta Physiol. Scand.* **109**, 283–296.

Cartaud, J., Sobel, A., Rousselet, A., Devaux, P., and Changeux, J. P. (1981). *J. Cell Biol.* **90**, 418–426.

Cartaud, J., Oswald, R., Clement, G., and Changeux, J.-P. (1982). *FEBS Lett.* **145**, 250–257.

Changeux, J. P. (1981). *Harvey Lect.* **75**, 85–254.

Changeux, J. P., Kasai, M., and Lee, C. Y. (1970). *Proc. Natl. Acad. Sci. U.S.A.* **67**, 1241–1247.

Chiu, A. Y., and Sanes, J. R. (1984). *Dev. Biol.* **103**, 456–467.

Chow, I., and Cohen, M. W. (1983). *J. Physiol.* **339**, 553–571.

Christian, C. N., Daniels, M. P., Sugiyama, H., Jaques, L., and Nelson, P. G. (1978). *Proc. Natl. Acad. Sci. U.S.A.* **75**, 4011–4014.

Claudio, T., Ballivet, M., Patrick, J., and Heinemann, S. (1983). *Proc. Natl. Acad. Sci. U.S.A.* **80**, 1111–1115.

Cohen, M. W. (1980). *J. Exp. Biol.* **89**, 43–56.

Cohen, M. W., and Weldon, P. R. (1980). *J. Cell Biol.* **86**, 388–401.

Cohen, S. A., and Pumplin, D. W. (1979). *J. Cell Biol.* **82**, 494–518.

Colquhoun, D., and Sakmann, B. (1983). *In* "Single Channel Recording" (B. Sakmann and E. Neher, eds.), pp. 345–364. Plenum Press, New York.

Connolly, J. A. (1984). *J. Cell Biol.* **99**, 148–154.

Courtney, J., and Steinbach, J. H. (1981). *J. Physiol.* **320**, 435–447.

Daniels, M. P., Vigny, M., Sonderegger, P., Bauer, H. C., and Vogel, Z. (1984). *Int. J. Dev. Neurosci.* **2**, 87–99.

Devreotes, P. N., and Fambrough, D. M. (1976). *Proc. Natl. Acad. Sci. U.S.A.* **73**, 161–164.

Diamond, J., and Miledi, R. (1962). *J. Physiol.* **162**, 393–408.

Dreyer, F., and Peper, K. (1974). *Pfluegers Arch.* **348**, 273–286.

Edwards, C., and Frisch, H. L. (1976). *J. Neurobiol.* **4**, 377–381.

Eldridge, L., Liebhold, M., and Steinbach, J. H. (1981). *J. Physiol.* **313**, 529–545.

Ellisman, M. H., Rash, J. E., Staehelin, L. A., and Porter, K. R. (1976). *J. Cell Biol.* **68**, 752–774.

Fairclough, R. H., Finer-Moore, J., Love, R. A., Kristofferson, D., Desmeules, P. J., and Stroud, R. M. (1983). *Cold Spring Harbor Symp. Quant. Biol.* **48**, 9–20.

Fambrough, D. M. (1979). *Physiol. Rev.* **59**, 165–227.

Fambrough, D. M., and Hartzell, H. C. (1972). *Science* **176**, 189–191.

Feldman, D. H., Sanes, J. R., and Lawrence, J. C. (1984). *Soc. Neurosci. Abstr.* **10**, 761.

Fertuck, H. C., and Salpeter, M. M. (1976). *J. Cell Biol.* **69**, 144–158.

Fischbach, G. D., and Cohen, S. A. (1973). *Dev. Biol.* **31**, 147–162.

Fischbach, G. D., and Schuetze, S. M. (1980). *J. Physiol.* **303**, 125–137.

Fischbach, G. D., Role, L. W., O'Brien, R., and Matossian, V. (1984). *Soc. Neurosci. Abstr.* **10**, 925.

Frank, E., and Fischbach, G. D. (1979). *J. Cell Biol.* **83**, 143–158.

Frank, E., Gautvik, K., and Sommerschild, H. (1975). *Acta Physiol. Scand.* **95**, 66–76.

Froehner, S. C., Gulbrandsen, V., Hyman, C., Yeng, A. Y., Neubig, R. R., and Cohen, J. B. (1981). *Proc. Natl. Acad. Sci. U.S.A.* **78**, 5230–5234.

Geiger, B. (1979). *Cell* **18**, 193–205.

Godfrey, E. W., Nitkin, R. M., Wallace, B. G., Rubin, L. L., and McMahan, U. J. (1984). *J. Cell Biol.* **99**, 615–627.

Gulley, R. L., and Reese, T. S. (1981). *J. Cell Biol.* **91**, 298–302.

Gysin, R., Wirth, M., and Flanagan, S. D. (1981). *J. Biol. Chem.* **256**, 11373–11376.

Hall, Z. W., Lubit, B. W., and Schwartz, J. H. (1981). *J. Cell Biol.* **90**, 789–792.

Hall, Z. W., Roisin, M. P., Gu, Y., and Gorin, P. D. (1983). *Cold Spring Harbor Symp. Quant. Biol.* **48**, 101–108.

Harris, A. J. (1981). *Philos. Trans. R. Soc. Lond. (Biol. Sci.)* **293**, 287–314.
Harris, A. J., Heinemann, S., Schubert, D., and Tarikas, H. (1971a). *Nature (London)* **231**, 296–301.
Harris, A. J., Kuffler, S. W., and Dennis, M. J. (1971b). *Proc. R. Soc. Lond. (Biol.)* **177**, 541–553.
Hartzell, H. C., and Fambrough, D. H. (1972). *J. Gen. Physiol.* **60**, 248–262.
Harvey, A. L., and van Helden, D. (1981). *J. Physiol.* **317**, 397–411.
Heuser, J., and Salpeter, S. R. (1979). *J. Cell Biol.* **82**, 150–173.
Hirokawa, N., and Heuser, J. E. (1982). *J. Neurocytol.* **11**, 487–510.
Ishikawa, Y., and Shimada, Y. (1982). *Dev. Brain Res.* **5**, 187–197.
Jacob, M., and Lentz, T. L. (1979). *J. Cell Biol.* **82**, 195–211.
Jacob, M. H., Berg, D. K., and Lindstrom, J. M. (1984). *Proc. Natl. Acad. Sci. U.S.A.* **81**, 3223–3227.
James, D. W., and Tresman, R. L. (1969). *Z. Zellforsch. Mikrosk. Anat.* **100**, 126–140.
Jessell, T. M., Siegel, R. E., and Fischbach, G. D. (1979). *Proc. Natl. Acad. Sci. U.S.A.* **76**, 5397–5401.
Jones, R., and Vrbova, G. (1974). *J. Physiol.* **236**, 517–538.
Jones, R., and Vyskocil, F. (1975). *Brain Res.* **88**, 309–317.
Kalcheim, C., Vogel, Z., and Duskin, D. (1982). *Proc. Natl. Acad. Sci. U.S.A.* **79**, 3077–3081.
Karlin, A. (1980). *In* "The Cell Surface and Neuronal Function" (C. W. Cotman, G. Poste, and G. L. Nicolson, eds.), pp. 191–260. North Holland Publ. Co., Amsterdam.
Katz, B., and Miledi, R. (1964). *J. Physiol.* **170**, 379–388.
Kelly, A. M., and Zachs, S. I. (1969). *J. Cell Biol.* **42**, 154–169.
Kelly, P. T., and Cotman, C. W. (1978). *J. Cell Biol.* **79**, 173–183.
Kidokoro, Y. (1980). *Dev. Biol.* **78**, 231–241.
Kidokoro, Y., and Patrick, J. (1978). *Brain Res.* **142**, 368–373.
Kidokoro, Y., Anderson, M. J., and Gruener, R. (1980). *Dev. Biol.* **78**, 464–483.
Knaack, D., and Podleski, T. (1984). *Soc. Neurosci. Abstr.* **10**, 1053.
Ko, P. K., Anderson, M. J., and Cohen, M. W. (1977). *Science* **196**, 540–542.
Kuffler, D., Thompson, W., and Jansen, J. K. S. (1977). *Brain Res.* **138**, 353–358.
Kuffler, S. W., and Yoshikami, D. (1975). *J. Physiol.* **244**, 703–730.
Kullberg, R. W., Lentz, T. L., and Cohen, M. W. (1977). *Dev. Biol.* **60**, 101–129.
Kullberg, R. W., Brehm, P., and Steinbach, J. H. (1981). *Nature (London)* **289**, 411–413.
Kuromi, H., and Kidokoro, Y. (1984a). *Dev. Biol.* **104**, 421–427.
Kuromi, H., and Kidokoro, Y. (1984b). *Dev. Biol.* **103**, 53–61.
Land, B. R., Podleski, T. R., Salpeter, E. E., and Salpeter, M. M. (1977). *J. Physiol.* **269**, 155–176.
Landis, D. M. D., and Reese, T. S. (1974). *J. Comp. Neurol.* **155**, 93–125.
Landis, D. M. D., Reese, T. S., and Raviola, E. (1974). *J. Comp. Neurol.* **155**, 67–92.
Landmesser, L., and Pilar, G. (1978). *Fed. Proc.* **37**, 2016–2022.
Lavoie, P. A., Collier, B., and Tenenhouse, A. (1977). *Exp. Neurol.* **54**, 148–171.
Levitt, T. A., Loring, R. H., and Salpeter, M. M. (1980). *Science* **210**, 550–551.
Linden, D. C., and Fambrough, D. M. (1979). *Neuroscience* **4**, 527–538.
Lo, M. M. S., Garland, P. B., Lamprecht, J., and Barnard, E. A. (1980). *FEBS Lett.* **111**, 407–412.
Lomo, T., and Rosenthal, J. (1972). *J. Physiol.* **221**, 493–513.
Lomo, T., and Slater, C. R. (1980a). *J. Physiol.* **303**, 191–202.
Lomo, T., and Slater, C. R. (1980b). *J. Physiol.* **303**, 173–189.
Matthews-Bellinger, J., and Salpeter, M. M. (1978). *J. Physiol.* **279**, 197–213.
Matthews-Bellinger, J. A., and Salpeter, M. M. (1983). *J. Neurosci.* **3**, 644–657.
McMahan, U. J., and Slater, C. R. (1984). *J. Cell Biol.* **98**, 1453–1473.
Merlie, J. P., Heinemann, S., and Lindstrom, J. M. (1979). *J. Biol. Chem.* **254**, 6320–6327.

Merlie, J. P., Sebbane, J., Gardner, S., and Lindstrom, J. (1983a). *Proc. Natl. Acad. Sci. U.S.A.*
    80, 3845–3849.
Merlie, J. P., Sebbane, R., Gardner, S., Olson, E., and Lindstrom, J. M. (1983b). *Cold Spring
    Harbor Symp. Quant. Biol.* **48**, 135–146.
Michler, A., and Sakmann, B. (1980). *Dev. Biol.* **80**, 1–17.
Miledi, R., and Zelena, J. (1966). *Nature (London)* **210**, 855–856.
Montal, M., Labarca, P., Fredkin, D. R., Suarez-Isla, B. A., and Lindstrom, J. (1984). *Biophys. J.*
    45, 165–174.
Moody-Corbett, F., and Cohen, M. W. (1982). *J. Embryol. Exp. Morphol.* **72**, 53–69.
Morrison-Graham, K. (1983). *Dev. Biol.* **99**, 298–311.
Nass, M. M., Lester, H. A., and Krause, M. E. (1978). *Biophys. J.* **22**, 135–155.
Neher, E., and Sakmann, B. (1976). *J. Physiol.* **258**, 705–730.
Neubig, R. R., Krodel, E. K., Boyd, N. D., and Cohen, J. B. (1979). *Proc. Natl. Acad. Sci. U.S.A.*
    76, 690–694.
Nghiem, H. O., Cartaud, J., Dubreuil, C., Kordeli, C., Buttin, G., and Changeux, J. P. (1983).
    Proc. Natl. Acad. Sci. U.S.A. **80**, 6403–6407.
Nitkin, R. M., Wallace, B. G., Spira, M. E., Godfrey, E. W., and McMahan, U. J. (1983). *Cold
    Spring Harbor Symp. Quant. Biol.* **48**, 653–665.
Noda, M., Furutani, Y., Takahashi, H., Toyosato, M., Tanabe, T., Shimizu, S., Kikyotani, S.,
    Kayano, T., Hirose, T., Inayama, S., and Numa, S. (1983a). *Nature (London)* **305**, 818–823.
Noda, M., Takahashi, H., Tanabe, T., Toyosato, M., Kikyotani, S., Furutani, Y., Hirose, T.,
    Takashima, H., Inayama, S., Miyata, T., and Numa, S. (1983b). *Nature (London)* **302**, 528–
    532.
Ocklind, C., Forsum, U., and Obrink, B. (1983). *J. Cell Biol.* **96**, 1168–1171.
Olek, A. J., and Robbins, N. (1983). *Neuroscience* **9**, 225–233.
Olek, A., Krikorian, J. G., and Daniels, M. P. (1984). *Soc. Neurosci. Abstr.* **10**, 36.
Olek, A. J., Pudimat, P. A., and Daniels, M. P. (1983). *Cell* **34**, 255–264.
Orkand, P. M., Orkand, R. K., and Cohen, M. W. (1978). *Neuroscience* **3**, 435–446.
Peng, H. B., and Cheng, P. C. (1982). *J. Neurosci.* **2**, 1760–1774.
Peng, H. B., and Phelan, K. A. (1984). *J. Cell Biol.* **99**, 344–349.
Peng, H. B., Nakajima, Y., and Bridgman, P. C. (1980). *Brain Res.* **196**, 11–31.
Peper, K., Dreyer, F., Sandri, C., Akert, K., and Moor, H. (1974). *Cell Tissue Res.* **149**, 437–455.
Podleski, T. R., Axelrod, D., Ravdin, P., Greenberg, I., Johnson, M. M., and Salpeter, M. M.
    (1978). *Proc. Natl. Acad. Sci. U.S.A.* **75**, 2035–2039.
Porter, K. (1984). *J. Cell Biol.* **99**(1): Part 2.
Porter, S., and Froehner, S. C. (1983). *J. Biol. Chem.* **258**, 10034–10040.
Pumplin, D. W., and Drachman, D. B. (1983). *J. Neurosci.* **3**, 576–584.
Pumplin, D. W., and Fambrough, D. M. (1982). *Annu. Rev. Physiol.* **44**, 319–335.
Reiness, C. G., and Weinberg, C. B. (1981). *Dev. Biol.* **84**, 247–254.
Rosenbluth, J. (1974). *J. Cell Biol.* **62**, 755–766.
Rosenbluth, J. (1975). *J. Neurocytol.* **4**, 697–712.
Salpeter, M. M., and Harris, R. (1983). *J. Cell Biol.* **96**, 1781–1785.
Salpeter, M. M., Spanton, S., Holley, K., and Podleski, T. R. (1982). *J. Cell Biol.* **93**, 417–425.
Sanes, J. R. (1982). *J. Cell Biol.* **93**, 442–451.
Sanes, J. R. (1983). *Annu. Rev. Physiol.* **45**, 581–600.
Sanes, J. R., and Chiu, A. Y. (1983). *Cold Spring Harbor Symp. Quant. Biol.* **48**, 667–678.
Sanes, J. R., and Lawrence, J. C., Jr. (1983). *Dev. Biol.* **97**, 123–136.
Sanes, J. R., and Merlie, J. P. (1984). *Soc. Neurosci. Abstr.* **10**, 924.
Sanes, J. R., Feldman, D. H., Cheney, J. M., and Lawrence, J. C., Jr. (1984). *J. Neurosci.* **4**, 464–
    473.

Schaffner, A. E., and Daniels, M. P. (1982). *J. Neurosci.* **2**, 623–632.
Schreiner, G. F., and Unanue, E. R. (1976). *In* "Advances in Immunology" (F. J. Dixon and H. G. Kunkel, eds.), Vol. 24, pp. 37–164. Academic Press, New York.
Schuetze, S. M., Frank, E. F., and Fischbach, G. D. (1978). *Proc. Natl. Acad. Sci. U.S.A.* **75**, 520–523
Sealock, R., Wray, B. E., and Froehner, S. C. (1984). *J. Biol. Chem.* **98**, 2239–2244.
Silberstein, L., Inestrosa, N., and Hall, Z. W. (1982). *Nature (London)* **295**, 143–145.
Sine, S. M., and Steinbach, J. H. (1984). *Biophys. J.* **45**, 175–185.
Sine, S. M., and Taylor, P. (1980). *J. Biol. Chem.* **255**, 10144–10156.
Singer, I. I. (1979). *Cell* **16**, 675–685.
Singer, I. I. (1982). *J. Cell Biol.* **92**, 398–408.
Slater, C. R. (1982a). *Dev. Biol.* **94**, 11–22.
Slater, C. R. (1982b). *Dev. Biol.* **94**, 23–30.
Smith, M. A., and Slater, C. R. (1983). *J. Neurocytol.* **12**, 993–1005.
Sobel, A., Heidmann, T., Hofler, J., and Changeux, J.-P. (1978). *Proc. Natl. Acad. Sci. U.S.A.* **75**, 510–514.
Steinbach, J. H. (1980). *In* "Cell Surface Reviews" (C. Cotman, G. Poste, and G. Nicolson, eds.), Vol. 6, pp. 120–156. Elsevier, Amsterdam.
Steinbach, J. H. (1981a). *J. Physiol.* **313**, 513–528.
Steinbach, J. H. (1981b). *Dev. Biol.* **84**, 267–276.
Steinbach, J. H., Harris, A. J., Patrick, J., Schubert, D., and Heinemann, S. F. (1973). *J. Gen. Physiol.* **62**, 255–270.
Steinbach, J. H., Merlie, J., Heinemann, S., and Bloch, R. (1979). *Proc. Natl. Acad. Sci. U.S.A.* **76**, 3547–3551.
Strader, C. D., Lazarides, E., and Raftery, M. A. (1980). *Biochem. Biophys. Res. Commun.* **92**, 365–373.
Stya, M., and Axelrod, D. (1983a). *J. Cell Biol.* **97**, 48–51.
Stya, M., and Axelrod, D. (1983b). *J. Neurosci.* **4**, 70–74.
Stya, M., and Axelrod, D. (1983c). *Proc. Natl. Acad. Sci. U.S.A.* **80**, 449–453.
Sytkowski, A. J., Vogel, Z., and Nirenberg, M. W. (1973). *Proc. Natl. Acad. Sci. U.S.A.* **70**, 270–274.
Vogel, Z., Christian, C. N., Vigny, M., Bauer, H. C., Sonderegger, P., and Daniels, M. P. (1983). *J. Neurosci.* **3**, 1058–1068.
Weinberg, C. B., and Hall, Z. W. (1979a). *Dev. Biol.* **68**, 631–635.
Weinberg, C. B., and Hall, Z. W. (1979b). *Proc. Natl. Acad. Sci. U.S.A.* **76**, 504–508.
Weinberg, C. B., Sanes, J. R., and Hall, Z. W. (1981a). *Dev. Biol.* **84**, 255–266.
Weinberg, C. B., Reiness, C. G., and Hall, Z. W. (1981b). *J. Cell Biol.* **88**, 215–218.
Weldon, P. R., Moody-Corbett, F., and Cohen, M. W. (1981). *Dev. Biol.* **84**, 341–350.
Young, S. H., and Poo, M. M. (1983). *J. Neurosci.* **3**, 225–231.
Ziskind-Conhaim, L., and Bennett, J. I. (1982). *Dev. Biol.* **90**, 185–197.
Ziskind-Conhaim, L., and Dennis, M. J. (1981). *Dev. Biol.* **85**, 243–251.
Ziskind-Conhaim, L., Geffen, I., and Hall, Z. W. (1984). *J. Neurosci.* **4**, 2346–2349.

8

The Molecular Basis of Intercellular Communication in Neuronal Development

RICHARD J. RIOPELLE AND KIMBERLY E. DOW

I. INTRODUCTION

Considered at its simplest level, the development of the neuron is the result of interactions between the phenotype of the neuron and the extracellular environment. While the general statement applies to all cells, the neuron is unique, since a novel feature of this cell is its ability to extend processes (axons or dendrites), often over long distances, such that at any one time the neuron is interacting with environments that may be exceedingly heterogeneous. Neuronal development can be understood fully only within the complexity of the nervous system; this complexity is defined by connectivity via synapse formation, which is based in a developmental context upon axonal growth and guidance toward appropriate

215

RECEPTORS IN CELLULAR RECOGNITION
AND DEVELOPMENTAL PROCESSES

targets of innervation. Thus, that part of the neuron of interest in a developmental sense is the neuronal process, and more specifically, the growth cone at the growing tip, and attempts to explain axonal growth and guidance must be accounted for by properties displayed by this growth cone. Ramon y Cajal (1904) recognized that the growth cone or ''cone d'acroissement'' possessed unique functional properties when he wrote in 1890 that it was ''. . . endowed with exquisite chemical sensitivity, with rapid ameboid movements, and with certain impulsive forces. . . .'' This chapter is devoted to a consideration of these functional properties of developing neurons predicted by Ramon y Cajal in 1890, and first demonstrated by Harrison (1910) 20 years later.

The concept of axonal guidance in development can be understood only within the context of molecular species in the extraneuronal environment to which the growth cone responds. Thus, to deal with the ''chemical sensitivity'' of growth cones, a review of those ligands within the extracellular environment and growth-cone receptors for these ligands will be undertaken. Current theories of patterning of axonal guidance and exquisite synaptic mapping in nervous system development based upon growth-cone interactions with the extracellular environment will then be reviewed with a bias towards explaining axon guidance and target recognition by related mechanisms.

II. GROWTH-CONE STRUCTURE AND MOTILITY AND AXONAL GROWTH

The growth cone consists of a varicosity or protoplasmic enlargement, the periphery of which is highly motile and composed of thin, fingerlike projections or filopodia (approximately 0.15 μm in diameter) and veil-like projections known as lamellipodia (Landis, 1983) (Fig. 1).

A relatively constant array of membranous organelles are found within the varicosity—a granular reticulum, clear vesicles and vacuoles, coated vesicles, large dense-core vesicles, mitochondria, and lysosomal structures (see Landis, 1983, for review). The cytoskeletal elements of the growth-cone neurofilaments, microfilaments, and microtubules have been identified by a number of techniques (Bray and Gilbert, 1981; Bunge, 1973; Goldman et al., 1979; Tennyson, 1970; Jockusch and Jockusch, 1981; Yamada et al., 1970, 1971; Kuczmarski and Rosenbaum, 1979; Letourneau, 1982). Within the neuronal process or neurite, microtubules are bundled together but splay out to form a fan when they extend to the growth-cone varicosity; microtubules do not enter the lamellipodia or filopodia. Intermediate or 10-nm neurofilaments appear to have a similar distribution. Actin microfilaments form a meshwork beneath the plasma membrane of the entire growth cone; some of the actin in the growth cone is bundled

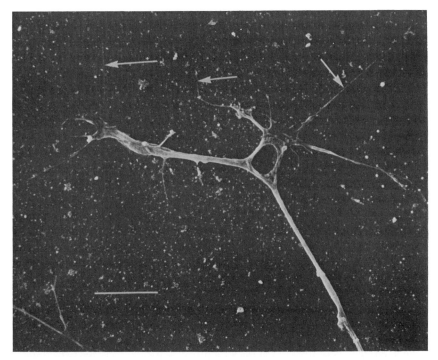

Fig. 1. Photomicrograph of growth cone of a neuron in culture. A varicosity at the tip of the neuronal process, the growth cone is a specialized structure bearing lamellipodia and filopodia. The lamellipodia are thin veil-like projections from the varicosity, while filopodia (designated by the arrows) are thin fingerlike projections that extend often over large distances. Bar represents 10 μm. Courtesy of S. Carbonetto.

and passes through the varicosity to the peripheral specializations, and some of the filaments appear to insert into the plasmalemma (Luduena and Wessells, 1973). Myosin immunoreactivity appears to colocalize with actin (Letourneau, 1981), and the actin filaments in filopodia are polarized such that new monomers are laid down distally (Isenberg and Small, 1978).

The cytoskeletal components of the growth cone are assumed to be responsible for extension, contraction, and motility of lamellipodia and filopodia, as well as the lengthening movements of the entire growth cone–neurite complex. The disruption of microtubule function by agents such as colchicine blocks neurite lengthening and results in retraction but has no influence on the morphology or activity of the growth-cone specializations (Yamada et al., 1970). Alternatively, microfilament disruption by cytochalasin causes rounding up of growth cones and inhibition of axonal lengthening (Yamada et al., 1970, 1971; Yamada and Wessells, 1973). These data are interpreted to suggest that structural support of the axon is carried out by microtubules, while motility is subserved by actin microfilaments.

Early observations of axonal elongation suggested that the addition of new plasma membrane occurs close to the growing tip (Harrison, 1910; Bray, 1970, 1973). These studies have been confirmed using a number of approaches, especially those relating to the distribution of identified molecular species of the membrane of neurons, axons, and the growth cone (see Landis, 1983 for review). The availability of lectins has perhaps contributed most to the understanding of membrane incorporation. Following labeling of growing neurons *in vitro* with a variety of lectins (concanavalin A, ricin I, wheat germ agglutinin), there is a time-dependent appearance of label-free membrane seen at the peripheral regions of the growth cone, especially on filopodia. These label-free regions, however, contain lectin-binding sites, since relabeling with the same lectin once again gives a uniform density of binding (Feldman *et al.*, 1981). These experiments suggest bulk addition of membrane at the growing tip.

III. DIRECTIONAL GUIDANCE OF THE AXON

The neuronal growth cone is a transient structure confined to a developmental time window between neuroblast generation and intimate target association. During this time, and during axonal regeneration where a growth cone is again recognized, the major function of the structure is to interact with the extracellular environment in order to guide the growing axon to its appropriate target. As has been pointed out by Jacobson (1978), the initial sprouting of a neuronal process is a genetically programed event. Subsequent axonal lengthening depends upon an interaction of growth-cone receptors with a host of ligands within the extraneuronal environment. Harrison (1910) has emphasized that the distance first neurons must navigate to interact with an appropriate milieu is relatively short, and direct observations of pioneering neurons in insects (Goodman *et al.*, 1984) would suggest that original axonal pathways may result from random filopodial outgrowth and adhesion to appropriate substrate in close proximity. To understand the directionality that the environment provides to axonal elongation, it is first necessary to consider the surface specializations the growth cone has at its disposal (Carbonetto, 1984). A filopodium can attach to the substrate of growth to form a firm adhesion site. This adhesion site is the result of an interaction of a substrate ligand with a receptor in the plasma membrane of the growth cone filopodium. By analogy with other cell systems, this interaction alters the association of the ligand–receptor complex with the bridging protein vinculin and actin filaments on the cytoplasmic face of the membrane, and the filopodium anchored to its attachment site then pulls on and extends the growth cone (Fig. 2). Subsequent increase in the attached filopodium diameter occurs, new filopodia are extended, some of which make initial attachment, and another adhe-

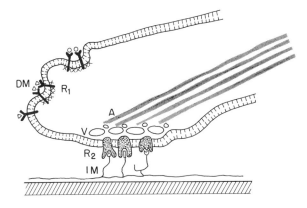

Fig. 2. Schematic diagram illustrating interaction of a filopodium of the neuronal growth cone with molecular species of the extracellular environment. Immobilized molecules (IM) and diffusible molecules (DM) interact with specific receptors (R_1 and R_2), which are integral membrane structures. By analogy with other cell systems, a diffusible neurotrophic molecule such as NGF is internalized by receptor-mediated endocytosis in clathrin-coated (xx) vesicles. This uptake process mediates the retrograde transport of NGF to the cell body of responsive neurons. Substrate adhesion is the result of an IM–R_2 interaction. The tethered filopodium is able to mediate lengthening of the neuronal process via cytoskeletal elements such as actin filaments (A), which are linked to the IM–R complex via the bridging molecular species vinculin (V).

sion site is formed. The orientation of the actin and myosin filaments within the growth cone and filopodia would appear to be appropriate to provide the contractile forces necessary for advancing the growth cone and neuronal process (Letourneau *et al.*, 1980).

Given this simple model of neuron-process lengthening, one can appreciate that similar phenomena occurring in a dynamic fashion within the filopodia of the growth cone might explain directionality of the growth cone if the extracellular milieu is oriented between less and more adhesive substrates, both in a qualitative (different substrates) sense and quantitatively (concentration, orientation). Letourneau (1983) has recently reviewed data demonstrating that neurons show a hierarchy of relative adhesion to various substrates; by patterning growth surfaces with these qualitatively different substrates, it has been demonstrated that growth cones always elongated on the more adhesive substrate, even though the filopodia sampled extensively and randomly. Similar observations on neuron-process orientation have been made for materials released from cultured cells and bound to culture surfaces using techniques that would form steep gradients— these experiments demonstrating the direction-promoting effect of a quantitative difference in the substrate (Collins and Garrett, 1980).

Other mechanisms have been proposed for axonal guidance *in vivo*. Directed axonal growth provided by the presence of a gradient of diffusible substance has been demonstrated most strikingly by the experiments of Gunderson and Barrett (1980) using nerve growth factor (NGF). Growing on an appropriate substrate in a milieu of low concentrations of NGF, it has been shown that neurites of sensory neurons consistently turned towards a micropipette containing high concentrations of NGF within 20 min of exposure. Mechanical factors and electrical fields also provide some orienting of axonal growth (Ebendal, 1976, Weiss, 1934; see Patel and Poo, 1984, for discussion). That these mechanisms are anything other than a favorable reorientation or polarization of the available substrate of growth has not been determined. Furthermore, no evidence has been obtained to suggest that endogenous electric fields within tissue serve as a guiding force for nerve growth.

IV. MOLECULAR SUBSTRATES OF AXONAL GUIDANCE

Reference has been made to molecular species within the extracellular milieu of developing neurons to which neurons respond by extending processes, and by which, depending on their disposition in that milieu, directionality of axonal elongation is achieved. For the purposes of the ensuing discussion, the extracellular milieu comprises the glycocalyx and extracellular matrix, the molecular species of which are synthesized and released by other cells (neurons and non-neurons) with which neurons interact. Neuronal interactions with immobilized molecular species are referred to as adhesive interactions. The property of adhesiveness between neurons and molecular species within the glycocalyx of other cells, or between neurons and molecular species within what is referred to as the extracellular matrix proper, likely governs many aspects of neuronal development, both in a permissive sense by facilitating access, and in an instructive sense by promoting exquisite targeting. However, neurons also interact with diffusible or semi-diffusible molecular species within the extracellular milieu. The role played by diffusible factors in development is less clear. A permissive effect (enhanced survival, accelerated differentiation) for at least one of these factors, NGF, has been clearly demonstrated (see Mobley et al., 1977, for review), but to date, despite observations on the guidance of neuritic growth by this diffusible neurotrophic factor in vitro, no instructive or guidance role in vivo has been determined.

Before discussing the interaction of receptors on neurons and growth cones with ligands within the extracellular milieu, it would be appropriate to review what is known about the ligands themselves. Much information is available on the biochemistry and mechanisms of action of the diffusible neurotrophic factor NGF. Since its discovery over 30 years ago, NGF has been the prototype neurotrophic protein, and a review of its mechanism of action and receptor interactions with responsive neurons will be presented.

The nondiffusible ligands within the extracellular environment that interact with neurons are found within the plasma membrane–glycocalyx (PM-GC) complex and within the extracellular matrix (ECM). One chooses to consider ECM and PM-GC complex ligands separately, recognizing that, in many respects, PM-GC complex and ECM form a structural and functional continuum. Much information on the structure and biochemistry of the ECM is available, although there is little information on the nature of the growth-cone and neuron receptor sites for the ECM ligands. Information is emerging, however, on the nature of adhesion molecules (receptors) on neuronal surfaces that interact with immobilized ligands on both neurons and glial cells.

A. Diffusible Ligands

1. Nerve Growth Factor—Biological Effects and Mechanisms of Action

Nerve growth factor (NGF), the only well-characterized neuronal trophic factor, was first discovered by Levi-Montalcini and Hamburger (1951). NGF is widely distributed, having been detected in a number of vertebrate species (Winick and Greenberg, 1965). A potent source of NGF has been found in the male mouse submaxillary gland (Cohen, 1960), and this finding led to isolation of the purified form of the NGF protein (Varon *et al.,* 1967). The biologically active NGF molecule known as βNGF is a dimer composed of two identical noncovalently linked peptide chains. Each polypeptide contains 118 amino acids and has a molecular weight of 13,259 (Greene *et al.,* 1971). This protein is contained in a high-molecular-weight complex known as 7SNGF, which also consists of α and γ subunits, the stoichiometry being 2 α subunits, 2 γ subunits, and one βNGF dimer.

NGF plays a critical role in the growth and development of sympathetic and certain sensory neurons and in the maintenance of the differentiated state of these neurons. Levi-Montalcini and Hamburger (1951) found that implantation of mouse sarcoma (a rich source of NGF) into day 3 chick embryos resulted in hypertrophy and hyperplasia of spinal sensory and sympathetic ganglia cells. This response was first seen at day 6 of incubation, at which time dorsal root ganglia (DRG) nerve fibers invaded the tumor. All the sympathetic ganglion cells increased in size, but only the mediodorsal (MD) cells of the sensory ganglia (DRG), not the ventrolateral (VL) cells, responded in this way to stimulation by the tumor. The hypertrophic response of sympathetic ganglia to NGF has been shown to persist into adult life (Levi-Montalcini and Angeletti, 1968). At the ultrastructural level, stimulated DRG and sympathetic neurons display and increase in the development of the Golgi apparatus and rough endoplasmic reticulum and increased numbers of neurofilaments and neurotubules (Angeletti *et al.*, 1971). Available evidence suggests that the increased number of sensory neurons noted after NGF treatment results from the maintenance of neuronal viability in MD cells (Banthorpe *et al.*, 1974) and also in VL cells (Hamburger and Yip, 1984), whereas the hyperplastic response of sympathetic neurons to NGF appears to be due to both an increase in the number of neurons surviving and accelerated maturation of undifferentiated "stem" cells (Hendry and Campbell, 1976). There is no evidence that NGF increases neuroblast mitotic activity in either sensory or sympathetic ganglia. In summary, NGF acts to maintain the viability and accelerate maturation of postmitotic sympathetic and sensory neurons.

When embryonic chick DRG were explanted to culture with sarcoma, a halo of neurites growing out of the ganglia was observed within 24 hr (Levi-Montalcini *et al.*, 1954). The measure of the size and density of neurite outgrowth from explanted ganglia in the presence of serial dilutions of NGF constitutes the classic biological assay for NGF (Levi-Montalcini, 1966). Dissociated cultures of chick embryo sensory ganglion neurons display enhanced survival, substrate attachment, and process formation in the presence of NGF. As with explanted ganglia, a single-cell biological assay for NGF has been developed based upon the percent of neurons bearing processes greater than 1 cell diameter at varying concentrations of NGF (Sutter *et al.*, 1979a; Riopelle and Cameron, 1981) (Fig. 3). NGF has also been shown to be essential for survival, substrate attachment, and process formation of sympathetic ganglion neurons (Levi-Montalcini and Angeletti, 1963; Cohen *et al.*, 1964).

Functions for NGF in the central nervous system (CNS) are suggested by the response to NGF by catecholaminergic neurons in the brain (Bjorklund and Stenevi, 1972; Bjerre *et al.*, 1973) and optic nerve (Turner and Glaze, 1977). Cholinergic septal neurons respond to NGF by an increase in choline acetyltransferase levels (Gnahn *et al.*, 1983). Shelton and Reichardt (1984) have recently detected mRNA for NGF in the brain with regional variations in its concentration.

NGF has been shown to cause rat pheochromocytoma cells (Tischler and Greene, 1975) and adrenal chromaffin cells (Unsicker and Chamley, 1977) to acquire neuronal characteristics. The PC12 rat pheochromocytoma cell line, when exposed to NGF, exhibits characteristic properties of sympathetic neurons, including neurite outgrowth, cessation of proliferation, increased electrical excitability, induction of choline acetyltransferase (Greene and Tischler, 1976; Dichter *et al.*, 1977; Tischler and Greene, 1978), and increased synthesis of the NGF-inducible large external (NILE) glycoprotein, a widespread neuronal marker (McGuire *et al.*, 1978). NGF-responsive cell lines, because of their ability to respond to but not require NGF, offer the opportunity to study as controls viable NGF-untreated cells and initial effects of NGF on cells without prior NGF exposure. This has led to considerable progress in the study of the mechanism of action of NGF.

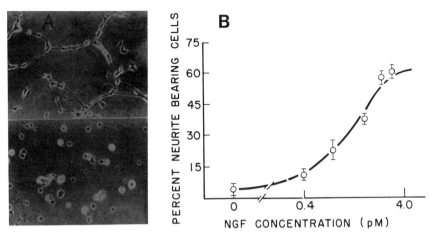

Fig. 3. (A) Appearance of dissociated day 8 chick embryo sensory neurons on a poly-D-lysine substrate, following incubation for 24 hr in the presence of 4.0 pM NGF (top) and in the absence of NGF (bottom). (B) Dose-response curve of a single cell bioassay for NGF. Sensory neurons are cultured for 24 hr in the presence of various concentrations of NGF. The number of neurons with processes is scored as a percentage of total viable (phase bright) cells, and the means and standard deviation of percent neurite-bearing cells are plotted as a function of NGF concentration.

Several investigators, using cultured chick embryo sympathetic and sensory ganglia, have suggested that NGF does not induce neuronal process formation by increasing RNA synthesis (Partlow and Larrabee, 1971; Burnham and Varon, 1974). However, Burstein and Greene (1978), using the PC12 line, found that NGF induction of process formation was suppressed in the presence of RNA synthesis inhibitors in cells not previously exposed to NGF. The induction by NGF of the NILE glycoprotein, HMW MAP1.2 (high-molecular-weight microtubule-associated protein), acetylcholinesterase, tyrosine hydroxylase, and neuron-specific enolase (NSE), and the appearance of electrical excitability are other examples of transcriptionally regulated processes. These transcription-dependent events have a latency period of at least 1 day and are referred to as delayed responses to NGF. In contrast, regeneration of neurites by PC12 cells preexposed to NGF is a rapidly onsetting response and is not blocked by RNA synthesis inhibitors. This response is similar to the early transcription-independent neurite outgrowth response demonstrated by ganglia *in vitro*. As these ganglia have had prior *in vivo* exposure, their *in vitro* behavior could represent regeneration rather than initiation of neurites.

These findings have led to the proposal of the "priming" model for the role of NGF in neurite outgrowth (Burstein and Greene, 1978)—the delayed response to NGF results in transcription-dependent changes, which give the cell potential for neurite outgrowth, but for process formation to occur, a rapid transcription-independent NGF response must also take place. Cultured neurons and PC12 cells pre-exposed or ''primed'' to NGF exhibit rapid transcription-independent neurite outgrowth because they have already undergone the delayed response. Thus, a pool of material whose synthesis is transcriptionally regulated by NGF is present to support neurite regeneration when transcription is blocked. Evidence is emerging to suggest that the rapidly onsetting response (approximately 20 min), which affects neurite outgrowth, may include local actions of NGF regulation on growth cone shape and motility (Gunderson and Barrett, 1980). Other examples of early responses to NGF include maintenance of neuronal survival (Greene, 1978), regulation of membrane Na^+/K^+ pump (Varon and Skaper, 1983), transport of nutrients (McGuire and Greene, 1979), and activation of tyrosine hydroxylase (Greene, 1984). Based on these considerations, Fig. 4 suggests that neurons might possess a mechanism to respond rapidly by virtue of

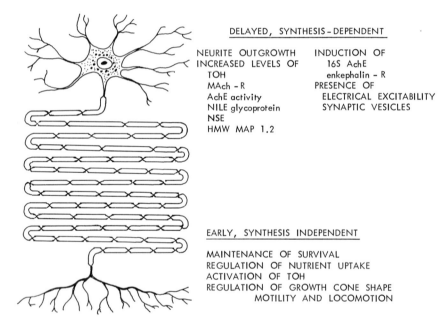

DELAYED, SYNTHESIS-DEPENDENT

NEURITE OUTGROWTH   INDUCTION OF
INCREASED LEVELS OF   16S AchE
   TOH   enkephalin - R
MAch - R   PRESENCE OF
AchE activity   ELECTRICAL EXCITABILITY
NILE glycoprotein   SYNAPTIC VESICLES
NSE
HMW MAP 1.2

EARLY, SYNTHESIS INDEPENDENT

MAINTENANCE OF SURVIVAL
REGULATION OF NUTRIENT UPTAKE
ACTIVATION OF TOH
REGULATION OF GROWTH CONE SHAPE
   MOTILITY AND LOCOMOTION

Fig. 4.   Schematic diagram summarizing delayed and early responses to NGF by a NGF receptor-bearing neuron. The long myelinated neuronal process serves to highlight the idea that the neuron could respond locally to NGF with short latency at a long distance from the cell body by having available a pool of material, thus precluding the requirement for a delayed transcription-dependent event. Synthesis-dependent (and -independent) refers to the requirement (and absence of requirement) for a long-latency transcription-dependent event. Abbreviations: TOH, tyrosine hydroxylase; MAch-R, muscarinic acetylcholine receptor; NILE, NGF-inducible large extracellular; NSE, neuron-specific enolase; HMW MAP 1.2, high-molecular-weight microtubule-associated protein 1.2; 16 S AchE, 16 S form of acetylcholinesterase; enkephalin-R, enkephalin receptor.

a permissive NGF effect to changes in the environment of the neuron even at a great distance from the cell body.

a. Retrograde Axonal Transport of NGF. NGF is internalized at peripheral sympathetic and peripheral and central sensory nerve terminals and retrogradely transported to the cell body (Hendry *et al.*, 1974; Stoeckel *et al.*, 1975; Johnson *et al.*, 1978; Richardson and Riopelle, 1984). Retrograde axonal transport of NGF has also been demonstrated in parasympathetic neurons of chick and rat ganglia (Max *et al.*, 1978) and in cholinergic septal–hippocampal neurons of the brain (Schwab *et al.*, 1979; Seiler and Schwab, 1984). This transport process involves a saturable, high-affinity uptake process, and the NGF that reaches the cell body in this way has been shown to be biologically active by its induction of tyrosine hydroxylase activity (Paravicini *et al.*, 1975). The administration of exogenous NGF prevents the degeneration of sympathetic neurons following axotomy (Hendry, 1975), and selective innervation of target organs that release large amounts of NGF has been demonstrated (Thoenen *et al.*, 1978). Recent evidence for the synthesis of NGF by target organs comes from the work of Shelton and Reichardt (1984), who found an increase in NGF mRNA content of irides grown *in vitro*. These data lend support to the hypothesis that NGF is an essential trophic factor transported from target organs to cell bodies. However, Richardson and Riopelle (1984) have demonstrated NGF uptake along peripheral and central axons of primary sensory neurons in the adult rat, and Shelton and Reichardt (1984) have detected NGF mRNA in peripheral nerve. These findings suggest that NGF, synthesized and released by Schwann cells or endoneurial fibroblasts (Riopelle *et al.*, 1981; Richardson and Ebendal, 1982; Thoenen *et al.*, 1983), may have local paracrine effects on neurons whose axons bear receptors in juxtaposition to the source of NGF.

2. Other Diffusible Neurotrophic Factors

Two recent additions to the family of diffusible neurotrophic factors are brain-derived neurotrophic factor (BDNF), a 12,300-dalton protein purified from pig brain (Barde *et al.*, 1982), and ciliary neurotrophic factor (CNTF), a 20,000-dalton species from eye tissues (Manthorpe *et al.*, 1982a, b). A number of other factors that support neuronal growth have been described (see Barde *et al.*, 1983, and Berg, 1984, for review), but for the majority, no details as to whether they are diffusible or immobilized, or whether they are related, are available.

Within a developmental framework, NGF, and likely other so-far uncharacterized diffusible neurotrophic factors, acts permissively to maintain viability and to accelerate the process of differentiation of responsive neuronal populations. Within the mature nervous system, NGF and related factors likely continue to play a role in maintenance of functional integrity of responsive cells.

B. Immobilized Ligands of the Extracellular Milieu

1. The Extracellular Matrix

a. Structural Components. In a restricted sense, the extracellular matrix (ECM) consists of a basal lamina (BL) and (with reference to cell membranes) a more external ground substance (GS) (Heathcote and Grant, 1981; Kefalides *et al.*, 1979; Hay, 1981). The more electron-dense layer of the BL is known as the lamina densa, is about 10 nm thick, and is separated from the cell membrane by a 2- to 5-nm electron-lucent layer known as the lamina rara. The GS contains striated and nonstriated fibrils embedded in an amorphous matrix. Where these fibrils are concentrated in a layer adjacent to the BL, a single structure identified by light microscopy is called the basement membrane (BM) as seen in schematic in Fig. 5. The BL is a feature of most nonmigratory cells, while migrating or mesenchymal cells do not have a BL but are embedded in other elements of ECM.

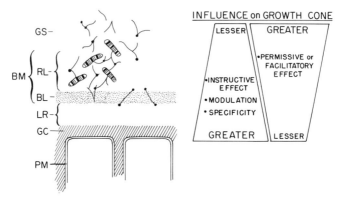

Fig. 5. Schematic diagram illustrating the extracellular milieu. Adjacent to the plasma membrane (PM) is the glycocalyx (GC) (hatched area). An electron-lucent layer, the lamina rara (LR), underlies an electron-dense layer of the basal lamina (BL), known as the lamina densa (dotted area). Collagen, as illustrated by the striated structures, and other fibrils are found within an amorphous matrix or ground substance (GS), which also contains diffusible growth factors. The basement membrane (BM) is a structure recognized by light microscopy; it consists of the BL and other fibrillary elements of the GS concentrated in a layer adjacent to the BL. The extracellular matrix is that part of the external milieu that does not include the PM-GC complex. It is suggested by the drawing indicating gradients that a growth cone of a neuron interacts with the ECM, which has a permissive influence on it, thereby facilitating access and passage by the growth cone. As the growth cone becomes more closely juxtaposed to the GC of adjacent cells (including neurons), this part of the extracellular milieu is better suited, by virtue of modulations of molecular species in the region, to effect instructive influences that could subserve exquisite pathfinding and specific connectivity. [(RL), reticular lamina.]

The Schwann cells of the peripheral nervous system deposit a sheet of ECM on their outer surfaces, and thus, Schwann cells and the axons with which they are associated are within a sheath of BM (Bunge and Bunge, 1983). Basement membrane also envelops skeletal muscle fibers. Within the central nervous system, BM is sparse; a BM is found beneath ependymal and meningeal cells and surrounds blood vessels within the CNS parenchyma. Following trauma, BM is found around the astrocytic scar. The internal limiting membrane between retina and vitreous body is BM. Most of the extracellular space of CNS is composed of amorphous matrix with no recognizable collagenous fibrous protein (Carbonetto, 1984).

b. Molecules of the ECM. Structures of the ECM are made of protein and carbohydrate molecules. The molecular species of the ECM have been difficult to isolate and thus difficult to characterize biochemically. Most information comes from work on connective tissue, a plentiful source of ECM. Generally speaking, however, the components of the ECM can be placed in three major categories as listed in Table I (adapted from Carbonetto, 1984).

Collagen molecules are unique in that they contain triple-helix segments and are rich in proline, hydroxyproline, and glycine in highly redundant amino acid sequences. Eight to 10 genetically distinct polypeptide chains are used to form the triple helixes, with the polypeptide makeup of the helixes identifying the collagen type. The molecular weight of collagen molecules approximates 3×10^6 daltons. Hydroxylysine residues within the polypeptide chains are glycosylated with glucosylgalactose. Collagen types I–V are found within the nervous system; in addition, there is a collagen-tailed form of acetylcholinesterase found on skeletal muscle BM, especially in synaptic regions. Types I, III, IV, and V collagen are Schwann cell-related, while type II is the major structural protein of the vitreous body. Collagens generally serve as the major structural components of the ECM, and the cellular interactions with collagen are mediated by other polyvalent proteins (connecting or "nectin" proteins, e.g., laminin and fibronectin), which bridge between collagen and receptors on the cell surface.

The noncollagenous glycoproteins differ from collagens because they are relatively deficient in hydroxyproline and hydroxylysine, insensitive to collagenases, and lack a triple helix structure. The glycoproteins are composed of asparagine-linked oligosaccharides.

Of the noncollagenous glycoproteins, fibronectin has been most extensively studied. Fibronectin exists in two forms with similar properties; each functional molecule is composed of two 220,000-dalton subunits, and each subunit has at least eight functional regions. Cellular receptors for fibronectin bind to a highly conserved tetrapeptide sequence, which is part of the 11.5-kDa cell attachment domain of the molecule; this sequence is present in at least five other proteins,

TABLE I

Molecules of the ECM[a]

Molecule	Nervous system distribution	Some characteristics
Collagens		$M_r \approx 3 \times 10^6$, triple helix, glycine-, hydroxyproline-, hydroxylysine-rich, glycosylated; 8–10 genetically distinct polypeptide chains
Type I	Schwann cells	2 α1 type I and 1 α2 type I chains
Type II	Vitreous body	3 α1 type II chains
Type III	Schwann cells	3 α1 type III chains
Type IV	Schwann cells	3 α1 type IV chains
Type V	Schwann cells	1 α type V and 2 α2 type V chains
Acetylcholinesterase	Skeletal muscle	Collagen-tailed form of acetylcholinesterase
Noncollagenous glycoproteins		Deficient in hydroxyproline and hydroxylysine, no triple helix, asparagine-linked heterosaccharides
Fibronectin	Radial glia developing brain; Schwann cells; neural crest ECM	2 $M_r \approx 220,000$ subunits, multiple binding domains cellular and secreted forms, 5% carbohydrate
Laminin	Schwann cells; skeletal muscle; probably all BM	Cruciform; $M_r \approx 1 \times 10^6$; subunits 2 × 220,000, 1 × 400,000, multiple binding domains, 12–15% carbohydrate
Entactin	Schwann cells	Sulfated glycoprotein; $M_r \approx 158,000$
Ligatin	Cerebrum; retina	$M_r \approx 10,000$, associated with phospholipid and cholesterol
β-Galactoside lectins	Brain; muscle	2 $M_r \approx 12,000–17,000$ subunits; binds to glycoconjugates
Glycosaminoglycans (GAGs)		Sugar polymers of alternating residues of uronic acid and glucosamine; negatively charged; includes heparan sulfate, chondroitin sulfate, dermatan sulfate, keratan sulfate, heparin
Proteoclycans		GAGs covalently bound to protein
Heparan sulfate	Brain; spinal cord; neuromuscular junction; Schwann cells	Binds to fibronectin and laminin; $M_r \approx 55,000$ (brain)
Chondroitin sulfate	Brain	Major proteoglycan in brain
Hyaluronic acid	Vitreous body	$M_r \approx 4 \times 10^3$ to 10^7; alternating residues of glucuronic acid and N-acetylglucosamine; not a proteoglycan

[a]Adapted from Carbonetto (1984).

including fibrinogen (Pierschbacher and Ruoslahti, 1984). Other domains bind to collagen, heparin, hyaluronic acid, acetylcholinesterase, fibrin transglutaminase, and fibronectin itself (Pierschbacher and Ruoslahti, 1984).

Immunocytochemical techniques show that fibronectin is a prominent component of the migratory pathway and BLs in the region of developing neural crest-derived cells, and levels of this glycoprotein increase in the region as migration begins. *In vitro* studies have demonstrated the dependence of crest cells on fibronectin for attachment and migration. Fibronectin has been found oriented in the same direction as radial glia in developing cerebral cortex (Pearlman *et al.*, 1984), and the protein is also found in Schwann cell and skeletal muscle basement membrane.

Laminin is a cruciform molecule with a molecular weight approximating 10^6 daltons, consisting of a 400,000-dalton subunit and two 220,000-dalton subunits. Laminin is also a multifunctional molecular species with domains that interact with cells, heparin, heparan sulfate, and type IV collagen. This glycoprotein is found in BMs of Schwann cells and skeletal muscle, is synthesized by Schwann cells, and is deposited in the lamina rara around nerve fibers. Laminin has been found to be associated with the neurite-promoting factors of conditioned media from a variety of cell cultures (Lander *et al.*, 1985).

A number of less well studied, noncollagenous glycoproteins related to the ECM of the nervous system have been described. These include entactin, a sulfated glycoprotein of molecular weight 158,000, ligatin, (molecular weight 10,000), which assembles into fibrils, and a series of β-galactoside lectins of 12,000–17,000 daltons.

Glycosaminoglycans (GAGs) are unbranched polymers of disaccharide units that are usually sulfated, highly negatively charged, and usually bound covalently to protein, in which form they are called proteoglycans. Examples of such GAGs are heparan sulfate, chondroitin sulfate, dermatan sulfate, keratan sulfate, and heparin. Hyaluronic acid differs from the other GAGs in that it is neither sulfated nor bound to a protein core. GAGs are found widely distributed in the nervous system where they act to facilitate neuroblast migration and adhesion. GAGs are (for the most part) nonimmunogenic but are stained by cationic dyes such as Alcian Blue and Ruthenium Red.

c. Localization of Molecules within the ECM. Immunocytochemical and histochemical analyses reveal where the various components of ECM are located (Sanes, 1983). Most BLs contain collagen type IV, laminin, and a heparan sulfate proteoglycan, while in addition, some contain fibronectin and collagen type V. Collagen type IV is highly enriched in the electron-dense layer of the BL, while laminin and proteoglycans are concentrated in the lamina rara. Large, striated fibers contain type I collagen, and smaller fibrils can contain types II, III, or V collagens, the distribution being somewhat tissue-dependent. The fibers and fibrils of collagen are coated with fibronectin and proteoglycans. GAGs (especially hyaluronic acid) are diffusely distributed throughout the ground substance. Increasingly sophisticated studies of ECM continue to add to our knowledge of the complexity of macromolecular networks within the ECM.

2. The Plasma Membrane–Glycocalyx Complex

The glycocalyx of the cell is that part of the extracellular environment that lies in contiguity to and coats the plasma membrane (Luft, 1976). Practically speaking, it is appropriate to consider the plasma membrane and glycocalyx as one functional unit (the receptor domain), and the glycocalyx and ECM as another functional unit (the ligand domain). For the purpose of this review, however, glycocalyx and ECM are considered separately. Even though plasma membrane receptors for various ligands of the extracellular environment make up some of the constituents of the glycocalyx, these are considered separately as well.

The glycocalyx coats the plasma membrane and includes external domains of integral plasma membrane proteins, as well as molecular species more loosely adherent to the cell surface (Luft, 1976). In addition to receptors for extracellular ligands, the PM-GC complex also contains ligands that can subserve intimate cell association. Cell–cell adhesion in nervous system development that mediates aggregations of neurons into various nuclei, as well as the adhesive phenomena that subserve neuron process bundling or fasciculation, are the result of molecular interactions within the PM-GC complex. Until recently, little was known of the molecular substrates of adhesive interactions between neurons or between neurons and other cells of the nervous system. Recently, however, a number of neuron-associated molecules identified by immunological techniques have been shown to play a role in neuronal adhesion. The best characterized of these molecules is the cell adhesion molecule (CAM) known as neuronal-CAM or N-CAM (Edelman, 1983). N-CAM is a bifunctional glycoprotein bearing both receptor and ligand domains, presumably constrained sufficiently to minimize intramolecular and to maximize intermolecular interactions; thus, the receptor domain of N-CAM on one neuron interacts with the ligand domain of N-CAM on the cell surface of a second neuron and subserves an adhesive interaction. Related molecular species include BSP-2, described by Goridis et al. (1983), glycoproteins bearing the epitopes L1 and L2, described by Schachner et al. (1983), and glycoproteins bearing the HNK-1 epitope, described by Riopelle et al. (1984). L1 would appear to be very similar in a functional sense to neuron-glia-CAM (Ng-CAM) (Edelman, 1984a,b; Grumet et al., 1983). All of the glycoproteins appear to be related in that they subserve adhesive interactions of neurons but, as will be detailed in a later section, these glycoproteins are also related antigenically.

Other properties of cell–cell adhesion glycoproteins are considered within a functional framework in a subsequent section.

C. Neuronal Receptors for Ligands

Using ^{125}I-labeled NGF, NGF receptors have been found on a variety of cell types of neural crest origin, including embryonic chick sympathetic and dorsal root ganglia neurons (Banerjee et al., 1973; Herrup and Shooter, 1973; Sutter et al., 1979b; Riopelle et al., 1980), Schwann cells (Zimmerman and Sutter, 1983), and on cultured human melanoma (Fabricant et al., 1977; Riopelle et al., 1983), mouse neuroblastoma (Revoltella et al., 1974) and PC12 cells (Herrup and Thoenen, 1979). Evidence for NGF receptors in brain has been reported by Frazier et al. (1974), Szutowicz et al. (1976) and Zimmerman et al. (1978). However, some NGF receptors detected in homogenates of CNS tissue may actually be derived from central projections of peripheral neurons (Richardson and Riopelle, 1984). Sutter et al. (1979b) have described two saturable binding sites on intact chick embryo DRG cells with dissociation equilibrium constants of $2 \times 10^{-11} M$ (site I) and $2 \times 10^{-9} M$ (site II). These two affinities of binding are not the result of negatively cooperative interactions (Sutter et al., 1979b; Riopelle et al., 1980) but, as will be reviewed below, may be due to conformational changes in a single receptor molecule. It is the interaction of NGF with site I receptors that has been strongly implicated in the mechanism responsible for the initiation of neurite outgrowth (Sutter et al., 1979a). The role of site II receptors in mediating a biological response to NGF is not yet clear. Specific binding of ^{125}I-NGF is not displaced by insulin and several other peptides (Sutter et al., 1979b; Herrup and Shooter, 1973), and is calcium-dependent and trypsin-sensitive (Banerjee et al., 1975). NGF receptors from DRG and SCG can be solubilized with Triton X-100 (Banerjee et al., 1976). Herrup and Shooter (1975) and Rohrer and Barde (1982) have reported a decrease in the number of NGF binding sites on chick embryo DRG neurons that parallels the loss in biological responsiveness to NGF observed with these cells between days 16 and 18 of incubation. However, retrograde transport of NGF in both sensory and sympathetic neurons persists in the adult, and the presence of NGF receptors along primary sensory axons in the adult rat has recently been demonstrated (Richardson and Riopelle, 1984), implying that sufficient receptors remain in adult to subserve and mediate an NGF effect. When membrane preparations from chick embryo DRG are used, internal binding sites for NGF (site III) are exposed, and these sites have a dissociation equilibrium constant of at least $10^{-6} M$ (Riopelle et al., 1980). As with site II receptors, the biological role of site III receptors has not yet been determined.

Available data suggest that binding sites I and II are two conformations of the same receptor molecule species. Cohen et al. (1980) observed that the binding affinity of NGF to both sites I and II was reduced to the same degree when NGF with oxidized tryptophan residues was used. It has recently been established that the NGF receptor can be converted from a trypsin-sensitive, rapidly dissociating state (site II) to a trypsin-resistant, slowly dissociating state (site I) by wheat germ agglutinin (Buxser *et al.*, 1983). By cross-linking to ^{125}I-NGF, followed by gel electrophoresis, the NGF receptor has been found to be a glycoprotein of approximately 85,000 daltons (Puma *et al.*, 1983; Grob *et al.*, 1983). Site III, as it is not seen when intact cells are used, is probably a separate site that does not participate with the other binding components (Riopelle *et al.*, 1980).

In contrast to the wealth of information that is available on the interaction of NGF with its cellular receptor site, very little is known of the interaction of other diffusible neurotrophic factors and ECM molecules with integral receptor molecules on the surface of growth cones. One of the problems encountered, however, is that the ECM molecules with which cells interact are large and are highly complexed, and it has been only relatively recently (compared to NGF) that homogeneous preparations of ECM molecules have been available for such analysis. In addition, the cellular binding domains for only one of these ECM molecules, fibronectin (Carbonetto *et al.*, 1983; Pierschbacher and Ruoslahti, 1984), have been characterized. The availability of synthetic peptide sequences of ECM molecules that bind to cells will aid in the characterization of cell and neuronal receptors for these species.

V. THEORIES OF PATTERNING AND EXQUISITE MAPPING IN NERVOUS SYSTEM DEVELOPMENT

In the previous sections, a review of the molecular substrates of neuronal interaction with the extracellular environment—diffusible factors and immobilized materials within the glycocalyx and the extracellular matrix—has been presented. In this section, we will review the data suggesting a role for these interactions in neural patterning—in other words, the cell–cell interactions that define pathway patterns and the complexity of synaptic connectivity of the nervous system. These interactions involving neurons must be understood within a framework that involves a sequential program of the primary processes of neuronal development: cell division and generation of many more neurons than will ultimately survive, neuronal differentiation, aggregation, adhesion, migration, connectivity, death of neurons failing to compete for target influence (Hamburger and Oppenheim, 1982), and synaptic elimination and consolidation.

Prior to the 1940s, the view of Weiss (1936) that neuronal specificity grew out of selection via activity (the resonance theory) was widely prevalent. These ideas were subjected to scrutiny by Sperry (1963), who perceived that theories of neural patterning must ultimately include a description of each of the key morphogenetic mechanisms at the molecular level. This led, in 1963, to Sperry's proposal to account for developmental patterning of the nervous system by the chemoaffinity hypothesis. According to the hypothesis, the cells and processes of the nervous system carry individual chemical identification markers distinguishable almost to the level of single cells. Complementary marker pairs would determine in precise detail the interconnection of the vast array of neurons and their processes, and thus, it was implied that a large number of marker pairs existed. This hypothesis also implied preexisting diversity and a static array of complementary markers that was not significantly influenced by the changing milieu encountered by the developing cells.

Since 1963, there has been little advance made in detecting those molecular species predicted to be markers unique to neurons that also play a functional role in recognition processes. Furthermore, a number of observations made by Sperry and his colleagues (1963), and subsequently by other investigators, could not be predicted by strict axon–target chemoaffinity theories. Attardi and Sperry (1963), using the regenerating retino–tectal system, found that regenerating retinal ganglion neurons reinnervated appropriate areas of tectum, thus supporting the idea of specific interactions between complementary marker pairs on retinal ganglion growth cones and tectal neurons. Since this was a regenerative paradigm, however, it remains conceivable that previous invasion of the tectum by retinal ganglion cell axons and the previously innervated tectal cells themselves could have imprinted the milieu into which the regenerating axons grew, such that the new growth cones recognized a substrate of growth that had been modified and now was highly favorable and provided specific cues. This interpretation would, in part, also suffice to explain the observation made in these experiments by Attardi and Sperry (1963), and subsequently by other investigators on guidance of regenerating axons (Fujisawa *et al.,* 1981; Scholes, 1979; Scholes, 1981), that the regenerating axons within the optic nerve were highly ordered and made appropriate decussations as they elongated toward the denervated tectum. That neurons and their axons can modify and imprint the milieu into which they grow has been demonstrated both *in vitro* and *in vivo,* and these data will be reviewed in a subsequent section. Pertinent to this discussion, however, is the observation that *in vivo* axons modify their matrix of growth in the region of a neuromuscular junction, such that a regenerating axon can be triggered to display presynaptic specializations when the matrix of growth previously in juxtaposition to a neuromuscular junction is reached during axonal elongation (Sanes and Chiu, 1983).

Two other observations that might not be predicted by strict growth cone–target chemoaffinity theories suggest that patterning and exquisite mapping may involve other mechanisms. During the development of the retino–tectal system, there is evidence to suggest that retinal ganglion axons are already topographically ordered as they leave the region of the eye and enter the optic nerve (Holt and Harris, 1983). This suggests that a certain degree of ordering intrinsic to the growth cones and independent of target influence is playing a role in pathway patterning. While at least one gradient of a molecular species that defines topography within the retina has been demonstrated by Trisler *et al.* (1981), the particular molecular species involved has not been shown to date to have a functional role in ordering of retinal ganglion cell axonal growth during development.

A second observation now made by a number of investigators (Gaze and Sharma, 1970; Yoon, 1971; Schmidt *et al.*, 1978) on regeneration of retinal ganglion cell axons from incomplete retinae also might not be predicted by the lock-and-key cell–cell interaction predicted by strict chemoaffinity theories. Although the regeneration of retinal ganglion axons to tectal targets is precise initially, with time the field of innervation spreads to include those targets previously innervated by retinal ganglion axons that are now absent. A lock-and-key arrangement with one-to-one matching predicted by strict chemoaffinity theories might not permit this plasticity to occur.

These observations suggest that to view neural specificity by the matching of recognition molecules on neuronal surfaces is inadequate. Evidence from a number of systems (see Purves and Lichtman, 1983, for review) suggests that the final connectivity pattern of neurons is the result of temporal and spatial opportunities, recognition at the level of axonal guidance and connectivity, and competition. Current theories of pattern formation and specific connectivity have begun to address these issues.

More parsimonious views of axonal pathway patterning and specificity of synaptic connectivity in nervous system development have arisen recently from two very different experimental approaches. In 1976, Edelman (1976) suggested that a relatively small number of surface molecules responsible for cell adhesion could generate exquisite patterning provided that modulatory changes consisting of number, distribution, or chemical alteration occurred in the adhesion molecules during development. According to the modulation or regulator hypothesis (Edelman, 1984b), local (versus global) cell-surface modulation of adhesion molecules would, by changing directly or indirectly their binding behaviour, influence directly or indirectly the dynamics and interactions of the primary processes of development alluded to above, thereby resulting in a change in pattern and form. In support of the hypothesis, Edelman and his colleagues (Edelman, 1984b; Edelman *et al.*, 1983a) have identified several different cell adhesion molecules in different tissues of a number of vertebrate species. A family of molecules that appears to be specific to the nervous system has been identified by immunological techniques and plays a role in neurite fasciculation (Edelman *et al.*, 1983a; Buskirk *et al.*, 1980) and neuromuscular interaction (Grumet *et al.*, 1982). In early embryogenesis at the time of neural induction, immunoreactivity to these molecules becomes restricted to neural plate structures (Edelman *et al.*, 1983b).

The nervous system cell adhesion molecule known as N-CAM is microheterogeneous with an apparent molecular weight of 200,000–250,000 in its embryonic form (Edelman *et al.*, 1983b). N-CAM is a glycoprotein containing large amounts of sialic acid (Edelman *et al.*, 1983a). During development there is a conversion to adult forms of the glycoprotein that migrate on polyacrylamide gels as bands of 180, 140, and 120 kDa (Hoffman *et al.*, 1982). The major age-dependent and region-dependent change in the glycoprotein is in the amount of sialic acid (Hoffman and Edelman, 1983).

N-CAM acts as an adhesion molecule by homophilic binding; a ligand domain of N-CAM on one cell binds to a "receptor" domain of N-CAM on another cell in a Ca^{2+}-independent manner (Hoffman and Edelman, 1983) via a region in the NH_2-terminal domain. The COOH-terminal domain is inserted into the plasma membrane, while the middle region contains the sialic acid (Cunningham *et al.*, 1983; Edelman *et al.*, 1983a). The rate of homophilic binding varies inversely with the amount of sialic acid present on N-CAM. In addition to the effect of this intrinsic qualitative change in N-CAM, quantitative changes in surface concentration of N-CAM lead to marked changes in rate of binding (Hoffman and Edelman, 1983). The mechanisms whereby N-CAMs are converted or regulated would appear to be under genetic control since they are time-, age-, and region-dependent, but whether sialidases or glycosyltransferases or other regulatory enzymes are involved remains to be determined. On a developmental framework, the implication from these observations of modulation of N-CAM prevalence and chemical structure and the related changes in binding is that cell-surface modulation would be expected to have primary effects on cell interaction and cell migration, possibly secondary and indirect effects on cell division, differentiation, and cell death, and ultimately, an influence on tissue pattern and form. Also implied is that the same mechanisms involved in patterning could be used to explain specific synaptic connectivity.

The modulation or regulator hypothesis suggested by Edelman (1984b) begins to address the fact that normal recognition during both axonal guidance and synaptic connectivity is relative. Recognition provides a bias rather than the absolute certainty implied by strict chemoaffinity phenomena.

Predictions arising from the modulation or regulator hypothesis are that the number of nervous system CAMs will be small and shared by vertebrate species. N-CAMs were initially defined in the chick nervous system (Thiery *et al.*, 1977) and subsequently found in rat and mouse nervous systems (Chuong *et al.*, 1982) and in the human nervous system (McClain and Edelman, 1982). Goridis and his colleagues (1983) have described a surface glycoprotein, BSP-2, in the mouse which shares the properties of N-CAM. Similarly, Schachner and her colleagues (1983) have detected two surface antigens of developing murine cerebellum denoted as L1 and L2. The L1 antigen appears to be responsible for the interaction of murine cerebellar granule neurons with Bergmann glia *in vivo* and functionally would appear to be the equivalent of neuron–glia CAM (Ng-CAM) described by Grumet *et al.* (1983). Recently, Riopelle *et al.* (1986) have demonstrated that a monoclonal antibody, HNK-1, which recognizes a carbohydrate epitope on myelin-associated glycoprotein (MAG) of the mature human nervous system, also recognizes neuronal surface proteins, some of which play a role in the interaction of neurons with certain immobilized molecular species of the extracellular matrix. Finally, Kruse *et al.* (1984) have made the important observation that the monoclonal antibodies L2 and HNK-1 recognize similar epitopes on BSP-2, N-CAM, L1, and MAG. These data would support the prediction that there is a highly conserved family of glycoproteins on neurons of the developing nervous system that share similar properties of acting as integral membrane molecules, and that there is intramolecular diversity of domains promoting various forms of neuronal adhesion. Implied, therefore, is that a restricted number of related glycoproteins could be involved in various aspects of adhesion, and that a series of posttranslational events take place by virtue of developmentally regulated cell–cell interactions that confer upon these glycoproteins dynamic heterofunctional adhesive properties.

Recent observations of cellular interaction in the developing nervous system of the grasshopper, coupled with monoclonal antibody techniques, have given rise to a hypothesis of selective fasciculation to explain the development of intricate connectivity within the nervous system. Goodman and his colleagues (1984) have observed that filopodia of developing grasshopper neurons mediate cell recognition by adhering to other cell surfaces, and that the selectivity displayed by these filopodia give rise to stereotyped patterns of selective fasciculation in which growth cones confronted with a scaffolding of axon bundles choose particular bundles along which to extend. Implied, therefore, is that cell recognition during neuronal development involves temporal and spatial expression of different molecules, and that the modulation of the growth-cone surface is due to phenotypic changes occurring as the result of previous interactions with the changing matrix of growth. Goodman and colleagues have described monoclonal antibodies that reveal cell-surface antigens whose temporal and spatial distribution in the embryo correlate with one of the predictions of the cellular studies, namely that neurons whose axons fasciculate together share common surface antigens.

There are a number of implied similarities to the hypotheses advanced by Edelman and colleagues and by Goodman and colleagues. The selective fasciculation model perhaps implies homophilic interactions that have been demonstrated for the N-CAMs. Both hypotheses imply modulation of cell-surface adhesion molecules resulting from cellular interaction. While the approach taken by Edelman and his colleagues demonstrates regional changes in N-CAMs and implies local modulatory phenomena to account for pattern, form, and precise synaptic mapping, the observations of Goodman and colleagues, while implying perhaps similar modulations, demonstrate temporal and spatial heterogeneity of antigens at the single-cell level.

Virtually all theories of patterning and exquisite mapping in neurogenesis are based upon cell–cell interactions. Ultimately, integral receptor molecules of growth cones interact with membrane proteins within the PM-GC complex. Within these models, roles for diffusible factors, or indeed other elements of the ECM, have not been given a prominent profile. Despite *in vitro* observations of directionality and guidance, the ubiquity of diffusible and immobilized factors within the extracellular milieu, excluding the glycocalyx, might suggest that they play a permissive role in patterning by facilitating axonal guidance and access to targets.

Nevertheless, *in vitro* observations on a chemotactic influence of NGF and of other non-specified trophic materials (diffusible or substrate-bound), as well as observations implicating the target of innervation as a source of neurotrophic materials, have given rise to theories of axonal guidance mediated by gradients of these factors (for example, see Bonhoeffer and Gierer, 1984). A number of constraints must be placed upon theories attempting to explain patterning in nervous system development on the basis of target-derived factors. First, the distances over which gradients have been shown to be effective are very small, and thus, if gradients act *in vivo* as directional cues, they would be more likely to be maximally effective in early neurogenesis where the distances between first-derived neurons and their potential targets are exceedingly small. At this stage of development, targets are primordial, and thus one would have to conclude that factors are released from primordial targets (Riopelle and Cameron, 1981); *in vitro* demonstrations of axonal guidance involving target-released factors have made use of mature tissue. In addition, except for NGF, virtually all target-derived influences involved in axonal guidance have not been completely specified. It is now accepted that many tissues produce both diffusible and substrate-attached materials, and that a common component to many is laminin (Lander *et al.*, 1985). Intuitively, one might expect that a substrate-attached factor produced by targets could form a fairly stable gradient; however, the same cannot be said for the so-called diffusible factors such as NGF. No experimental data are available at this time to determine whether the so-called diffusible factors are rendered semidiffusible by interactions with molecular species of the ECM, such that they function similarly to laminin and fibronectin as nectins or trophic influences bridging between the growth cone and the substrate. If NGF or similar factors were partially immobilized, it could set the stage for establishment of stable gradients that might act over relatively large distances.

Finally, as reviewed earlier, evidence is emerging to suggest that the importance of the target as a source of trophic influence for neurons may have been overestimated. Recent experiments (Riopelle *et al.*, 1981; Richardson and Ebendal, 1982) have demonstrated that peripheral nerve contains both NGF and non-NGF neurotrophic factors. Using a cDNA probe for NGF, NGF messenger RNA has been detected within endoneurial tissue (Shelton and Reichardt, 1984), and receptors for this trophic factor have been found to be displayed along peripheral axons of primary sensory neurons (Richardson and Riopelle, 1984). These findings suggest that endoneurial cells that are intimately associated with axons throughout their length even during development can provide neurotrophic material, and that axonal receptors for these materials are strategically juxtaposed to the source in order to permit uptake and subsequent retrograde transport of the material to the neuronal cell body.

VI. NEURONAL MODIFICATION OF THE EXTRACELLULAR MILIEU

If parsimonious theories of patterning in nervous system development are valid, then the number of adhesive proteins will likely be restricted, possibly bifunctional (i.e., containing both ligand and receptor domains), and it will be possible to explain the diversity demanded by exquisite mapping only by invoking a series of modulations of the adhesive proteins occurring as a result of previous interactions of cells bearing the proteins—these modulations thereby altering the biases within the recognition process. Implied is that there must be both temporal and spatial structural diversity within the family of adhesion molecules acting as substrates for regulatory gene products that will function as enzymes. Observations to date would support these predictions. The best characterized of the adhesion proteins, N-CAM, appears to be a receptor/ligand bifunctional species; furthermore, all of the molecular species of the nervous system found to have adhesive properties are glycoproteins, and as reviewed above, related glycoproteins subserve different adhesive phenomena.

Even within single glycoproteins, oligosaccharides exhibit considerable microheterogeneity (Berger *et al.,* 1982). Glycoproteins are formed by the sequential action of glycosyltransferases, and thus the product of one glycosyltransferase is utilized as an acceptor substrate by another glycosyltransferase; no template, as in the case of protein biosynthesis, is utilized in the elongation process. Thus, oligosaccharide structure will likely be determined by a host of factors, including the timing of expression, type, and number of glycosyltransferases produced by a cell, the relative amounts of various glycosyltransferases that will determine which biosynthetic pathway will predominate, and the amounts of glycosyltransferase that will determine whether a reaction will go to completion. These considerations, coupled with the concurrent processing of oligosaccharides by glycosidases, begin to set in place a framework whereby considerable structural diversity can be developed. Thus, it is conceivable that relatively few glycoproteins bear, on carbohydrate and polypeptide residues, a large number of adhesion domains—either as receptor or as ligand—as well as a diverse array of substrates for regulatory enzymes that could modify these glycoproteins directly at the receptor or ligand sites, or indirectly at other sites which could influence receptor and ligand conformation. The observations of Edelman and colleagues (1983a) demonstrating an influence of sialic acid residues on N-CAM binding rates are examples of indirect influences on the functional properties of this adhesion molecule. The regulatory gene products that can alter the adhesive properties of glycoproteins both directly and indirectly will likely turn out to be neuron-produced glycosidases, glycosyltransferases (Edelman, 1984a), or "creative" proteolytic enzymes (Loh et al., 1984). Similar regulatory enzymes will likely be found to subserve modulation of neuron–glia adhesive interactions. Evidence is emerging to suggest that neurons can modify their growth milieu enzymatically. Krysostek and Seeds (1981) have demonstrated that neurons release to their microenvironment protease activity that is a fibrinogen activator. Similar observations have been made by Kalderon (1979) and Moonen et al. (1982). More recently, Pittman (1985) has identified two protein kinases released by cultured neurons: one a 62,000-dalton metalloprotease, and the other a 51,000-dalton urokinase.

Other neuronal modifications of the milieu of growth also occur; Riopelle and Cameron (1984) have demonstrated that neurons can release to their microenvironments nondiffusible factors that promote neurite extension. In addition, evidence from a regenerating frog neuromuscular system referred to earlier suggests that the BL in the region of a previous neuromuscular junction is imprinted by previous axonal association, such that the regenerating axon specializes its presynaptic region when this part of the BL is contacted by the growth cone (Sanes and Chiu, 1983). Both by contributing to the milieu of growth and by altering the milieu through which they pass, neurons and their axons could be modifying their environment in ways that begin to address the diversity of interactions demanded to establish the normal patterns of nervous system development and connectivity.

VII. A UNIFYING HYPOTHESIS

At the present time there is little evidence *in vivo* to support the notion that specificity of connectivity involves instructive signals provided by extracellular molecular species that are not closely associated with the PM-GC complex. More plausible is the idea that these extracellular molecular species act permissively to permit or facilitate access, as would fuel stations (diffusible factors) and highways (ECM), and thus, play a major role in axonal guidance and pathway patterning, but that modulations of adhesive interactions of molecular species more closely associated with the PM-GC complex are responsible for instructive signals that subserve exquisite pathway patterning and specific synaptic connectivity. Once connectivity has been established, the diffusible and immobilized non-PM-GC complex molecules derived from targets and glial elements may serve to protect and maintain the functional integrity of the neurons and axons with which they are intimately associated.

Figure 5 attempts to schematize these ideas in reference to the plasma membrane-glycocalyx-ECM of the neuron and its environment by suggesting that, in relation to neuronal connectivity, as growth-cone interactions with the environment become more intimately associated with the membrane of a contiguous neuron, less permissive but more modulatory, and thus more instructive and specific, influences upon the neuronal growth cone are effected.

The authors recognize that the vast and rapidly expanding areas of study devoted to the molecular substrates of neuronal development have only been highlighted in this chapter because of the constraints imposed by the present format. A number of recent reviews covering various aspects of neuronal development are available. To facilitate literature perusal by the interested reader, reference citations have been oriented towards these reviews where possible (Barde *et al.*, 1983; Berg, 1984; Berger *et al.*, 1982; Carbonetto, 1984; Edelman, 1983, 1984a and b; Edelman *et al.*, 1983a; Goodman *et al.*, in press; Greene and Shooter, 1980; Greene, 1984; Landis, 1983; Letourneau, 1983; Letourneau *et al.*, 1980; Loh *et al.*, 1984; Mobley *et al.*, 1977; Purves and Lichtman, 1983).

ACKNOWLEDGMENTS

The authors wish to acknowledge the support of the following agencies: Medical Research Council of Canada (R.J.R.), The Physicians' Services Incorporated Foundation of Ontario (K.E.D. and R.J.R.), The Hospital for Sick Children Foundation (K.E.D. and R.J.R.), and the Clare Nelson Bequest of the Kingston General Hospital (R.J.R.). We are indebted to S. McCaughey for typing the manuscript.

REFERENCES

Angeletti, P. U., Levi-Montalcini, R., and Caramia, F. (1971). *J. Ultrastruct. Res.* **36**, 24–36.
Attardi, D. G., and Sperry, R. W. (1963). *Exp. Neurol.* **7**, 46–64.
Banerjee, S. P., Snyder, S. H., Cuatrecasas, P., and Greene, L. A. (1973). *Proc. Natl. Acad. Sci. U.S.A.* **70**, 2519–2523.
Banerjee, S. P., Cuatrecasas, P., and Snyder, S. H. (1975). *J. Biol. Chem.* **250**, 1427–1433.
Banerjee, S. P., Cuatrecasas, P., and Snyder, S. H. (1976). *J. Biol. Chem.* **251**, 5680–5685.
Banthorpe, D. V., Pearce, F. L., and Vernon, C. A. (1974). *J. Embryol. Exp. Morphol.* **31**, 151–167.
Barde, Y.-A., Edgar, D., and Thoenen, H. (1982). *EMBO J.* **1**, 549–553.

Barde, Y.-A., Edgar, D., and Thoenen, H. (1983). *Annu. Rev. Physiol.* **45**, 601–612.

Berg, D. K. (1984). *Annu. Rev. Neurosci.* **7**, 149–170.

Berger, E. G., Buddecke, E., Kamerling, J. P., Kobata, A., Paulson, J. C., and Vliegenthart, J. F. G. (1982). *Experientia* **38**, 1129–1258.

Bjerre, B., Bjorklund, A., and Stenevi, U. (1973). *Brain Res.* **60**, 161–176.

Bjorklund, A., and Stenevi, U. (1972). *Science* **175**, 1251–1253.

Bonhoeffer, F., and Gierer, A. (1984). *Trends Neurosci.* **7**, 378–381.

Bray, D. (1970). *Proc. Natl. Acad. Sci. U.S.A.* **65**, 905–910.

Bray, D. (1973). *J. Cell Biol.* **56**, 702–712.

Bray, D., and Gilbert, D. (1981). *Annu. Rev. Neurosci.* **4**, 505–523.

Bunge, M. B. (1973). *J. Cell Biol.* **56**, 713–735.

Bunge, R. P., and Bunge, M. B. (1983). *Trends Neurosci.* **6**, 499–505.

Burnham, P. A., and Varon, S. (1974). *Neurobiology* **4**, 57–70.

Burstein, D. E., and Greene, L. A. (1978). *Proc. Natl. Acad. Sci. U.S.A.* **75**, 6059–6063.

Buskirk, D. R., Thiery, J.-P., Rutishauser, U., and Edelman, G. M. (1980). *Nature (London)* **285**, 488–489.

Buxser, S. E., Kelleher, D. J., Watson, L., Puma, P., and Johnson, G. L. (1983). *J. Biol. Chem.* **258**, 3741–3749.

Carbonetto, S. (1984). *Trends Neurosci.* **7**, 382–387.

Carbonetto, S., Gruver, M. M., and Turner, D. C. (1983). *J. Neurosci.* **3**, 2324–2335.

Chuong, C.-M., McClain, D. A., Streit, P., and Edelman, G. M. (1982). *Proc. Natl. Acad. Sci. U.S.A.* **79**, 4234.

Cohen, A. I., Nicol, E. C., and Richter, W. (1964). *Proc. Soc. Exp. Biol. Med.* **116**, 784–789.

Cohen, P. A., Sutter, G., Landreth, A., Zimmerman, A., and Shooter, E. M. (1980). *J. Biol. Chem.* **255**, 2949–2954.

Cohen, S. (1960). *Proc. Natl. Acad. Sci. U.S.A.* **46**, 302–311.

Collins, F., and Garrett, J. E. (1980). *Proc. Natl. Acad. Sci. U.S.A.* **77**, 6226–6228.

Cunningham, B. A., Hoffman, S., Rutishauser, U., Hemperly, J. J., and Edelman, G. M. (1983). *Proc. Natl. Acad. Sci. U.S.A.* **80**, 3116–3120.

Dichter, M. A., Tischler, A. S., and Greene, L. A. (1977). *Nature (London)* **268**, 501–504.

Ebendal, T. (1976). *Exp. Cell Res.* **98**, 159–169.

Edelman, G. M. (1976). *Science* **192**, 218–226.

Edelman, G. M. (1983). *Science* **219**, 450–457.

Edelman, G. M. (1984a). *Trends Neurosci.* **7**, 78–84.

Edelman, G. M. (1984b). *Proc. Natl. Acad. Sci. U.S.A.* **81**, 1460–1464.

Edelman, G. M., Gallin, W. J., Delouvee, A., Cunningham, B. A., and Thiery, J.-P. (1983a). *Proc. Natl. Acad. Sci. U.S.A.* **80**, 4384–4388.

Edelman, G. M., Hoffman, S., Chuong, C.-M., Thiery, J.-P., Brackenbury, R., Gallin, W. J., Grumet, M., Greenberg, M., Hemperly, J. J., Cohen, C., and Cunningham, B. A. (1983b). *Cold Spring Harbour Symp. Quant. Biol.* **48**, 515–526.

Fabricant, R. N., DeLarco, J. E., and Todaro, G. J. (1977). *Proc. Natl. Acad. Sci. U.S.A.* **74**, 565–569.

Feldman, E. L., Axelrod, D., Schwartz, M., Heacock, A. M., and Agranoff, B. W. (1981). *J. Neurobiol.* **12**, 591–598.

Frazier, W. A., Boyd, L. F., Pulliam, M. W., Szutowicz, A., and Bradshaw, R. A. (1974). *J. Biol. Chem.* **249**, 5918–5923.

Fujisawa, H., Watanabe, K., Tani, N., and Ibata, Y. (1981). *Brain Res.* **206**, 21–26.

Gaze, R. M., and Sharma, S. C. (1970). *Exp. Brain Res.* **10**, 171–181.

Gnahn, H., Hefti, F., Heumann, R., Schwab, M. E., and Thoenen, H. (1983). *Brain Res.* **285**, 45–52.

Goldman, R. D., Milsted, A., Schloss, J. A., Starger, J., and Yerna, M.-J. (1979). *Annu. Rev. Physiol.* **41**, 703–722.

Goodman, C. S., Bastiani, M. J., Doecq, duLac S., Helfand, S., Kuwada, J. Y., and Thomas, J. B. (1984). *Science* **225**, 1271–1279.

Goridis, C., Deagostini-Bazin, H., Hirn, M., Hirsch, M.-R., Rougon, G., Sadoul, R., Langley, O., Gombos, G., and Finne, J. (1983). *Cold Spring Harbour Symp. Quant. Biol.* **48**, 527–537.

Greene, L. A. (1978). *J. Cell Biol.* **78**, 747–755.

Greene, L. A. (1984). *Trends Neurosci.* **7**, 91–94.

Greene, L. A., and Shooter, E. M. (1980). *Annu. Rev. Neurosci.* **3**, 353–402.

Greene, L. A., and Tischler, A. S. (1976). *Proc. Natl. Acad. Sci. U.S.A.* **73**, 2424–2428.

Greene, L. A., Varon, S., Piltch, A., and Shooter, E. M. (1971). *Neurobiology.* **1**, 37–48.

Grob, P. M., Berlot, C. H., and Bothwell, M. A. (1983). *Proc. Natl. Acad. Sci. U.S.A.* **80**, 6819–6823.

Grumet, M., Rutishauser, U., and Edelman, G. M. (1982). *Nature (London)* **295**, 693–695.

Grumet, M., Rutishauser, U., and Edelman, G. M. (1983). *Science* **222**, 60–62.

Gunderson, R., and Barrett, J. (1980). *J. Cell Biol.* **87**, 546–554.

Hamburger, V., and Oppenheim, R. W. (1982). *Neurosci. Commentaries* **1**, 39–55.

Hamburger, V., and Yip, Y. W. (1984). *J. Neurosci.* **4**, 767–774.

Harrison, R. G. (1910). *J. Exp. Zool.* **9**, 787–846.

Hay, E. D. (1981). *J. Cell Biol.* **91**, 205–223.

Heathcote, J. G., and Grant, M. E. (1981). *Int. Rev. Connect. Tissue Res.* **9**, 191–264.

Hendry, I. A. (1975). *Brain Res.* **94**, 87–97.

Hendry, I. A., and Campbell, J. (1976). *J. Neurocytol.* **5**, 351–360.

Hendry, I. A., Stoeckel, K., Thoenen, H., and Iversen, L. L. (1974). *Brain Res.* **86**, 103–121.

Herrup, K., and Shooter, E. M. (1973). *Proc. Natl. Acad. Sci. U.S.A.* **70**, 3384–3388.

Herrup, K., and Shooter, E. M. (1975). *J. Cell Biol.* **67**, 118–125.

Herrup, K., and Thoenen, H. (1979). *Exp. Cell Res.* **121**, 71–78.

Hoffman, S., and Edelman, G. M. (1983). *Proc. Natl. Acad. Sci. U.S.A.* **80**, 5762–5766.

Hoffman, S., Sorkin, B. C., White, P. C., Brackenbury, R., Mailhammer, R., Rutishauser, U., Cunningham, B. A., and Edelman, G. M. (1982). *J. Biol. Chem.* **257**, 7720–7729.

Holt, C., and Harris, W. (1983). *Nature (London)* **301**, 150–152.

Isenberg, G., and Small, J. V. (1978). *Cytobiologie* **16**, 326–344.

Jacobson, M. (1978). *In* "Developmental Neurobiology," 2nd ed., pp. 253–307. Plenum Press, New York.

Jockusch, H., and Jockusch, B. N. (1981). *Exp. Cell Res.* **131**, 345–352.

Johnson, E. M., Jr., Andres, R. Y., and Bradshaw, R. A. (1978). *Brain Res.* **150**, 319–331.

Kalderon, N. (1979). *Proc. Natl. Acad. Sci. U.S.A.* **76**, 5992–5996.

Kefalides, N. A., Alper, R., and Clark, C. C. (1979). *Int. Rev. Connect. Tissue Res.* **61**, 167–228.

Kruse, J., Mailhammer, R., Wernecke, H., Faissner, A., Sommer, I., Goridis, C., and Schachner, M. (1984). *Nature (London)* **311**, 153–155.

Krysostek, A., and Seeds, N. (1981). *Science* **213**, 1532–1534.

Kuczmarski, E. R., and Rosenbaum, J. L. (1979). *J. Cell Biol.* **80**, 356–371.

Lander, A. D., Fujii, D. K., and Reichardt, L. F. (1985). *Proc. Natl. Acad. Sci. U.S.A.* **82**, 2183–2187.

Landis, S. C. (1983). *Annu. Rev. Physiol.* **45**, 567–580.

Letourneau, P. C. (1981). *Dev. Biol.* **85**, 113–122.

Letourneau, P. C. (1982). *J. Neurosci.* **2**, 806–814.

Letourneau, P. C. (1983). *Trends Neurosci.* **6**, 451–455.

Letourneau, P. C., Ray, P. N., and Bernfield, M. R. (1980). *In* "Biological Regulation and Development" (R. Goldberg, ed.), Vol. 2, pp. 339–376. Plenum Press, New York.

Levi-Montalcini, R. (1966). *Harvey Lect.* **60**, 217–259.
Levi-Montalcini, R., and Angeletti, P. U. (1963). *Dev. Biol.* **7**, 653–657.
Levi-Montalcini, R., and Angeletti, P. U. (1968). *Physiol. Rev.* **48**, 534–569.
Levi-Montalcini, R., and Hamburger, V. (1951). *J. Exp. Zool.* **116**, 321–362.
Levi-Montalcini, R., Meyer, H., and Hamburger, V. (1954). *Cancer Res.* **14**, 49–57.
Loh, Y. P., Brownstein, M. J., and Gainer, H. (1984). *Annu. Rev. Neurosci.* **7**, 189–222.
Luduena, M. A., and Wessells, N. K. (1973). *Dev. Biol.* **30**, 427–440.
Luft, J. H. (1976). *Int. Rev. Cytol.* **45**, 291–382.
McClain, D. A., and Edelman, G. M. (1982). *Proc. Natl. Acad. Sci. U.S.A.* **79**, 6380.
McGuire, J. C., and Greene, L. A. (1979). *J. Biol. Chem.* **254**, 3362–3367.
McGuire, J. C., Greene, L. A., and Furano, A. V. (1978). *Cell* **15**, 357–365.
Manthorpe, M., Barbin, G., and Varon, S. (1982a). *J. Neurosci. Res.* **8**, 233–239.
Manthorpe, M., Skaper, S. D., Barbin, G., and Varon, S. (1982b). *J. Neurochem.* **38**, 415–421.
Max, S. R., Schwab, M., Dumas, M., and Thoenen, H. (1978). *Brain Res.* **159**, 411–415.
Mobley, W. C., Server, A. C., Ishii, D. N., Riopelle, R. J., and Shooter, E. M. (1977). *N. Engl. J. Med.* **297**, 1096–1104, 1149–1158, 1211–1218.
Moonen, G., Grau-Wagemans, M. P., and Selak, I. (1982). *Nature (London)* **298**, 753–755.
Paravicini, U., Stoeckel, K., and Thoenen, H. (1975). *Brain Res.* **84**, 279–291.
Partlow, L. M., and Larabee, M. G. (1971). *J. Neurochem.* **18**, 2101–2118.
Patel, N. B., and Poo, M. M. (1984). *J. Neurosci.* **4**, 2939–2947.
Pearlman, A. L., Stewart, G. R., and Cohen, J. P. (1984). *Soc. Neurosci. Abstr.* **10**, 39.
Pierschbacher, M. D., and Ruoslahti, E. (1984). *Nature (London)* **309**, 30–33.
Pittman, R. (1985). *Dev. Biol.* **110**, 91–101.
Puma, P., Buxser, S. E., Watson, L., Kelleher, D. J., and Johnson, G. L. (1983). *J. Biol. Chem.* **258**, 3370–3375.
Purves, D., and Lichtman, J. W. (1983). *Annu. Rev. Physiol.* **45**, 553–565.
Ramon y Cajal, S. (1904). *In* "Histologie du Systeme Nerveux de L'Homme et des Vertebres", Vol. I., pp. 597–598, 603, 604, 609. Consejo Superior de Investigaciones Cientificas, Madrid.
Revoltella, R., Bertolini, L., Pediconi, M., and Vignetti, E. (1974). *J. Exp. Med.* **140**, 437–451.
Richardson, P. M., and Ebendal, T. (1982). *Brain Res.* **246**, 57–64.
Richardson, P. M., and Riopelle, R. J. (1984). *J. Neurosci.* **4**, 1683–1689.
Riopelle, R. J., and Cameron, D. A. (1981). *J. Neurobiol.* **12**, 175–186.
Riopelle, R. J., and Cameron, D. A. (1984). *Dev. Brain Res.* **15**, 265–274.
Riopelle, R. J., Klearman, M., and Sutter, A. (1980). *Brain Res.* **199**, 63–77.
Riopelle, R. J., Boegman, R. J., and Cameron, D. A. (1981). *Neurosci. Lett.* **25**, 311–316.
Riopelle, R. J., Haliotis, T., and Roder, J. C. (1983). *Cancer Res.* **43**, 5184–5189.
Riopelle, R. J., McGarry, R. C., Mirski, S., and Roder, J. C. (1986). *Brain Res.* (in press).
Rohrer, H., and Barde, Y.-A. (1982). *Dev. Biol.* **89**, 309–315.
Sanes, J. R. (1983). *Annu. Rev. Physiol.* **45**, 581–600.
Sanes, J. S., and Chiu, A. Y. (1983). *Cold Spring Harbor Symp. Quant. Biol.* **48**, 667–678.
Schachner, M., Faissner, A., Kruse, J., Lindner, J., Meier, D. H., Rathjen, F. G., and Wernecke, H. (1983). *Cold Spring Harbor Symp. Quant. Biol.* **48**, 557–568.
Schmidt, J. T., Cicerone, C. M., and Easter, S. S. (1978). *J. Comp. Neurol.* **177**, 257–278.
Scholes, J. H. (1979). *Nature (London)* **278**, 620–624.
Scholes, J. H. (1981). *In* "Development in the Nervous System" (D. R. Garrod and J. D. Feldman, eds.), pp. 181–214. Cambridge University Press, Cambridge.
Schwab, M. E., Otten, U., Agid, Y., and Thoenen, H. (1979). *Brain Res.* **168**, 473–483.
Seiler, M., and Schwab, M. E. (1984). *Brain Res.* **300**, 33–39.
Shelton, D. L., and Reichardt, L. F. (1984). *Soc. Neurosci. Abstr.* **10**, 369.
Sperry, R. W. (1963). *Proc. Natl. Acad. Sci. U.S.A.* **50**, 703–710.

Stoeckel, K. M., Schwab, M., and Thoenen, H. (1975). *Brain Res.* **89,** 1–14.

Sutter, A., Riopelle, R. J., Harris-Warrick, R. M., and Shooter, E. M. (1979a). *J. Biol. Chem.* **254,** 5972–5982.

Sutter, A., Riopelle, R. J., Harris-Warrick, R. M., and Shooter, E. M. (1979b). *In* "Progress in Clinical and Biological Research 31: Transmembrane Signalling" (M. Bitensky, R. J. Collier, D. F. Steiner, and C. F. Fox, eds.), pp. 659–667. Alan R. Liss, New York.

Szutowicz, A., Frazier, W. A., and Bradshaw, R. A. (1976). *J. Biol. Chem.* **251,** 1516–1523.

Tennyson, V. M. (1970). *J. Cell Biol.* **44,** 62–79.

Thiery, J.-P., Brackenbury, R., Rutishauser, U., Edelman, G. M. (1977). *J. Biol. Chem.* **252,** 6841.

Thoenen, H., Schwab, M., and Otten, U. (1978). *Symp. Soc. Dev. Biol.* **35,** 101–118.

Thoenen, H., Korsching, S., Barde, Y.-A., and Edgar, D. (1983). *Cold Spring Harbor Symp. Quant. Biol.* **48,** 679–684.

Tischler, A. S., and Greene, L. A. (1978). *Lab Invest.* **39,** 77–89.

Trisler, G. D., Schneider, M. D., and Nirenberg, M. (1981). *Proc. Natl. Acad. Sci. U.S.A.* **78,** 2145–2149.

Turner, J. E., and Glaze, K. A. (1977). *Exp. Neurol.* **57,** 687–697.

Unsiker, K., and Chamley, J. H. (1977). *Cell Tissue Res.* **177,** 247–268.

Varon, S., and Skaper, S. D. (1983). *Trends Biochem. Sci.* **8,** 22–25.

Varon, S. S., Nomura, J., and Shooter, E. M. (1967). *Biochemistry* **6,** 2022–2029.

Weiss, P. (1934). *J. Exp. Zool.* **68,** 393–448.

Weiss, P. (1936). *Biol. Rev.* **11,** 494–531.

Winick, M., and Greenberg, R. E. (1965). *Pediatrics* **35,** 221–228.

Yamada, K. M., and Wessells, N. K. (1973). *Dev. Biol.* **31,** 413–420.

Yamada, K. M., Spooner, B. S., and Wessells, N. K. (1970). *Proc. Natl. Acad. Sci. U.S.A.* **66,** 1206–1212.

Yamada, K. M., Spooner, B. S., and Wessells, N. K. (1971). *J. Cell Biol.* **49,** 614–635.

Yoon, M. (1971). *Exp. Neurol.* **33,** 395–411.

Zimmerman, A., and Sutter, A. (1983). *EMBO J.* **2,** 879–885.

Zimmerman, A., Sutter, A., Samuelson, J., and Shooter, E. M. (1978). *J. Supramol. Struct.* **9,** 351–361.

III (CHAPTERS 9–12)

Biochemical Analysis of Cell-Surface Glycoproteins— The Search for a Structure/Function Relationship

REGINALD M. GORCZYNSKI

What is known of the biochemistry of cell surface molecules and cell association, communication, and differentiation? One of the most thoroughly explored mammalian biological differentiative systems is the lympho-hemopoietic system. This system delivers to the periphery functional blood cells to maintain the body's physiological state (homeostasis) and responds flexibly to signals that call for specific changes in representation of those peripheral cells. Rarely does deviant behavior occur (resulting in, e.g., aplasia, leukemia) in this hierarchy of cells arranged in a continuum of differentiation and commitment. Where deviance does occur it is frequently manifest at the plasma membrane, e.g., the aberrant plasma membrane found in granulocytes and erythrocytes of CML patients (a neoplasm arising in a stem cell common to both granulocytes and erythrocytes) (Kumar and Gupta, 1983). For a detailed review of differentiation and development within the hemopoietic system, the reader is referred elsewhere (Till *et al.*, 1977). However, it is important to realize at this stage that two

245

distinct forms of signal have been documented to operate within this hierarchy from pluripotential stem cells to mature end-stage cells (e.g., erythrocytes). The one, intrinsic to the process of differentiation itself, leads to the (genetic) determination of limited specialized functions available to commited progeny. Thus, analysis of the W gene in the mouse indicates that expression of an allele of this locus leads to an anemia originating in the pluripotential stem cell despite the preservation of normal function in many committed progenitors (McCulloch *et al.*, 1964). A second form of signal is of an extrinsic nature and takes the form of long-range hormonal-like messages associated with membrane-derived mediators (Krantz and Jacobson, 1970; Price and McCulloch, 1978). Interestingly, structural data on these latter high-molecular-weight stimulating factors suggest a link between them and MHC-derived molecules.

The section that follows concentrates on erythropoiesis and provides insight into the approaches used to study the relevance of surface macromolecules for intracellular communication in hemopoiesis. Following this is a discussion by Raz and Fidler detailing some of the biochemical properties that seem to be of relevance in determining metastatic potential in malignant cells.

Much of our contemporary knowledge of biological membrane structure per se comes from analysis of erythrocytes, and Gahmberg and Karhi (Chapter 9) have provided a review of the general biochemistry of some of the surface molecules at the cell surface of erythroid cells, and their possible functional significance in erythroid differentiation and development. By a variety of techniques, including radiolabeling *in situ* with galactose oxidase/NaB^3H_4 and subsequent digestion with endogalactosidase, or by the use of blood-group-specific lectins, it has been possible to demonstrate ABH activity in band 3 glycoproteins (Steck, 1974) as well as in the sialoglycoprotein, glycophorin A (Tomita and Marchesi, 1975). The majority of ABH activity in red-cell membranes may not be associated with glycolipids, as was the earlier belief, but with these glycoprotein molecules (Schenkel-Brunner, 1980). It has been established for some time that terminal N-acetylgalactosamine specifies the blood-group A antigen, and galactose the blood-group B antigen (Kabat, 1956). More recent work with AB cells suggests that there is a topographical separation in the membrane of the glycosyltransferase responsible for the biosynthesis of the terminal portions of the oligosaccharides responsible for these specificities. Internal repeating saccharide residues in these erythroglycans are responsible for the i/I antigens (Fukuda *et al.*, 1979), which are characteristic of the ontogenetic development of erythroid cells.

In contrast to the saccharide-determined specificity of A and B blood-group antigens, the MN blood-group specificities, carried on glycophorin A, are apparently a function of differences in the glycophorin polypeptide of MM or NN cells, with the same glycosylation, N-acetylneuraminic acid, $\alpha2{\rightarrow}3$-galactose(N-acetylneuraminic acid, $\alpha2{\rightarrow}6)\beta1{\rightarrow}3$-$N$-acetylgalactosamine, $\alpha1{\rightarrow}0$-Ser/Thre, occurring in each case. However, a minor sialoglycoprotein, glycophorin B, whose amino-terminal residues are identical to glycophorin A from NN cells over the first 23 amino acids, is similarly glycosylated and is present in MM and NN cells, so that MM cells invariably carry some N activity (Furthmayer, 1978). The blood-group Ss activity is determined by the amino acid interaction sequence of glycophorin B (Dahr *et al.*, 1980). The early appearance of glycophorin and blood-group A antigens during erythroid differentiation, at a stage similar to hemoglobin synthesis and their restriction to the erythroid line of hemopoiesis (Anderson *et al.*, 1979) suggest a possible physiological role for these molecules as receptors or mediators of intercellular communication during erythropoiesis.

Since glycophorin A is specific for cells of the erythroid lineage it has already proven of use in clinical diagnosis of early erythroleukemia (Ekblom *et al.*, 1983). In addition, Springer (1984) has used markers of the precursors *O*-glycosidic oligosaccharides of glycophorin A as markers of malignancy in a number of disease types.

An interesting ''spin-off'' of our detailed knowledge of the structure of the red-cell membrane and its many polymorphisms is that we are now in a position to understand at the molecular level receptor–ligand interaction in a situation where the ligand is a living (and disease-causing) entity, i.e., the malarial parasite. Recent evidence suggests that the glycophorins represent receptor molecules for the merozoite form of the parasite, with particular relevance being attached to the N-terminal sugar *N*-acetylglucosamine (Howard *et al.*, 1982; Weiss *et al.*, 1981). Building upon the foundations of studies in other fields we may thus soon be led closer to the development of reagents suitable for use in the treatment and eradication of this crippling disease.

The chapter which follows from Trowbridge (Chapter 10) provides a feel for much of the excitement concerning the putative importance of the comparative expression of cell surface determinants on the related differentiated functions of cells in the hematopoietic lineage. Many (if not all) of the cell surface antigens, now mainly detected using monoclonal antibody reagents (Kohler and Milstein, 1975), are glycoproteins, though as yet there is scant data concerning the possible functional relevance of the majority of these molecules in development and differentiation. Structural information may provide a clue in this regard (e.g., the sequence homology between Thy-1 and immunoglobulin variable-region domains suggests a likely ''recognition role'' for Thy-1), though the importance of comparative tissue distributions (amongst species) should not be overlooked. Again, citing Thy-1 as an example, the evidence for its apparent universal appearance in brain tissue (there is an homologous molecule apparently in squid brain) suggests that the brain (and not lymphocytes per se) might be the most suitable tissue to investigate to look for a physiological role for Thy-1 (Williams *et al.*, 1977). We might also make a case for the power of structure/function analysis from the recent success of such studies in defining the role of the so-called T3 molecule on human T lymphocytes. Antibodies to this glycoprotein of 20,000 daltons, which is first expressed on cells in the thymus late in development and is also present on peripheral T cells, inhibit T-cell function and can act as T-cell mitogens. More recently it has been shown that the T3 molecule co-modulates with a molecule recognized by a clonotypic antiserum directed to the putative specific recognition receptor on T cells (Meuer *et al.*, 1983). An interpretation of these data would thus envisage T3 and the specific receptor as being noncovalently bound in the functional recognition complex at the T-cell surface.

A question related to the functional role of cell-surface markers in differentiation concerns the genetic regulation of their expression. As yet this is relatively unexplored territory, though some preliminary information is available, again for the Thy-1 antigen, and possible means of approach using somatic cell hybridization techniques and gene cloning are discussed. The importance of these questions to the practice of clinical medicine is already evident from the data we have to date concerning the usefulness of monoclonal reagents to lymphocyte cell-surface antigens in the diagnosis of leukemias and lymphomas (Sallan *et al.*, 1980)—the possibility still exists for a more practical role in the therapy of malignant disease.

One key physiological role for cell-surface oligosaccharides has long been thought to be in the regulation of the cell–cell adhesion necessary for morphogenesis and organogenesis. This issue has been discussed in earlier sections of this volume. However, we should note at this stage the detailed evidence from recent studies in the immune system that specialized acceptor/receptor sites exist on endothelial cells and lymphocytes, respectively, that seem to allow for specific trafficking of lymphocytes. Monoclonal antibodies recognizing lymphocyte-derived glycoproteins (Gallatin *et al.*, 1983) that affect lymphocyte adhesion to endothelial surfaces and inhibit lymphocyte trafficking in a predictable and specific fashion have been described (Chin *et al.*, 1982). We do not yet know answers to questions such as: What is the nature of cellular interaction at the endothelial surface? Are there specialized endothelial sites that permit (and may explain) selective patterns of lymphocyte flow? Are there characteristic molecular changes in the endothelial surface in disease processes (which in turn may explain alteration in lymphocyte trafficking in disease)?

It is a small step to consider then that if normal differentiation and development is regulated by information transfer mediated by communication via cell-surface molecules, one might anticipate that the failure to regulate appropriately the local growth of a given tissue is itself a reflection of a biochemical alteration at the cell surface of cells comprising that tissue type. There are many crucial events that contribute to the development of a metastatic growth from a localized neoplasm, any or all of which may in turn be seen as causally related to changes at the tumor cell membrane. Perhaps the most dramatic connection between growth control and tumorigenesis per se that has been made in recent years has been the evidence that products of known viral oncogenes are closely related to well-described growth factors or their receptors. Thus the product of v-*sis* is related to platelet-derived growth factor (PDGF) (Deuel *et al.*, 1983) and there has now been demonstrated a remarkable similarity between the product of the v-*erb-B* oncogene and the receptor for epidermal growth factor (EGF) (Downward *et al.*, 1984).

A variety of other cell-surface changes may (secondarily) affect tumor cells to cause metastatic growth. As described by Raz and Fidler, alteration in membrane-degradative enzymes, especially fibrolytic and collagenolytic activities, may enhance release of tumor cells into the circulation (Dobrossy *et al.*, 1980). The relative loss of adhesiveness of tumor cells to one another, or to the substratum, possibly due to a decrease in the cell surface glycoprotein fibronectin, may also be a causal event (Pearlstein, 1979).

To date there is still no one universal feature that has been recognized as inextricably associated with tumor metastasis. In addition to characteristic changes in the cell-surface oligosaccharides on tumor cells themselves (Dennis *et al.*, 1984) and in localized anionic sites in the cell membrane (Raz *et al.*, 1980), there are studies that suggest the importance of changes in host tissue, the "bed" of metastatic growth (Grabel *et al.*, 1979). There is too a growing body of evidence that points to a role for genes mapping in the major histocompatibility complex as playing a role in tumor metastasis (Kerbel *et al.*, 1978). Perhaps these genes regulate the expression of cell surface glycosyltransferases, important in intercellular communication (e.g., Chapter 3). One could then consider that the initiating event in secondary tumor growth may be associated with a failure to develop the MHC-regulated appropriate cell surface communication with the nearest neighbors. Note that in this discussion by Raz and Fidler there is not an a priori assumption made concerning the genetic basis for the production of metastatic variants or the stability of the variants so produced. A discussion of the generation and stability of metastatic variants is included, however (see for instance Ling *et al.*, 1984).

REFERENCES

Anderson, L. C., Gahmberg, C. G., Siimes, M. A., Teerenhavi, L., and Vuopio, P. (1979). *Int. J. Cancer* **23**, 306.

Chin, Y. H., Carey , G. D., and Woodruff, J. J.(1982). *J. Immunol.* **129**, 1911.

Dahr, W., Gielen, W., Beyreuter, K., and Kruger, J. (1980). *Physiol. Chem.* **361**, 145.

Dennis, J. W., Carver, J. R., and Schachter, H. (1984). *J. Cell. Biol.* **99**, 1034.

Deuel, T. F., Huang, J. S., Huang, S. S., Stroobant, P., and Waterfield, M. (1983). *Science* **221**, 1348.

Dobrossy, L., Pavelic, L. P., Vaughan, M., Porter, N., and Bernacki, R. J. (1980). *Cancer Res.* **40,** 3281.

Downward, J., Yarden, Y., Mayes, E., Scrace, C., Tatty, N., Stockwell, P., Ullrich, A., Schlessinger, J., and Waterfield, M. D. (1984). *Nature (London)* **307,** 521.

Ekblom, M., Borgstrom, G., Von Willebrand, E., Gahmberg, C. G., Vuopio, P., and Anderson, L. C. (1983). *Blood* **62,** 591.

Fukuda, M., Fukuda, M. N., and Hakamori, S. I. (1979). *J. Biol. Chem.* **254,** 3700.

Furthmayer, H. (1978). *Nature (London)* **271,** 519.

Gallatin, W. M., Weissman, I. L., and Butcher, E. C. (1983). *Nature (London)* **304,** 30.

Grabel, L. B., Rosen, S. D., and Martin, G. R. (1979). *Cell* **17,** 477.

Howard, R. J., Barnwell, J. W., Kao, V., Daniel, W. A., and Aley, S. B. (1982). *Mol. Biochem. Parasitol.* **6,** 303.

Kabat, E. A. (1956). "Blood Group Substances: Their Chemistry and Immunochemistry." Academic Press, New York.

Kerbel, R. S., Twiddy, R. R., and Robertson, D. M. (1978). *Int. J. Cancer* **22,** 583.

Kohler, G., and Milstein, C. (1975). *Nature (London)* **256,** 495.

Krantz, S. B., and Jacobson, L. O. (1970). "Erythropoietin and the Regulation of Erythropoiesis." University of Chicago Press, Chicago.

Kumar, A., and Gupta, C. M. (1983). *Nature (London)* **303,** 632.

Ling, V., Chambers, A. F., Harris, J. F., and Hill, R. P. (1984). *J. Cell. Physiol.* **3,** 99.

McCulloch, E. A., Siminovitch, L., and Till, J. E. (1964). *Science* **144,** 844.

Meuer, S. C., Fitzgerald, K. A., Hussey, R. E., Hodgdon, J. C., Schlossman, S. F., and Reinherz, E. L. (1983). *J. Exp. Med.* **157,** 705.

Pearlstein, E. (1976). *Nature (London)* **262,** 497.

Price, G. B., and McCulloch, E. A. (1978). *Semin. Hemol.* **15,** 283.

Raz, A., Bucana, C., McLellan, W., and Fidler, I. J. (1980). *Nature (London)* **284,** 363.

Sallan, S. E., Ritz, J., Pesando, J., Gelber, R., O'Brien, C., Hitchcock, S., Coral, F., and Schlossman, S. F. (1980). *Blood* **55,** 395.

Schenkel-Brunner, H. (1980). *Eur. J. Biochem.* **104,** 529.

Springer, G. F. (1984). *Science* **224,** 1193.

Steck, T. L. (1974). *J. Cell Biol.* **62,** 1.

Till, J. E., Price, G. B., and Senn, J. S. (1977). *In* "Growth Kinetics and Biochemical Regulation of Normal and Malignant Cells," 29th Annual Symposium on Fundamental Cancer Research, (B. Drewinko and R. M. Humphrey, eds.), p. 223. Williams and Wilkins, Baltimore.

Tomita, M., and Marchesi, V. T. (1975). *Proc. Natl. Acad. Sci. U.S.A.* **72,** 2964.

Weiss, M. M., Oppenhein, J. D., and Vanderberg, J. P. (1981). *Exp. Parasitol.* **51,** 400.

Williams, A. F., Galfre, G., and Milstein, C. (1977). *Cell* **12,** 663.

9

Chemistry of ABH/Ii, MN/Ss, and Rh$_o$(D) Blood Group-Active Proteins of the Human Red-Cell Membrane

CARL G. GAHMBERG AND KIMMO K. KARHI

In this review we summarize recent progress in the identification and characterization of blood group-active proteins in the human red-cell membrane. These include (1) the classical ABH and Ii antigens, specified by carbohydrate structures, which in part are confined to membrane glycoproteins, (2) the MN/Ss active glycoprotein system, the antigenic sites of which are formed by carbohydrates and protein on the sialoglycoproteins, and (3) the Rh$_o$(D) antigen, which is confined to a nonglycosylated cell-surface protein.

RECEPTORS IN CELLULAR RECOGNITION
AND DEVELOPMENTAL PROCESSES

I. THE ABO BLOOD-GROUP SYSTEM

A. Introduction

The human red-cell blood-group antigens constitute receptors for different antibodies present in human plasma, and for various carbohydrate-binding lectins from plants and lower animals. When Landsteiner discovered the ABO blood-group system (Landsteiner, 1900), the importance of such genetically determined specific surface receptors was realized. The elucidation of the molecular nature of these and other red-cell antigens has, however, been extremely difficult and slow, mainly due to lack of techniques applicable for separation of hydrophobic membrane proteins. Therefore the original work on the specificities and structures of ABH antigens was done using soluble blood-group substances, which were obtained from ovarian cyst fluids and other body fluids. Through the elegant work of Morgan and Watkins in London (Watkins and Morgan, 1952; Morgan and Watkins, 1953) and Kabat in New York (Kabat, 1956) in the 1940s and 1950s, it was shown that α-N-acetylgalactosamine specifies the blood-group A antigen and α-galactose the B antigen.

But for several years, the molecules carrying the ABH specificities in the red cell remained unclear. Important progress was made when ABH-active glycolipids of relatively low molecular weights were isolated from red cells and their structures were determined (Yamakawa and Suzuki, 1952; Kościelak and Zakrzewski, 1960; Hakomori and Jeanloz, 1961; Stellner et al., 1973; Wherrett and Hakomori, 1973). Later it was shown that ABH-active glycolipids of high molecular weights (macroglycolipids) containing 20–60 monosaccharides existed in the erythrocyte membrane (Gardas and Kościelak, 1973; Kościelak et al., 1976). Due to their amphilic nature these macroglycolipids easily contaminated various protein fractions. Therefore it was considered that ABH antigens in the red cell are exclusively glycolipids.

B. Identification and Characterization of ABH-Active Glycopeptides from the Red-Cell Membrane

The erythrocyte membrane has been extensively studied and much of our present knowledge of membrane structure derives from studies of this membrane. On its outer surface it contains two major glycoproteins that are band 3 (Steck, 1974), or the anion transport protein (Cabantchik and Rothstein, 1974; Ho and Guidotti, 1975) and the major sialoglycoprotein, glycophorin A (Tomita and Marchesi, 1975), the physiological function of which is unknown. Many relatively poorly characterized minor glycoproteins also exist in this membrane

(Gahmberg, 1976). For review see Steck (1974), Marchesi *et al.* (1976), and Gahmberg (1977).

During studies on its structure, band 3 found to contain an N-linked oligosaccharide with exceptional features. The molecular weight of the oligosaccharide(s) was 7000–11,000 (Gahmberg *et al.*, 1976). This is an unusually high value because *N*-glycosidic oligosaccharides commonly found in glycoproteins have molecular weights of 2000–4000. Subsequent studies using Pronase digestion of whole erythrocyte membranes showed that a major portion of the released glycopeptides were of high molecular weights, and, most important, they contained ABH activity (Finne *et al.*, 1978; Järnefelt *et al.*, 1978). Significant contamination by glycosphingolipids could be excluded after assaying for sphingosine.

Structural studies of the isolated glycopeptides showed the presence of a repeating internal disaccharide galactose-β1→4-*N*-acetylglucosamine-β1→3 unit in these erythroglycans or lactosaminoglycans (Järnefelt *et al.*, 1978; Krusius *et al.*, 1978). Such structures are not normally found in glycoproteins but in keratan sulfate (Lindahl and Höök, 1978). The structure of the band 3 lactosaminoglycans has recently been elucidated (M. Fukuda *et al.*, 1984). The repeating linear carbohydrate sequence occurring in the internal regions of ABH glycoconjugates is recognized by i antibodies, whereas the I antigen contains a branched internal structure consisting of galactose-β1→4-*N*-acetylglucosamine-β1→3 (R-6) galactose (Watanabe *et al.*, 1979; M. Fukuda *et al.*, 1979; M. N. Fukuda *et al.*, 1979; Wood and Feizi, 1979; Zdebska and Kościelak, 1978). The i antigen is found on red cells from fetuses and newborns and disappears soon after birth. It is replaced by the I antigen, but in exceptional cases i antigens are found in red cells from adults.

Interestingly, the same i-type repeating linear carbohydrate unit has recently been found on the human erythroleukemia cell line K562 (Andersson *et al.*, 1979c), which contains i activity (Turco *et al.*, 1980; Fukuda, 1980). This cell line can be induced to further erythroid differentiation (Andersson *et al.*, 1979b). During differentiation the i antigen activity increased but it was not replaced by the I antigen (Benz *et al.*, 1980). The i antigen structure is also found on erythroblasts from fetuses and newborns, whereas erythroblasts from adults seem to contain the I structure (Fukuda *et al.*, 1980). Thus the I/i activity of erythroglycans seems to depend primarily on the ontogenic stage of the erythroblast progenitors. Furthermore, different leukocytes seem to contain such structures (Childs and Feizi, 1981; Childs *et al.*, 1983).

Only a fraction of the erythroglycan type of oligosaccharides, even from blood-group AB cells, contains α-*N*-acetylgalactosamine/α-galactose at the nonreducing terminals. These glycopeptides have substantially higher molecular weights than the ABH-negative erythroglycans (Viitala *et al.*, 1981). This cannot simply be accounted for by the ABH-specific determinants occurring at the non-reducing terminals, but other internal structural differences exist.

There is now good evidence that band 3 and glycophorin A normally are closely associated in the normal red-cell membrane and evidently constitute part of the intramembrane particles seen after freeze-fracturing of the membrane. Antibodies to glycophorin A decreased the rotational movement of band 3 (Nigg *et al.*, 1980), and absence of glycophorin A in En(a−) red cells (Gahmberg *et al.*, 1976; Tanner *et al.*, 1976; Dahr *et al.*, 1976; Furthmayr, 1978) leads to increased suspectibility to clustering of the particles at decreased pH (Gahmberg *et al.*, 1978b). In En(a−) red cells the oligosaccharide of band 3 is larger than that of normal cells (Gahmberg *et al.*, 1976). We think that the absence of glycophorin A makes glycosylation of band 3 by sugar transferases more efficient due to decreased steric hindrance.

Individual ABH-active glycopeptides from normal AB red cells contain either the A or the B antigenic determinants but not both, although the branched structures of the oligosaccharides would make this possible (Viitala *et al.*, 1981). This raises the interesting possibility that the glycosyl transferases responsible for the biosynthesis of the terminal portions of the oligosaccharides are topographically separated.

Taken together, these data show that the regulation of the size, branching, and terminal specificities of the erythroglycan oligosaccharides is extremely complex and subject to different types of regulatory mechanisms, which essentially are not understood.

C. Identification of ABH-Active Red-Cell Glycoproteins

A large amount of work has been done on the isolation of different red-cell integral membrane glycoproteins followed by assay for the presence of blood-group antigens in the isolated fractions. However, definite proof for ABH antigen-active glycoproteins has been difficult to obtain because contamination by blood group-active glycolipids has been difficult to exclude. Therefore, other means to approach the intriguing question of whether ABH-active red-cell glycoproteins really exist and, if so, which proteins are antigenically active have become necessary.

Blood-group A and B glycosyltransferases from gastric mucosa, plasma, milk, and salivary glands have been purified (Schenkel-Brunner and Tuppy, 1970; Nagai *et al.*, 1978a,b; Takasaki and Kobata, 1976; Schwyzer and Hill, 1977a) and used for the transfer of N-[^{14}C]acetylgalactosamine or [^{14}C]galactose from the UDP sugars to blood group O cells (Schenkel-Brunner and Tuppy, 1970; Takasaki and Kobata, 1976; Schwyzer and Hill, 1977b). Kobata and co-workers obtained evidence that primarily the sialoglycoproteins (PAS1–3, or glycophor-

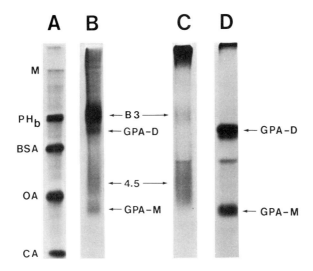

Fig. 1. Fluorography pattern of a polyacrylamide slab gel of erythrocyte membranes labeled with [^{14}C]N-acetylgalactosamine using N-acetylgalactosaminyltransferase. (A) Standard proteins: M, myosin; PH$_b$, phosphorylase b; BSA, bovine serum albumin; OA, ovalbumin; CA, carbonic anhydrase. (B) Pattern of N-[^{14}C]acetylgalactosamine-labeled O erythrocyte membranes. B3, band 3; GPA-D, glycophorin A dimer; 4.5, glycoproteins in the 4.5 region of the gel; GPA-M, glycophorin A monomer. The labeled membranes were solubilized in Triton X-100 and fractionated on a *Vicia cracca* lectin–Sepharose column. (C) The lectin-bound and sugar-eluted fraction. (D) The fraction not bound to the lectin.

ins A and B) gained A activity (Takasaki and Kobata, 1976). Schwyzer and Hill (1977b) used a similar approach, but the assignment of H-active receptors to well-characterized erythrocyte glycoproteins remained unclear.

We have purified α-N-acetylgalactosaminyltransferase from human A$_1$ plasma and used this for transfer of N-[^{14}C]acetylgalactosamine from UDP-N-acetyl-[^{14}C]galactosamine to blood-group O cells. After this the labeled glycoproteins were studied by polyacrylamide slab gel electrophoresis followed by fluorography. The introduction of fluorography enabled a much better resolution than previously obtained. It is obvious that H-active acceptors are present mainly on band 3 but also in some other glycoproteins, mainly in the band 4.5 region [Fig. 1(B)]. A significant amount of radioactivity was incorporated into glycoprotein in the glycophorin A region of the gel [GPA-D and GPA-M, Fig. 1(B)]. However, glycophorin A did not bind to the blood group A-specific *Vicia cracca* lectin [Fig. 1(D)]. The acceptor activity of glycophorin A was ascertained by immunoprecipitation of ^{14}C-labeled glycophorin A [Fig. 2(C)] using specific antiglycophorin A antiserum (Karhi, 1982).

It is important to point out that the results obtained with external blood-group

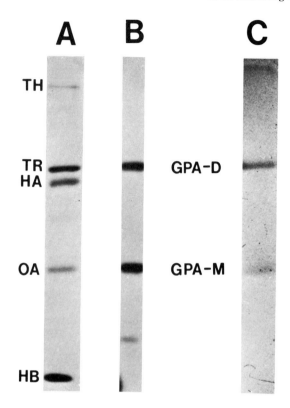

Fig. 2. Visualization of A-like activity in glycophorin A after *in vitro* transfer of N-[^{14}C]-acetylgalactosamine. (A) Standard proteins: TH, thyroglobulin; TR, transferrin; HA, human albumin; OA, ovalbumin; HB, hemoglobin. (B) Pattern of periodate/NaB^{3}H$_4$-labeled erythrocyte membranes. (C) Immunoprecipitate obtained with anti-glycophorin A antiserum from O cells incubated with UDP–N-[^{14}C]acetylgalactosamine. GPA-D, glycophorin A dimer; GPA-M, glycophorin A monomer.

transferases may not correspond to the situation during normal biosynthesis. The molecular details of blood-group antigen biosynthesis are poorly understood. The glycosyltransferases, the availability of acceptors, and their topographical organization in the internal membranes of the erythroblast may well be different than in the mature red-cell plasma membrane. In fact, we have shown that by using serum blood-group A transferase with B cells as acceptors, a substantial fraction of the incorporated N-[^{14}C]acetylgalactosamine was found in B-active oligosaccharides (C. G. Gahmberg and K. K. Karhi, unpublished), although, as noted above, most A- and B-specific terminal carbohydrate occurs in separate glycopeptides in normal AB cells.

An elegant approach to identify Ii/ABH-active glycoproteins was taken by M. N. Fukuda *et al.* (1979) and Mueller *et al.* (1979). They surface-labeled red cells by the galactose oxidase/NaB^{3}H$_4$ technique (Gahmberg and Hakomori, 1973) and digested part of the labeled cells with *endo*-β-galactosidase. This enzyme is specific for internal galactose β1→4-*N*-acetylglucosamine linkages (Fukuda and Matsumura, 1975, 1976; Fukuda *et al.*, 1978). The labeled and digested glyco-proteins were studied by polyacrylamide gel electrophoresis. Concomitantly with the disappearance of the labeled band 3 and band 4.5 glycoproteins from the gels, the ABH and Ii activities went down. The glycophorin A and B bands were not affected by the enzyme treatment.

The same technique has been used to study erythroglycan-containing glyco-

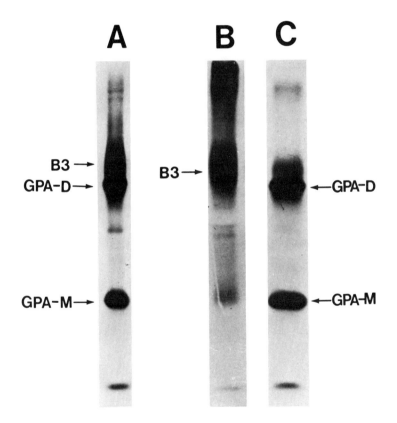

Fig. 3.   Polyacrylamide slab gel electrophoresis of blood-group A-active red cell glycoproteins. (A) Pattern of lactoperoxidase–^{125}I-labeled red-cell membranes. Abbreviations as in Figs. 1 and 2. (B) ^{125}I-Labeled red-cell glycoproteins adsorbed and eluted from *Vicia cracca* lectin–Sepharose. (C) ^{125}I-Labeled red-cell glycoproteins not interacting with *Vicia cracca* lectin–Sepharose.

proteins of K562 erythroleukemic cells and erythroid progenitor cells (Turco *et al.*, 1980; Fukuda *et al.*, 1980). The major glycoprotein band partially digested by *endo*-β-galactosidase had an apparent molecular weight of 105,000 and was absent from mature red cells (Turco *et al.*, 1980; Fukuda, 1980).

A third approach to identify ABH-active red-cell glycoproteins has been the use of blood-group-specific lectins. ABH-specific lectins are well known, and A- and H-specific lectins especially are routinely used in blood group typing. We isolated the blood group A-specific lectin from *Vicia cracca* seeds (Karhi and Gahmberg, 1980a) and coupled it to Sepharose. We then labeled red cells of different ABO blood groups by the galactose oxidase/NaB^3H_4 and the lactoperoxidase ^{125}I techniques, and after solubilization in detergent the extract was passed through the affinity column. Radioactive material adsorbed only from A cells, and after elution with the monosaccharide hapten, the radioactive A-active glycoproteins were identified by polyacrylamide gel electrophoresis followed by fluorography (Karhi and Gahmberg, 1980b). The band 3 and band 4.5 glycoproteins were the major components showing A activity [Fig. 3(B)]. Glycophorin A or B did not bind significantly to the affinity column [Fig. 3(C)].

Finne used a different approach (Finne, 1980). He separated unlabeled red cell membrane glycoproteins of different ABH blood groups by polyacrylamide gel electrophoresis and after fixation he incubated the gels with ^{125}I-labeled blood-group-specific lectins from *Bandeiraea simplicifolia* and *Lotus tetragonolobus*. Specific adsorption occurred to band 3 and band 4.5 proteins, but surprisingly glycophorin A also bound an appreciable amount. It is possible that internal α-*N*-acetyl galactosamines linked to Ser/Thr in glycophorin A were exposed during the acid fixation due to liberation of *N*-acetylgalactosamine-linked sialic acids. Terminal α-*N*-acetylgalactosamines would then bind blood group A-specific lectins.

It is obviously important to determine the quantitative contribution of low-molecular-weight glycolipids, macroglycolipids, and glycoproteins to the total ABH activity of red cells. This has been attempted by Schenkel-Brunner (1980) and Wilczyńska *et al.* (1980). They used blood-group transferase and measured the incorporation of *N*-[^{14}C]acetylgalactosamine into the three major antigenic molecular groups. Both arrived at similar conclusions: about 5% of the total activity was associated with low-molecular-weight glycolipids, 10–20% with macroglycolipids, and the rest with protein. The results must, however, be interpreted with caution.

II. THE MN/Ss BLOOD GROUP-ACTIVE RED-CELL GLYCOPROTEINS

In 1928 Landsteiner and Levine reported that after injection of rabbits with human red cells followed by appropriate absorptions, they obtained antisera that

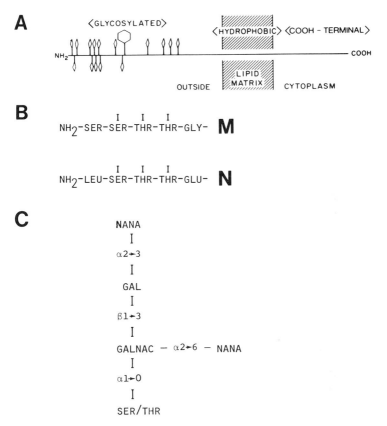

Fig. 4. Structure of glycophorin A from MM and NN cells. (A) Schematic structure of glycophorin A. (B) The amino-terminal sequences of glycophorin A, s from MM and NN cells; I indicates attachment sites of *O*-glycosidic oligosaccharides. (C) Structure of the major *O*-glycosidic oligosaccharide of glycophorin A.

discriminated between red cells of different individuals, irrespective of the ABO blood group. This blood-group system was given the name MN, and it was shown to be composed of the simple alleles M and N. Important progress in the chemistry of the MN blood group system occurred when Mäkelä and Cantell (1958) and Springer and Ansell (1958) showed that neuraminidase treatment abolished MN activity. It was later shown that the antigenic activity is associated with glycoprotein, and it has now been firmly established that the major sialoglycoprotein, glycophorin A, carries the activity. En(a−) cells, which lack glycophorin A (Gahmberg *et al.*, 1976; Tanner and Anstee, 1976; Dahr *et al.*, 1976), do not contain M activity although a small amount of N antigen is found. The minor sialoglycoprotein, glycophorin B, is always N-active irrespective of the MN blood group (see below), and this protein is present in En(a−) red cells.

Glycophorin A was the first integral membrane protein that was sequenced (Tomita and Marchesi, 1975), and it has no doubt served as a classical model for membrane protein structure [see Fig. 4(A)] and for biosynthesis (Jokinen *et al.*, 1979; Gahmberg *et al.*, 1980; Jokinen *et al.*, 1981). Springer and co-workers claimed that the distinction between M and N antigens was due to differences in the degree of sialylation and that the M antigen could be converted to N by limited acid treatment (Springer and Desai, 1975).

However, Lisowska and Duk (1975) showed that modification of the polypeptide portion of glycophorin A with amino-reactive reagents abolished the MN specificity, and soon it was established that the amino acid sequence in the NH_2-terminal portion of glycophorin A was different depending on whether the protein was isolated from MM or NN cells (Furthmayr, 1978; Waśniowska *et al.*, 1977; Dahr *et al.*, 1977; Kordowicz and Lisowska, 1978). The amino-terminal sequence of glycophorin A from MM cells is Ser-Ser-Thr-Thr-Gly-, whereas that from NN cells is Leu-Ser-Thr-Thr-Glu- [Fig. 4(B)]. Residues two to four are glycosylated containing the oligosaccharide N-acetylneuraminic acid-$\alpha 2 \rightarrow 3$-galactose (N-acetylneuraminic acid-$\alpha 2 \rightarrow 6$)$\beta 1 \rightarrow 3$-N-acetylgalactosamine-$\alpha 1 \rightarrow$ 0-Ser/Thr [Fig. 4(C)].

Definite proof for the importance of the polypeptide portion in MN specificity was obtained when Sadler *et al.* (1979) were able to regain the MN activity of desialylated erythrocytes with purified sialyl transferases. When the desialylated cells had been of M type, either the transferase adding sialic acid to the non-reducing galactosyl residue or the transferase sialylating the N-acetylgalactosamine residue restored M activity but N activity was never obtained. On the other hand, cells of N origin only gave N activity after resialylation. Thus it now seems clear that sialic acid residues are necessary for the MN antigen activity but the polypeptide sequence is responsible for the MN specificity.

The minor sialoglycoprotein glycophorin B (PAS3) is always N-active regardless of the MN phenotype. The sequence of the first 23 amino acid residues of glycophorin B is identical to that of glycophorin A from N cells (Furthmayr, 1978). This explains why MM cells always contain a low N acticity.

Glycophorin B carries the blood group Ss activity. Rare red-cell variants have been found that lack glycophorin B or contain a decreased level of carbohydrate on the protein. These lack the Ss antigens (Dahr *et al.*, 1975; Tanner *et al.*, 1977). It has been reported that the Ss specificity is due to replacement of the methionine residue at position 29 in SS cells with threonine in ss cells (Dahr *et al.*, 1980).

III. THE $Rh_o(D)$-ACTIVE PROTEIN

The $Rh_o(D)$ antigen (Levine and Stetson, 1939; Landsteiner and Wiener, 1940) is one of the clinically most important blood group antigens. Isoimmuniza-

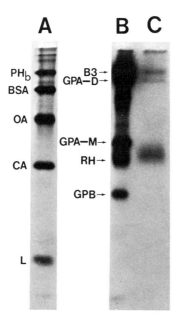

Fig. 5. Polyacrylamide slab gel electrophoresis of ^{125}I-labeled -D-/-D- red-cell membranes and immune precipitate obtained with anti-Rh_o(D) antiserum. (A) Standard proteins. See legend to Fig. 1; L, lysozyme. (B) Pattern of ^{125}I-labeled -D-/-D- membranes. RH, Rh_o(D) antigen. (C) Pattern of immune precipitate obtained with anti-Rh_o(D) antiserum.

tion of the pregnant woman may result in hemolytic disease of the newborn. Fortunately, this can often be prevented by passive transfer of anti-Rh_o(D) antibodies (Clarke, 1982). The molecular nature of the Rh_o(D) antigen has remained controversial and development in the field has been slow. Most earlier results indicated that it is a protein, and lipids may be needed for full activity (Green, 1972; Plapp *et al.*, 1980). However, the molecular weight estimates ranged from 7000 to 174,000 (Abraham and Bakerman, 1975; Folkerd *et al.*, 1977; Plapp *et al.*, 1979; Victoria *et al.*, 1981). Recently, Moore *et al.* (1982) and Gahmberg (1982) independently showed that a polypeptide with an apparent molecular weight of 28,000–33,000 was specifically immune precipitated from ^{125}I-surface-labeled cells with anti-Rh_o(D) antibodies and protein A.

The availability of -D-/-D- red cells containing several times the normal amount of the Rh_o(D) antigens greatly facilitated its immune precipitation (Fig. 5) and further characterization (Gahmberg, 1983). The protein was found to be unusual. Most cell-surface proteins are glycoproteins (Gahmberg, 1976), but the Rh_o(D) antigen seemed to lack carbohydrate. It was not labeled using carbohydrate-specific surface-labeling techniques (Gahmberg and Hakomori, 1973; Gahmberg and Andersson, 1977) and did not bind to any of several lectin–Sepharose columns tested. Furthermore, it was not degraded by glycosidases.

Further work showed that the $Rh_o(D)$ polypeptide is not easily solubilized by neutral detergents. Whereas most cell-surface proteins are easily soluble in Triton X-100 and similar detergents, 80% of the $Rh_o(D)$ polypeptide remained firmly associated with the membrane skeleton (Gahmberg and Karhi, 1984). The disposition of the $Rh_o(D)$ polypeptide in the red-cell membrane is not exactly known. Our present knowledge supports, however, a model in which the polypeptide contains a $Rh_o(D)$-specific amino acid sequence on the cell surface, penetrates the lipid bilayer, and interacts with cytoskeletal proteins at the cytoplasmic aspect of the membrane.

IV. GENERAL DISCUSSION

The extreme molecular complexity of the blood-group substances, the generation of which involves several specific biosynthetic enzymes, would make one believe that these are functionally important. Nature seldom wastes. However, no specific physiological functions have yet been found either for the ABO blood group substances or for the glycophorins. In fact, individuals of the Bombay blood group lack not only A and B antigens but also the H precursor molecules, and this abnormality does not seem to have any adverse effects. Similarly, persons with the En(a−) blood group who lack glycophorin A do not, as far as is known, have hematologic disorders.

It should, however, be pointed out that in En(a−) red cells, the other major surface glycoproteins, band 3, is overglycosylated (Gahmberg et al., 1976), indicating some sort of compensation due to lack of glycophorin A. In contrast, absence of all Rh antigens (Rh_{null} cells) leads to increased fragility of the red cells and changed ionic composition (Ballas et al., 1984).

The presence of ABH activity both in glycoproteins and glycolipids in the red cell has some important general implications. It shows that one must be careful when ascribing some receptor activity to a certain type of glycoconjugate. For example, some glycolipids have been claimed to carry specific receptor functions (for review, see Gahmberg, 1981). However, it may well be that glycoproteins carrying similar terminal oligosaccharide structures primarily are responsible for the physiological functions of the receptors.

Another interesting question concerns the biosynthesis of the carbohydrate in glycoproteins and glycolipids. Are there transferases that discriminate between glycoproteins and glycolipids? Obviously the same transferase is involved in the biosynthesis of the blood-group A antigen (or the B antigen) both in glycoproteins and glycolipids, but the synthesis of the inner regions of the oligosaccharide remains rather unclear. On the other hand, the core regions of the oligosaccharides are obviously synthesized by different mechanisms in glycoproteins and glycolipids (Parodi and Leloir, 1979; Hubbard and Ivatt, 1981). Only after purification of the different types of acceptors and transferases may it become possible to answer these questions.

Both glycophorin A (Gahmberg *et al.*, 1978a) and the blood-group A antigens (probably also the H and B antigens) (Karhi *et al.*, 1981) appear during early erythroid differentiation in normal human bone marrow. Glycophorin A appears somewhat before, and A antigens in blood-group A individuals approximately at the same stage as hemoglobin synthesis is observed, that is, in basophilic nor-moblasts. Both antigenic activities are specific for the erythroid cells in bone marrow, and the other cell lineages are negative.

It has recently been shown that glycophorin A from immature erythroid cells is less O-glycosylated than the glycophorin A molecule from mature red cells (Gahmberg *et al.*, 1984). This was seen both in normal bone marrow erythroid cells and in the erythroid cell lines K562 and HEL induced to differentiation.

One could hypothesize that if the ABO and MN antigens have any physiologi-cal functions these could operate during the earlier stages of erythroid differentia-tion, for example, in cell–cell interactions as receptors.

Glycophorin A is known to act as a receptor for various microbes. Influenza virus uses glycophorin A as a receptor and so does *Plasmodium falciparum* (Perkins, 1981; Pasvol, 1981). Immature erythroid cells are not infected by the merozoites (Tanner, 1982), which probably is due to incomplete glycosylation of the molecule (Gahmberg *et al.*, 1984). In addition, we have recently shown that certain *Escherichia coli* strains use glycophorin A as a receptor and specifically recognize the M type of glycophorin A (Jokinen *et al.*, 1985). The receptor structure seemed to involve both the NH_2-terminal polypeptide portion and the O-glycosidic carbohydrate.

The finding that glycophorin A is specific for the erythroid lineage has proven very useful for the clinical diagnosis of early erythroleukemias (Andersson *et al.*, 1979a; Ekblom *et al.*, 1983). Such malignant cells lack or have very low levels of hemoglobin and are therefore difficult to diagnose as erythroid. The malignant cells, however, often contain glycophorin A, and using specific rabbit anti-glycophorin A antiserum several positive cases were recorded. Some monoclonal antisera were less effective in detecting early erythroid cells (Greaves *et al.*, 1983).

Springer and co-workers have found that precursors to the O-glycosidic oligosaccharide of glycophorin A, the T and T_n antigens, function as carcinoma markers in various malignant diseases (Springer, 1984). These incompletely glycosylated oligosaccharides are found in different solid tumors and are useful in diagnosis of the diseases.

Studies on the blood group-active red-cell proteins are still in their infancy. But the increased knowledge now available on the structures of some blood group-active proteins and the advances in the methodology of their isolation and characterization should facilitate research in this fascinating field of studies on structural–functional relationships.

ACKNOWLEDGMENTS

This study was supported by the Academy of Finland, the Finska Läkaresällskapet, the Sigrid Jusélius Foundation, and National Cancer Institute grant 2 RO1 CA 26294-04. We thank Ulla Katajarinne for assistance and Barbara Björnberg for secretarial help.

REFERENCES

Abraham, C. V., and Bakerman, S. (1975). *Clin. Chim. Acta* **60**, 33–43.

Andersson, L. C., Gahmberg, C. G., Siimes, M. A., Teerenhovi, L., and Vuopio, P. (1979a). *Int. J. Cancer* **23**, 306–311.

Andersson, L. C., Jokinen, M., and Gahmberg, C. G. (1979b). *Nature (London)* **278**, 364–365.

Andersson, L. C., Nilsson, K., and Gahmberg, C. G. (1979c). *Int. J. Cancer* **23**, 143–147.

Ballas, S. K., Clark, M. R., Mohandas, N., Colfer, H. F., Caswell, M. S., Bergren, M. D., Perkins, H. A., and Shohet, S. B. (1984). *Blood* **63**, 1046–1055.

Benz, E. J., Murnane, M. J., Tonkonow, B. L., Berman, B. W., Mazur, E. M., Cavallesco, C., Jenko, T., Snyder, E. L., Forget, B. G., and Hoffman, R. (1980). *Proc. Natl. Acad. Sci. U.S.A.* **77**, 3509–3513.

Cabantchik, Z. I., and Rothstein, A. (1974). *J. Membrane Biol.* **15**, 207–226.

Childs, R. A., and Feizi, T. (1981). *Biochem. Biophys. Res. Commun.* **102**, 1158–1164.

Childs, R. A., Dalchau, R., Scudder, P., Hounsell, E. F., Fabre, J. F., and Feizi, T. (1983). *Biochem. Biophys. Res. Commun.* **110**, 424–431.

Clarke, C. (1982). *Br. J. Haematol.* **52**, 525–535.

Dahr, W., Uhlenbruck, G., Issitt, P. D., and Allen, F. M. (1975). *J. Immunogenet.* **2**, 249–251.

Dahr, W., Uhlenbruck, G., Leikola, J., Wagstaff, W., and Landfried, K. (1976). *J. Immunogenet.* **3**, 329–346.

Dahr, W., Uhlenbruck, G., Janssen, E., and Schmalisch, R. (1977). *Hum. Genet.* **35**, 335–343.

Dahr, W., Gielen, W., Beyreuther, K., and Krüger, J. (1980). *Hoppe Seylers Z. Physiol. Chem.* **361**, 145–152.

Ekblom, M., Borgström, G., von Willebrand, E., Gahmberg, C. G., Vuopio, P., and Andersson, L. C. (1983). *Blood* **62**, 591–596.

Finne, J. (1980). *Eur. J. Biochem.* **104**, 181–189.

Finne, J., Krusius, T., Rauvala, H., Kekomäki, R., and Myllylä, G. (1978). *FEBS Lett.* **89**, 111–115.

Folkerd, E. J., Ellory, J. C., and Hughes-Jones, N. C. (1977). *Immunochemistry* **14**, 529–531.

Fukuda, M. (1980). *Nature (London)* **285**, 405–407.

Fukuda, M., Fukuda, M. N., and Hakomori, S.-I. (1979). *J. Biol. Chem.* **254**, 3700–3703.

Fukuda, M., Fukuda, M. N., Papayannopoulou, T., and Hakomori, S.-I. (1980). *Proc. Natl. Acad. Sci. U.S.A.* **77**, 3474–3478.

Fukuda, M., Dell, A., Oates, J. E., and Fukuda, M. N. (1984). *J. Biol. Chem.* **259**, 8260–8273.

Fukuda, M. N., and Matsumura, G. (1975). *Biochem. Biophys. Res. Commun.* **64**, 465–471.

Fukuda, M. N., and Matsumura, G. (1976). *J. Biol. Chem.* **251**, 6218–6225.

Fukuda, M. N., Watanabe, K., and Hakomori, S.-I. (1978). *J. Biol. Chem.* **253**, 6814–6819.

Fukuda, M. N., Fukuda, M., and Hakomori, S.-I. (1979). *J. Biol. Chem.* **254**, 5458–5465.

Furthmayr, H. (1978). *Nature (London)* **271**, 519–524.

Gahmberg, C. G. (1976). *J. Biol. Chem.* **251**, 510–515.

Gahmberg, C. G. (1977). *In* "Dynamic Aspects of Cell Surface Organization" (G. Poste and G. L. Nicolson, eds.), pp. 371–421. North Holland, Amsterdam.

Gahmberg, C. G. (1981). "Comprehensive Biochemical Membrane Structure" (J. B. Finean and R. H. Michell, eds.), pp. 127–160. Elsevier, Amsterdam.

Gahmberg, C. G. (1982). *FEBS Lett.* **140**, 93–97.

Gahmberg, C. G. (1983). *EMBO J.* **2**, 223–227.

Gahmberg, C. G., and Andersson, L. C. (1977). *J. Biol. Chem.* **252**, 5888–5894.

Gahmberg, C. G., and Hakomori, S. (1973). *J. Biol. Chem.* **248**, 4311–4317.

Gahmberg, C. G., and Karhi, K. K. (1984). *J. Immunol.* **133**, 334–337.

Gahmberg, C. G., Myllylä, G., Leikola, A., Pirkola, A., and Nordling, S. (1976). *J. Biol. Chem.* **251**, 6108–6116.

Gahmberg, C. G., Jokinen, M., and Andersson, L. C. (1978a). *Blood* **52**, 379–387.

Gahmberg, C. G., Taurén, G., Virtanen, J., and Wartiovaara, J. (1978b). *J. Supramol. Struct.* **8**, 337–387.

Gahmberg, C. G., Jokinen, M., Karhi, K. K., and Andersson, L. C. (1980). *J. Biol. Chem.* **255**, 2169–2175.

Gahmberg, C. G., Ekblom, M., and Andersson, L. C. (1984). *Proc. Natl. Acad. Sci. U.S.A.* **81**, 6752–6756.

Gardas, A., and Kościelak, J. (1973). *Eur. J. Biochem.* **32**, 178–187.

Greaves, M. F., Sieff, C., and Edwards, P. A. W. (1983). *Blood* **61**, 645–651.

Green, F. A. (1972). *J. Biol. Chem.* **247**, 881–887.

Hakomori, S., and Jeanloz, R. W. (1961). *J. Biol. Chem.* **236**, 2827–2834.

Ho, M. K., and Guidotti, G. (1975). *J. Biol. Chem.* **250**, 675–683.

Hubbard, S. C., and Ivatt, R. J. (1981). *Annu. Rev. Biochem.* **50**, 555–583.

Järnefelt, J., Rush, J., Li, Y.-T., and Laine, R. A. (1978). *J. Biol. Chem.* **253**, 8006–8009.

Jokinen, M., Gahmberg, C. G., and Andersson, L. C. (1979). *Nature (London)* **279**, 604–607.

Jokinen, M., Ehnholm, C., Väisänen-Rhen, V., Korhonen, T. K., Pipkorn, R., Kalkkinen, N., and Gahmberg, C. G. (1985). *Eur. J. Biochem.* **147**, 47–52.

Jokinen, M., Ulmanen, I., Andersson, L. C., Kääriäinen, L., and Gahmberg, C. G. (1981). *Eur. J. Biochem.* **114**, 393–397.

Kabat, E. A. (1956). "Blood Group Substances: Their Chemistry and Immunochemistry." Academic Press, New York.

Karhi, K. K. (1982). *FEBS Lett.* **142**, 203–206.

Karhi, K. K., and Gahmberg, C. G. (1980a). *Biochim. Biophys. Acta* **622**, 337–343.

Karhi, K. K., and Gahmberg, C. G. (1980b). *Biochim. Biophys. Acta* **622**, 344–354.

Karhi, K. K., Andersson, L. C., Vuopio, P., and Gahmberg, C. G. (1981). *Blood* **57**, 147–151.

Kordowicz, M., and Lisowska, E. (1978). *Arch. Immunol. Ther. Exp. (Warsz)* **26**, 127–132.

Kościelak, J., and Zakrzewski, K. (1960). *Nature (London)* **187**, 516–517.

Kościelak, J., Miller-Podraza, H., Krauze, R., and Piasek, A. (1976). *Eur. J. Biochem.* **71**, 9–18.

Krusius, T., Finne, J., and Rauvala, H. (1978). *Eur. J. Biochem.* **92**, 289–300.

Landsteiner, K. (1900). *Zentralbl. Bakteriol. (Orig.)* **28**, 357–362.

Landsteiner, K., and Levine, P. (1928). *J. Exp. Med.* **47**, 757–775.

Landsteiner, K., and Wiener, A. S. (1940). *Proc. Soc. Exp. Biol. Med.* **43**, 223.

Levine, P., and Stetson, R. E. (1939). *J. A. M. A.* **113**, 126–127.

Lindahl, U., and Höök, M. (1978). *Annu. Rev. Biochem.* **47**, 385–417.

Lisowska, E., and Duk, M. (1975). *Eur. J. Biochem.* **54**, 469–474.

Mäkelä, O., and Cantell, K. (1958). Ann. Med. Exp. Biol. Fenn. 36, 366–374.
Marchesi, V. T., Furthmayr, H., and Tomita, M. (1976). Annu. Rev. Biochem. 45, 667–698.
Moore, S., Woodrow, C. F., and McClelland, D. B. L. (1982). Nature (London) 295, 529–531.
Morgan, W. T. J., and Watkins, W. M. (1953). Br. J. Exp. Pathol. 34, 94–103.
Mueller, T. J., Li, Y.-T., and Morrison, M. (1979). J. Biol. Chem. 254, 8103–8106.
Nagai, M., Davè, V., Kaplan, B. E., and Yoshida, A. (1978a). J. Biol. Chem. 253, 377–379.
Nagai, M., Davè, V., Muensch, H., and Yoshida, A. (1978b). J. Biol. Chem. 253, 380–381.
Nigg, E. A., Bron, C., Girardet, M., and Cherry, R. J. (1980). Biochemistry 19, 1887–1893.
Parodi, A. J., and Leloir, L. F. (1979). Biochim. Biophys. Acta 559, 1–37.
Pasvol, G., Wainscoat, J. S., and Weatherall, D. J. (1981). Nature (London) 297, 64–66.
Perkins, M. (1981). J. Cell Biol. 90, 563–567.
Plapp, F. V., Kowalski, M. M., Tilzer, L., Brown, P. J., Evans, J., and Chiga, M. (1979). Proc. Natl. Acad. Sci. U.S.A. 76, 2964–2968.
Plapp, F. V., Kowalski, M. M., Evans, J. P., Tilzer, L. L., and Chiga, M. (1980). Proc. Soc. Exp. Biol. Med. 164, 561–568.
Sadler, J. E., Paulson, J. C., and Hill, R. L. (1979). J. Biol. Chem. 254, 2112–2119.
Schenkel-Brunner, H. (1980). Eur. J. Biochem. 104, 529–534.
Schenkel-Brunner, H., and Tuppy, H. (1970). Eur. J. Biochem. 17, 218–222.
Schwyzer, M., and Hill, R. L. (1977a). J. Biol. Chem. 252, 2338–2345.
Schwyzer, M., and Hill, R. L. (1977b). J. Biol. Chem. 252, 2346–2355.
Springer, G. F. (1984). Science 224, 1198–1206.
Springer, G. F., and Ansell, N. J. (1958). Proc. Natl. Acad. Sci. U.S.A. 44, 182–189.
Springer, G. F., and Desai, P. D. (1975). Carbohydrate Res. 40, 183–192.
Steck, T. L. (1974). J. Cell Biol. 62, 1–19.
Stellner, K., Watanabe, K., and Hakomori, S. (1973). Biochemistry 12, 656–661.
Takasaki, S., and Kobata, A. (1976). J. Biol. Chem. 251, 3610–3615.
Tanner, M. J. (1982). Trends Biochem. Sci. 7, 231.
Tanner, M. J. A., and Anstee, D. J. (1976). Biochem. J. 153, 271–277.
Tanner, M. J. A., Jenkins, R. E., Anstee, D. J., and Clamp, J. R. (1976). Biochem. J. 155, 701–703.
Tanner, M. J. A., Anstee, D. J., and Judson, P. A. (1977). Biochem. J. 165, 157–161.
Tomita, M., and Marchesi, V. T. (1975). Proc. Natl. Acad. Sci. U.S.A. 72, 2964–2968.
Turco, S. J., Rush, J. S., and Laine, R. A. (1980). J. Biol. Chem. 255, 3266–3269.
Victoria, E. J., Mahan, L. C., and Masouredis, S. P. (1981). Proc. Natl. Acad. Sci. U.S.A. 78, 2898–2902.
Viitala, J., Karhi, K. K., Gahmberg, C. G., Finne, J., Järnefelt, J., Myllylä, J., and Krusius, T. (1981). Eur. J. Biol. Chem. 113, 259–265.
Waśniowska, K., Drzeniek, Z., and Lisowska, E. (1977). Biochem. Biophys. Res. Commun. 76, 385–390.
Watanabe, K., Hakomori, S.-I., Childs, R. A., and Feizi, T. (1979). J. Biol. Chem. 254, 3221–3228.
Watkins, W. M., and Morgan, W. T. J. (1952). Nature (London) 169, 825–826.
Wherrett, J., and Hakomori, S. (1973). J. Biol. Chem. 248, 3046–3051.
Wilczyńska, Z., Miller-Podraza, H., and Kościelak, J. (1980). FEBS Lett. 112, 277–279.
Wood, E., and Feizi, T. (1979). FEBS Lett. 104, 135–140.
Yamakawa, T., and Suzuki, S. (1952). J. Biochem. (Tokyo) 39, 393.
Zdebska, E., and Kościelak, J. (1978). Eur. J. Biochem. 91, 517–525.

10

Cell-Surface Receptors
and Differentiation

I. S. TROWBRIDGE

I. INTRODUCTION

The relationship of the cell surface to cellular differentiation can be conveniently divided into three different problems. First, there is a substantial body of evidence to support the notion that during differentiation along various lineages, changes occur in the array of molecules that are found on the cell surface. The most straightforward question is therefore to define in quantitative terms the molecular structure of the cell surface of particular cell types at a specific stage of differentiation. The second problem that arises as an obvious consequence of the fact that changes in the expression of individual surface components occur during

267

cellular differentiation is the nature of the genetic regulation that programs these changes. Finally, it is almost axiomatic that the cell surface itself plays a role in regulating cellular differentiation.

Current ideas about each of these aspects of the cell surface in relationship to differentiation owe a great deal to the classical work on lymphoid surface antigens begun in the 1960s by Boyse, Old, and their colleagues (Boyse and Old, 1969; Boyse, 1973; Bennett *et al.*, 1972; Old and Stockert, 1977). Using quantitative serological techniques based on the cytotoxic assay developed by Gorer and O'Gorman in 1956 and classical genetical analysis, they succeeded in establishing the lymphoid system as the experimental model of choice for studying cell-surface differentiation. This is still true today, and for this reason I have chosen to restrict the scope of this review to the cell-surface molecules of the hematopoietic system. Many of the problems encountered within the hematopoietic lineages in relating cell-surface structure to differentiation are germane to other differentiated tissues, even though hematopoietic cells are somewhat atypical in spending a fraction of their lifetime free in the circulation.

Since the early immunogenetic analysis of lymphocyte cell-surface antigens, the direction of cell-surface studies has changed in two major respects. First, the emphasis has moved from serological analysis to biochemical characterization of surface molecules, as exemplified by the purification and structural characterization of Thy-1 glycoprotein by Williams and his co-workers (Williams and Gagnon, 1982). Second, the development of the hybridoma technique by Kohler and Milstein (1975) has led to the widespread use of xenogeneic monoclonal antibodies to identify and study cell-surface molecules (Williams *et al.*, 1977; Milstein and Lennox, 1980). This technological advance has provided the antibody reagents with which to study the surface molecules of cells from species other than the mouse, and now almost as much is known about the cell surfaces of hematopoietic cells of the rat and human as the mouse.

It is intended in this chapter to review the current status of our knowledge of the differentiation antigens of hematopoietic cells and to discuss the direction future experimental work in the area might take.

II. GENERAL REMARKS

The term differentiation antigen was introduced by Boyse and Old (1969) to describe cell-surface antigens that are restricted to particular cell lineages or to specific stages of differentiation within a lineage. It is now clear that there are no clear-cut distinctions between antigens that are widely distributed on many tissues and those that are highly restricted, and it is known that many cell components show quantitative rather than qualitative changes in expression during differentiation. Furthermore, in at least one case, the transferrin receptor, there is a situation in which a surface molecule has properties of a differentiation antigen within one lineage, the erythroid series, and yet in most other tissues is associated with cell proliferation (see Section III,F). An additional complexity is introduced at a molecular level because different cell types may express related but structurally distinct cell-surface glycoproteins that may have common, cross-reactive and unique antigenic determinants (see Sections III,B and III,D). Despite these complications, the term differentiation antigen is a useful term provided it is used in a less restricted sense than was probably originally envisaged, to encompass all surface molecules that are expressed in different amounts on differentiated cells.

Table I gives a summary of the known cell-surface differentiation antigens of hematopoietic cells of the mouse, rat, and man, as of 1982. The criteria for inclusion of an antigen in the table are that it should be well defined serologically (in practice, this implies that a monoclonal antibody against the antigen has been obtained), show a quantitative or qualitative difference in expression on different cell types, and be biochemically characterized at least with defined apparent molecular weight by sodium dodecyl sulfate (SDS)–polyacrylamide gel electrophoresis. The decision not to include antigens such as W3/25 in the rat (Williams *et al.,* 1977; White *et al.,* 1978) that have been demonstrated empirically to define functional subsets of cells, but that have not been characterized biochemically, is based on the view that although such antigens may provide useful markers, they are not accessible to detailed analysis without a knowledge of their molecular structure. Antigens associated with the major histocompatibility locus have been arbitrarily excluded.

Progress made in the past 2 years in defining hematopoietic cell-surface differentiation antigens can be gauged by comparing the number of antigens listed in Table I to the relatively few cell-surface antigens reviewed previously in 1980 (Milstein and Lennox, 1980). All the differentiation antigens listed are cell-surface glycoproteins, and although glycolipid differentiation antigens exist (Springer *et al.,* 1978; Stern *et al.,* 1978), it is likely that glycoproteins are the class of molecules that mediate most of the important functions associated with the plasma membrane.

Some generalizations have emerged from the detailed data on the hematopoietic cell-surface glycoproteins that have been identified on the basis of their antigenicity. First, almost all differentiation antigens exhibit quantitative rather than qualitative differences in expression on different cell types, and few cell-surface glycoproteins have been defined with xenogeneic monoclonal antibodies that are restricted to a particular cell lineage. Even Lyt-1, a differentiation antigen *par excellence,* which was originally thought to be expressed on the surface of a T cell subset on the basis of studies with alloantisera, has now been detected using monoclonal antibodies in lesser amounts on peripheral T cells

TABLE I

Differentiation Antigens of Hematopoietic Cells[a]

Molecule	Species	Alternative designation	$M_r \times 10^{-3}$	Tissue distribution	References
T200	Mouse, man, rat	L-C, leukocyte common antigen	170–220	All leukocytes but not other cell types	Trowbridge, 1978; Omary et al., 1980; Sunderland et al., 1979; Coffman and Weissman, 1981; Dalchau and Fabre, 1981
Thy-1	Mouse, man, rat and others	Previously θ	18[b]	In mouse predominantly brain, thymus and peripheral T cells	Milstein and Lennox, 1980
Lyt-1	Mouse, man	In man, Leu-1	70	Predominantly T-cell subset but also some B cells	Cantor and Boyse, 1977; Lanier et al., 1981; Ledbetter et al., 1981
Lyt-2	Mouse, man	In man, Leu-2a; OKT5, OKT8?	S–S subunits = 30 + 35	T-cell subset, maybe some dendritic cells	Cantor and Boyse, 1977; Ledbetter et al, 1981; Terhorst et al., 1980; Nussenzweig et al., 1981
Lyt-3	Mouse, man	In man, Leu-2b	S–S subunits = 30 + 35	T-cell subset	Cantor and Boyse, 1977 Ledbetter et al., 1981
Mac-1	Mouse, man[c]	None	95, 170	Predominantly phagocytes	Sunderland et al., 1979; Coffman and Weissman, 1981; Dalchau and Fabre, 1981; Springer et al., 1978, 1979; Trowbridge and Omary, 1981b
LFA-1[d]	Mouse	None	95, 170	Related to Mac-1, many hematopoietic cells	Trowbridge and Omary, 1981b; Davignon et al., 1981a,b; Kurzinger et al., 1982
Pgp-1	Mouse, man[c]	None	95	Predominantly myeloid cells, fibroblasts	Hughes et al., 1981; Trowbridge et al., 1982b; Lesley and Trowbridge, 1982
Ly-9	Mouse	Lgp-100	100	Various lymphoid cells	Ledbetter et al., 1979; Hogarth et al., 1980
Transferrin receptor	Mouse, man	In man, B3/25, OKT9, 5E9	S–S subunits, 95 + 95	Actively proliferating cells	Trowbridge and Omary, 1981a; Sutherland et al., 1981; Haynes et al., 1981; Haynes et al., 1981; God-ing and Burns, 1981; Trowbridge and Domingo, 1981
CALLA	Man	None	95	Some leukemias, 0–20% bone-marrow cells, various normal tissues	Greaves and Janossy, 1978; Ritz et al., 1980; Metzgar et al., 1981

Antigen	Species	Other designation	Apparent $M_r (\times 10^{-3})$[b]	Cellular distribution	References
OKT1	Man	None	69	10% Thymocytes, 100% T cells	Reinherz et al., 1979; Agthoven et al., 1981; Reinherz and Schlossman, 1980
OKT3	Man	Leu-4	19	10% Thymocytes, 100% T cells	Agthoven et al., 1981; Reinherz and Schlossman, 1980; Kung et al., 1979; Chang et al., 1980
OKT4	Man	Leu-3 may be identical	62	75% Thymocytes, 55–60% T cells	Ledbetter et al., 1981; Reinherz and Schlossman, 1980; Kung et al., 1979; Tehorst et al., 1980
HTA-1	Man	OKT6	49 + 12	70% Thymocytes	Reinherz et al., 1979; Tehorst et al., 1980; Tehorst et al., 1981
OKT10	Man		45	Various lymphoid cells	Tehorst et al., 1981; Minowada et al., 1981
T65	Man	10.2, 17F12	65	100% Thymocytes, 100% T cells	Royston et al., 1980; Wang et al., 1981; Martin et al., 1980; Haynes, 1981
4F2	Man	None	S–S subunits, 80 + 40	Monocytes, hematopoietic and non-hematopoietic cell lines	Haynes, 1981
T28	Man	None	28	Cortical thymocytes	Levy et al., 1979
P24	Man	None	24	3% Bone marrow mononuclear cells, subset of all; no data on non-hematopoietic tissues	Kersey et al., 1981
F4/80	Mouse	None	160	Macrophages	Hirsch et al., 1981; Austyn and Gordon, 1981
gp IIb + IIIa	Man	None	140, 90	Platelets	McEver et al., 1980
63D3	Man	None	200	Monocytes	Ugolini et al., 1980
W3/13	Rat	None	95	Thymocytes, T cells, plasma cells, neutrophils, brain, not B cells	Brown et al., 1981
MRC OX2	Rat	None	60	100% Thymocytes, B lymphocytes, not T lymphocytes, brain, liver, or kidney	McMaster and Williams, 1979

[a] All the cell-surface antigens included in the table are defined by monoclonal antibodies and have been characterized by SDS–polyacrylamide gel electrophoresis (SDS-PAGE). Cell-surface antigens of the major histocompatibility complex have not been included.

[b] Apparent M_r on SDS–PAGE 25,000–30,000.

[c] Rat monoclonal antibody against mouse molecule cross-reacts with human antigen.

[d] LFA-1 is structurally related to Mac-1.

thought to be Lyt-1⁻2⁺ and on a small subpopulation of Thy-1⁻ cells in the primary follicles and germinal centers (Ledbetter *et al.*, 1980) as well as on B-cell tumors (Lanier *et al.*, 1981). Moreover, there is a report suggesting a subpopulation of dendritic cells express Lyt-2 (Nussenzweig *et al.*, 1981). Since many of the monoclonal antibodies against human lymphoid differentiation antigens have not been tested extensively for reactivity with nonhematopoietic tissues, it is likely some of the antigens currently believed to be restricted to subsets of hematopoietic cells will prove to be more broadly distributed (see for example Metzgar *et al.*, 1981).

Second, the importance of biochemical characterization of differentiation antigens defined by monoclonal antibodies is emphasized by the existence of families of related glycoproteins such as T200 and Mac-1 whose members share common structural and antigenic features but can be distinguished serologically and biochemically (Trowbridge, 1978; Coffman and Weissman, 1981; Dalchau and Fabre, 1981; Trowbridge and Omary, 1981b). Another example of this type of structural microheterogeneity may be the human CALLA antigen which has been reported to be a determinant restricted to one of a family of related glycoproteins defined by a conventional xenoantiserum (Pesando *et al.*, 1980). In these circumstances, it is extraordinarily difficult, if not impossible, to assess the functional significance of cell-surface antigens with restricted tissue distribution without information about the structural basis of the antigenic specificity.

Third, there are now more than 10 human lymphocyte cell-surface glycoproteins (Table I) that have been defined by means of monoclonal antibodies with no known counterparts in other species. Several explanations are possible. It is known that even cell-surface molecules that exist as homologues in different species may not have the same tissue distribution. This is best documented for Thy-1 glycoprotein (Williams *et al.*, 1977; Dalchau and Fabre, 1979). It is possible therefore that some human lymphocyte differentiation antigens either do not have counterparts in other species or that, if they exist, they are not expressed on hematopoietic cells. However, it may be that cell-surface molecules with similar properties do exist in other species but that more effort has been put into identifying human lymphocyte differentiation antigens because of their potential clinical significance, and that homologues in other species will be identified in due course. Finally, the current library of antigens may represent a distortion of the immune response in that monoclonal antibodies against the most immunogenic cell surface components are most readily obtained. It could be argued that the murine counterparts of some human lymphocyte differentiation antigens are not as immunogenic in the rat (at the moment the only species available for producing xenogeneic monoclonal antibodies against murine differentiation antigens) as their human homologues are in the mouse. The contribution of each of these factors to the high proportion of lymphocyte cell-surface antigens listed in Table I known to exist only in humans remains to be determined.

Finally, the list of differentiation antigens shown in Table I is sufficiently large that it is now evident that the identification of cell-surface molecules is no longer the major obstacle in understanding the role of the plasma membrane in differentiation and growth regulation. Indeed, from what is known about lymphocyte membrane glycoproteins purified by conventional methods, it appears likely that monoclonal antibodies have been obtained against most if not all of the major murine and rat lymphocyte surface glycoproteins. As discussed later (Section IV), what is needed most urgently is insight into the biological functions of known differentiation antigens, and this continues to be a difficult problem to approach experimentally.

III. CELL-SURFACE GLYCOPROTEINS OF HEMATOPOIETIC CELLS

A. Thy-1 Glycoprotein

Thy-1 glycoprotein is the most extensively characterized lymphocyte differentiation antigen, due in large part to the efforts of Williams and his colleagues at Oxford. Thy-1 was originally identified as a murine thymocyte alloantigen by Reif and Allen (1964), occurring in two allotypic forms, Thy-1.1 and Thy-1.2. Despite suggestions to the contrary (Vitetta *et al.*, 1973; Esselman and Miller, 1974; Milewicz *et al.*, 1976), Thy-1 is a glycoprotein, and the alloantigenic determinant is almost certainly the result of an Arg(Thy-1.1)–Gln(Thy-1.2) interchange at residue 89 of the polypeptide chain (Williams *et al.*, 1977; Cotmore *et al.*, 1981). Thy-1 glycoprotein is a small (M_r = 17,000–18,000), heavily glycosylated molecule and in the mouse is a major component of the thymocyte cell surface (10^6 molecules/cell). In the mouse, within the hematopoietic system it is restricted to thymocytes and T cells. In other species, however, homologues of Thy-1 glycoprotein show different cellular distributions (see Table 1, Williams *et al.*, 1977). In the rat, Thy-1 is found on B-cell precursors and CFU-s in the bone marrow (Hunt *et al.*, 1978; Hunt, 1979), while in humans and the chicken, Thy-1 is not found on either thymocytes or peripheral T cells(Williams *et al.*, 1977; Dalchau and Fabre, 1979; Arndt *et al.*, 1978). However, one invariant feature of the tissue distribution of Thy-1 in all species is its abundance in brain tissue, and there is evidence that a Thy-1 homologue is present in squid brain (Williams *et al.*, 1977). This would therefore appear to be the most promising tissue in which to begin the search for a physiological role for Thy-1.

Thy-1 has a number of interesting structural features. Foremost is the sequence homology of Thy-1 glycoprotein to the *V* region of immunoglobulin and β_2-microglobulin, which has led to suggestions that these proteins may carry out related functions (presently ill-defined) as cell-surface receptors (see Marchalonis *et al.*, Chapter 2, this volume). Another interesting point is that Thy-1 glycoprotein does not seem to have the hydrophobic amino acid sequence typical of integral membrane proteins, even though it binds detergent in the same way as such molecules (Campbell *et al.*, 1981; Cohen *et al.*, 1981; Kuchel *et al.*, 1978). The key to how Thy-1 is anchored in the cell membrane may lie in further studies of a non-protein hydrophobic group, presumably lipid, that appears to be covalently attached to the C-terminus of the molecule. Finally, studies of Thy-1$^-$ mutant lymphoma cells has shown that the expression of Thy-1 glycoprotein is dependent upon appropriate glycosylation, and it is possible that the glycoprotein's conformation is unusually sensitive to the oligosaccharides it displays (Trowbridge and Hyman, 1979; Chapman *et al.*, 1980).

B. T200 Glycoprotein

The T200 glycoprotein is another major glycoprotein of thymocytes (50,000 molecules/cell) and on a weight basis is present in about the same amount as Thy-1 glycoprotein. First identified in the mouse (Trowbridge, 1978; Trowbridge *et al.*, 1975), homologous molecules have now been identified in the rat and in humans (Omary *et al.*, 1980: Sunderland *et al.*, 1979; Dalchau *et al.*, 1980). The tissue distribution and structural properties of the glycoprotein are similar in all three species. In the mouse and rat, T200 glycoprotein (also known as leukocyte common antigen) exists in two alternative allotypic forms defined in the mouse by Ly-5 alloantisera. T200 glycoprotein also exhibits nonallelic microheterogeneity, which can be detected by SDS–polyacrylamide gel electrophoresis and serologically. The B-lymphocyte variant of the glycoprotein (apparent M_r = 220,000) appears larger than the glycoprotein found on thymocytes and T cells (apparent M_r = 170,000–190,000), and these variants of the glycoprotein can be distinguished by means of monoclonal antibodies (Trowbridge, 1978; Coffman and Weissman, 1981; Dalchau and Fabre, 1981; Trowbridge *et al.*, 1975; Sarmiento *et al.*, 1982). The glycoprotein is found on all hematopoietic cells, except mature cells of the erythroid series, but is not present on cells of nonhematopoietic tissues. For this reason, monoclonal antibodies against human T200 glycoprotein are likely to find clinical use in the differential diagnosis of malignant lymphoma (Battifora and Trowbridge, 1982). There is strong evidence that T200 glycoprotein spans the plasma membrane and that several serine residues in the cytoplasmic portion of the molecule are phosphorylated (Omary and Trowbridge, 1980). T200-negative murine lymphoma cell lines have been obtained by cytotoxic selection, and from one of these lines an interesting revertant has been obtained (Hyman *et al.*, 1982). This revertant expresses a truncated form of T200 glycoprotein (M_r = 105,000) that appears to lack the cytoplasmic portion of the glycoprotein. A comparison of wild-type and revertant cell lines may eventually lead to an understanding of the interactions of T200 glycoprotein with intracellular proteins such as cytoskeletal components.

C. W3/13 Glycoprotein

The W3/13 glycoprotein (35,000 molecules/cell) is the third major cell-surface glycoprotein of rat thymocytes (W. R. A. Brown *et al.*, 1981). It is likely that homologous glycoproteins are present in other species, although none have been clearly identified. The glycoprotein has an apparent M_r of 95,000 as determined by SDS–polyacrylamide gel electrophoresis. In addition to thymocytes, it is found on peripheral T cells, plasma cells, neutrophils, and brain. The composition of W3/13 glycoprotein is 60% (w/w) carbohydrate, in the form of O-glycosidically linked oligosaccharides composed exclusively of galactose, galactosamine, and sialic acid. As noted by W. R. A. Brown *et al.* (1981), this is similar to the oligosaccharide structure of the erythrocyte glycoprotein, glycophorin. However, since the function of glycophorin is also unknown (Chapter 9), this does not give any insight as to the function of W3/13 glycoprotein.

D. Mac-1

Mac-1, a differentiation antigen selectively expressed on murine phagocytes (Springer *et al.*, 1978, 1979), is an antigenic determinant of a glycoprotein complex consisting of two glycoprotein components (M_r = 170,000 and 95,000, respectively). Recently it has been shown that other murine hematopoietic cells express a related glycoprotein complex that consists of a similar 95,000 molecular weight polypeptide complexed with a different high-molecular-weight component (Trowbridge and Omary, 1981b). Thus, there seems to be a family of MAC-1 glycoproteins that in some respects is similar to that T200 glycoprotein family. Interestingly, monoclonal antibodies against the glycoprotein complex found on nonphagocytic hematopoietic cells referred to by Springer and his colleagues as LFA-1 (Davignon *et al.*, 1981a,b; Kurzinger *et al.*, 1981) inhibit T-cell-mediated cytotoxicity (Davignon *et al.*, 1981b; Sarmiento *et al.*, 1982). There is evidence that these monoclonal antibodies block the interaction between cytotoxic T lymphocytes and their target cells (Davignon *et al.*, 1981b), and it appears likely that the Mac-1 glycoprotein family plays some role in cell–cell interactions within the hematopoietic system. A rat monoclonal antibody against Mac-1 has been shown to react with human monocytes and natural killer cells (Ault and Springer, 1981), implying that homologues of the Mac-1 glycoproteins probably exist in humans.

E. Pgp-1 Glycoprotein

Pgp-1 glycoprotein was first described as a major polymorphic cell surface glycoprotein of murine fibroblasts (M_r = 80,000) (Hughes and August, 1981). Subsequently it was shown also to be a major surface component of hemotopoietic cells of the myeloid series (Hughes *et al.*, 1981; Trowbridge *et al.*, 1982a; Lesley and Trowbridge, 1982). By SDS–polyacrylamide gel elec-trophoresis, the apparent M_r of the glycoprotein found on hematopoietic cells is 95,000, somewhat larger than the fibroblast form of the antigen. Genetic map-ping studies show that the allelic genes controlling the expression of the poly-morphism map to chromosome 2 and are closely linked to the antigenic determi-nant Ly-mll (Lesley and Trowbridge, 1982). Although present in higher amounts on myeloid cells, Pgp-1 glycoprotein is also expressed in detectable amounts on other hematopoietic cell types, including prothymocytes (Trowbridge *et al.*, 1982). Pgp-1 glycoprotein is not, however, found on the majority of thymocytes. Only about 5% of thymocytes express very low amounts of the glycoprotein. The molecule is found in abundance on several Thy-1$^+$ T-cell lymphomas, and this raises the possibility that the antigen may be a marker for either a specific lineage of T cells or a particular stage of differentiation. As a monoclonal antibody against a nonpolymorphic determinant of the murine glycoprotein reacts with human peripheral blood leukocytes, a similar glycoprotein probably exists in humans (Trowbridge *et al.*, 1982a).

F. Transferrin Receptor

The transferrin receptor is a striking example of the difficulties that arise if attempts are made to define the term differentiation antigen in a narrow sense. There is no question that the transferrin receptor fulfills many of the criteria of a differentiation antigen. Its expression on certain hematopoietic tissues changes during development (see later), and transferrin receptors are found on erythroid cells at a particular stage of their maturation. On the other hand, in most tissues transferrin receptors are not expressed on specific cell types but are associated with growing cells in general, and the receptor can be considered to be a pro-liferation-associated or activation antigen. In our own studies of the transferrin receptor, it was not immediately obvious whether we were studying a differentia-tion antigen, a tumor-associated antigen, or a proliferation antigen, and indeed in practical terms considerable work is required to distinguish between these classes of surface antigen. A similar situation to that of the transferrin receptor probably exists for the cell-surface receptor for T-cell growth factor. This receptor is associated with proliferating T cells and perhaps also with activated B lympho-cytes (Robb, 1984) but as far as is known is not found on other cell types (Morgan *et al.*, 1976; Gootenberg *et al.*, 1981; Robb *et al.*, 1981). It too, therefore, has characteristics of a proliferation antigen and differentiation antigen.

The transferrin receptor is virtually unique among the presently identified cell-surface glycoproteins in that it is reasonably well characterized structurally and its function is known. Although receptors for transferrin, the major serum iron transport protein, had been identified on immature erythroid cells in 1963 (Jandl and Katz, 1963), the receptor was first purified and adequately characterized from placenta in 1979 (Seligman *et al.*, 1979). About this time, it was also realized that transferrin receptors were abundant on proliferating cells *in vitro* (Larrick and Cresswell, 1979). More recently, several monoclonal antibodies obtained by immunizing mice with various hematopoietic cells were shown to react with the transferrin receptor of human cells (Trowbridge and Omary, 1981a; Sutherland *et al.*, 1981; Haynes *et al.*, 1981; Goding and Burns, 1981). There is general agreement that the transferrin receptor is a glycoprotein that consists of two similar disulfide-bonded subunits with an apparent M_r of 95,000. The receptor appears to be associated with covalently bound fatty acid (Omary and Trowbridge, 1981a,b). A monoclonal antibody against the transferrin receptor of mouse cells has also been obtained, and the murine receptor has a similar chemical structure (Trowbridge *et al.*, 1982a). More recent studies (Kuhn *et al.*, 1984) report the cloning of the receptor using a DNA-transfer approach and selection of transferrin-receptor-expressing cells by fluorescence-activated cell sorting.

The receptor is of biological interest for several reasons. Continued characterization of the receptor should give some clues as to the regulation of iron metabolism in mammalian cells, which is currently poorly understood. Blocking of the transferrin-receptor function with monoclonal antibodies to the binding site leads to an arrest in cell proliferation (Trowbridge and Lopez, 1982), a result consistent with the idea that endocytosis (by which transferrin receptor:diferric transferrin enters the cell) is required for iron uptake in most cells (Hopkins and Trowbridge, 1983). Second, the transferrin receptor may be useful as a marker for stem cells within the hematopoietic system and possibly in other tissues. In the adult mouse less than 1% of spleen cells and thymocytes and only about 5% of bone-marrow cells express detectable amounts of transferrin receptor, whereas in fetal liver and neonatal spleen, both sites of active hematopoiesis, about half the nucleated cells are positive when stained with monoclonal anti-receptor antibodies and analyzed by flow cytometry (Trowbridge *et al.*, 1982a). It will be of some interest to determine the proportion of CFU-s, CFU-c, and other hematopoietic stem-cell populations that express transferrin receptors and whether this varies after treatments that stimulate hematopoiesis. One curious fact is that although the thymus is believed to be a site of intense cellular proliferation, few adult thymocytes have detectable numbers of receptors on their surface. This apparent contradiction is partially explained by the fact that even in regenerating thymus only about 10% of the thymocyte population is actively cycling (Bryant, 1972; Boersma *et al.*, 1981). Clearly, how the proliferation of thymocytes is regulated is an extremely important question, and it is possible transferrin receptors could be involved in the process.

The receptor for transferrin is readily detectable on dividing cells, including tumor cells and normal cells proliferating in tissue culture, but not on fully differentiated, nondividing cells (Trowbridge et al., 1982a). Since transferrin receptors may be selectively expressed on certain tumors in vivo, such as T-cell leukemias, it is possible that anti-receptor monoclonal antibodies may be useful as therapeutic agents. Preliminary experiments have shown, for example, that the growth of human melanoma cells as subcutaneous tumors in nude mice can be inhibited by intravenous injection of anti-transferrin-receptor antibodies (Trowbridge and Domingo, 1981). More importantly, an anti-receptor antibody has been obtained that blocks transferrin binding and has a direct pharmacological effect on the growth of a human T leukemia cell line in vitro (Trowbridge and Lopez, 1982). At present, however, much more work is required to determine whether the therapeutic use of monoclonal antibodies in general, and anti-transferrin-receptor antibody in particular, is feasible.

There is also considerable interest in the use of the ligand (IL-2) for the T-cell growth-factor receptor in therapy for some diseases. IL-2 is a lymphokine synthesized and secreted by some T cells after activation in the presence of antigen (mitogen) and another lymphokine, interleukin 1 (IL-1) (Morgan et al., 1976). The gene encoding IL-2 has been cloned and is located as a single copy on human chromosome 4 (Taniguchi et al., 1983; Clark et al., 1984; Holbrook et al., 1984; Robb, 1984). IL-2 allows for long-term growth in tissue culture of T-cell lines with helper and suppressor functions, as well as promoting growth and/or activity of natural killer cells (Watson, 1979; Coutinho et al., 1979; Gillis et al., 1978; Shaw et al., 1978). IL-2 administration itself is currently under intense investigation as a possible therapy in some neoplasms and in acquired immunodeficiency syndrome. Note that IL-2 only exerts its effect on T cells after interactions with specific membrane receptors (Robb et al., 1981), which, like the interacting ligand, are only expressed on T cells after activation. The receptors (55,000 glycoprotein) are expressed at high numbers on human T-cell leukemia/lymphoma virus (HTLV-1) infected T cells and may be associated with uncontrolled growth in these cells (Poiesz et al., 1981; Kolyanaraman et al., 1982). Separate groups have recently repeated the molecular cloning of cDNA encoding human interleukin-2 receptor, which should advance our understanding of its structure/function and the way in which control of its expression regulates early events in T-cell activation and membrane signal transduction (Leonard et al., 1984; Nikaido et al., 1984).

G. OKT3

OKT3 is a human lymphocyte cell-surface glycoprotein that has no known counterpart in other species. The major molecular species detected with anti-OKT3 monoclonal antibody after metabolic labeling or surface iodination of cells and immunoprecipitation is a glycoprotein with an apparent M_r of 19,000. There are additional polypeptides associated with this moiety (Van Agthoven *et al.*, 1981). The antigen is present on 10–20% of thymocytes, the majority of which are found in the thymic cortex (Bhan *et al.*, 1980), and all peripheral T cells. The initial primary interest of OKT3 glycoprotein centered around the fact that the monoclonal antibody defining this antigen is an extremely potent mitogen. Maximum proliferation of peripheral T cells is obtained at a concentration of 1–10 ng/ml (10^{-11} to $10^{-12} M$) (Chang *et al.*, 1981; Van Wauwe *et al.*, 1980), which is within the concentration range at which many polypeptide hormones are active. It has been estimated that there are about 5×10^4 molecules of OKT3 per peripheral T cell, and mitogenesis appears to be triggered when only 10–20% of these sites are occupied (Van Wauwe *et al.*, 1980). At higher concentrations ($10^{-8} M$), OKT3 monoclonal antibody blocks target-cell lysis by allogeneic cytotoxic T cells (Chang *et al.*, 1981).

Subsequent studies have shown that monoclonal antibodies defining clonotypic markers on T cells (putative markers to the T cell receptor, TcR) co-migrated with OKT3 (Meuer *et al.*, 1983; Acuto *et al.*, 1983). More recent studies in both mouse and human have proceeded apace to define the gene(s) encoding the α and β chain of the clonally distributed TcR (Hedrick *et al.*, 1984; Yanagi *et al.*, 1984). Considerable homology with genomic arrangement of the genes for heavy chains of immunogobulin has become apparent. In the context of the present discussion, however, most pertinent is the singular failure to date, from what we know of the structure of α and β chains of the TcR, to understand such phenomena as major histocompatibility complex (MHC) restriction of T-cell recognition (i.e., the mechanism by which T cells are restricted to see antigen only in association with other gene products of the histocompatibily locus). The ability of anti-OKT3, anti-clonotype, and anti-OKT4 (OKT8) antibodies to inhibit stimulation of antigen-specific T cells may point towards a resolution of this problem (Meuer *et al.*, 1983). It seems likely that the receptor on the T-cell surface may represent a complex of a number of molecules that contribute differentially to the antigen, class, and MHC restriction specificity of the cells.

IV. FUTURE DIRECTIONS

A. Functions of Differentiation Antigens

Although the title of this chapter is ''Cell-Surface Receptors and Differentiation,'' little has been said about the functions of most of the cell-surface molecules discussed. This is, of course, because at present we still have little idea as to the roles most of these glycoproteins play. One general point can be made, however. If one accepts the definition of differentiation antigen used in this discussion, then it is clear that most of these molecules are unlikely to play a direct role in the process that has been termed morphogenetic differentiation (Bennett *et al.*, 1972) or morphogenesis (Boyse and Old, 1969), whereby the cells of a developing metazoan become divided into separate groups or tissues according to their function (Bennett *et al.*, 1972). Differentiation antigens are likely to include molecules that are involved in quite general cellular processes, while others probably perform roles related to the specialized functions of the differentiated cells upon which they are expressed.

The challenge will be to sort out which cell-surface molecules are involved in which of these processes. While it is not immediately obvious how this can be done, some potential strategies can be outlined. First, structural information about specific cell-surface molecules may give clues about their function. Thy-1 glycoprotein is one example in this regard as it now is reasonable to pursue the hypothesis that Thy-1 is involved in molecular recognition given its structural homology with immunoglobulin variable region domains. Perhaps an even more striking example is that of p97, a glycoprotein selectively expressed on human melanoma cells (Woodbury *et al.*, 1980; J. P. Brown *et al.*, 1981). Sequence studies of the N-terminus of the glycoprotein revealed a strikingly homology with transferrin and lactoferrin and led to the demonstration that the glycoprotein shared the functional property of these secreted glycoproteins of binding Fe^{3+} (Brown *et al.*, 1982).

Second, there is little doubt that the use of somatic genetic techniques to study cell-surface molecules will make a major contribution to understanding their function. The feasibility of such an approach has been clearly demonstrated by the genetic studies of Thy-1, TL, and T200 glycoproteins carried out over the last decade by Hyman and his co-workers (Hyman, 1973; Hyman and Stallings, 1977, 1978; Hyman *et al.*, 1980a,b; Hyman and Trowbridge, 1981), and in the past 2 years this approach has been extended to a genetic analysis of the roles of cell surface molecules of functional T-cell lines (Dialynas *et al.*, 1981; Nabholz, 1982). Such genetic studies, in principle, can also be used to analyze the role of cell-surface molecules in the regulation of differentiation (Hyman *et al.*, 1980a).

Finally, in the event that all else fails, one may be forced to rely from time to time upon serendipity. In our own studies of the transferrin receptor, the initial observations that led to the experiments showing that the cell-surface glycoprotein we were studying bound transferrin occurred fortuitously (Trowbridge and Omary, 1981a). The main point is, of course, that it is not possible to predict the technological and conceptual advances that may occur in the future.

B. Genetic Regulation of the Expression of Differentiation Antigens

As mentioned in the introduction, one of the major questions concerning differentiation antigens is how their expression is genetically regulated. The best studied case is Thy-1 glycoprotein, where it has been shown that at least four

genes other than the structural gene for Thy-1 are required for the expression of
the glycoprotein on the cell surface (Hyman and Trowbridge, 1978). While it is
probable that all these genes control posttranslational modifications of Thy-1
glycoprotein, such as glycosylation, there is also evidence for a regulatory
gene(s) that extinguished the expression of Thy-1 in myeloma–lymphoma
hybrids (Hyman and Stallings, 1978). However, it is clear that there are limita-
tions to the information that can be obtained by somatic genetic techniques alone,
and there is a need to develop the necessary tools to analyze these questions at a
molecular level. The recent successes in the molecular cloning of cDNA coding
for HLA and β_2-microglobulin (Ploegh et al., 1980; Parnes et al., 1981) suggest
that this will prove to be an appropriate route of approach to isolate many of the
more abundant membrane glycoproteins within a few years (e.g., the recent
cloning of genes for the transferrin receptor, T-cell growth-factor receptor, referred to already).

C. Clinical Applications of Monoclonal Antibodies against Differentiation Antigens

The knowledge that has been acquired in the past several years about human
differentiation antigens and the availability of monoclonal antibodies that detect
them has opened up new possibilities in clinical diagnosis and therapy. While
this is of little direct relevance to understanding the role of cell-surface receptors
and differentiation, the clinical significance of monoclonal antibodies fuels
efforts to detect new cell-surface molecules and, indirectly, information accrued
in clinical and preclinical studies is likely to contribute useful basic information
about human differentiation antigens.

The most clear-cut clinical use of monoclonal antibodies against differentiation antigens is in the differential diagnosis of tumors, and it is already established that anti-T200 monoclonal antibody is useful in the diagnosis of lymphoma (Battifora and Trowbridge, 1982; Pizzolo et al., 1980). Monoclonal antibodies against human lymphocyte differentiation antigens are also being used to
classify different forms of acute lymphoblastic leukemia (Minowada et al., 1981;
Haynes, 1981; Sallan et al., 1980). Another possible use of monoclonal antibodies against differentiation antigens is in the detection of small tumor masses
by localization of radiolabeled antibody using photoscanning if technical obsta-
cles can be overcome. Finally, a major effort is being put into establishing
whether monoclonal antibodies against differentiation antigens or proliferation
antigens may be of therapeutic value in the treatment of cancer, either by targeting toxic agents to tumor cells or by direct effects on tumor cell growth (Trowbridge and Domingo, 1981; Trowbridge and Lopez, 1982, and references there-
in). There is already a growing body of experimental data to suggest that this
approach may have validity for tumors of mature cells in the immune system that
express clonotypic surface markers (idiotypes), e.g., lymphomas (Miller et al.,

1982). It is too early to say what the final outcome of these experiments will be, but the prospects are good that such approaches will be useful in the treatment of some kinds of tumor.

ACKNOWLEDGMENTS

I thank my colleagues, particularly Robert Hyman, Bishr Omary, Jayne Lesley, Catherine Mazauskas, Derrick Domingo, Roberta Schulte, Hector Battifora, and Amy Koide for their contributions to my own work described herein. The work was supported in part by American Cancer Society grant CH-175 and grant CA 17733 from the National Cancer Institute.

REFERENCES

Acuto, O., Meuer, S. C., Hodgdon, J. C., Schlossman, S. F., and Reinherz, E. L. (1983). *J. Exp. Med.* **158**, 1368.
Arndt, R., Stark, R., Klein, P., and Thiele, H.-G. (1978). *Immunology* **35**, 95.
Ault, K. A., and Springer, T. A. (1981). *J. Immunol.* **126**, 359.
Austyn, J. M., and Gordon, S. (1981). *Eur. J. Immunol.* **11**, 805.
Battifora, H., and Trowbridge, I. S. (1982). *Cancer* **51**, 816.
Bennett, D., Boyse, E. A., and Old, L. J. (1972). *In* "Cell Interactions" (L. Silvestri, ed.), pp. 247–263. Third Lepetit Colloquium, North Holland, Amsterdam.
Bhan, A. K., Reinherz, E. L., Poppema, S., McCluskey, R. T., and Schlossman, S. F. (1980). *J. Exp. Med.* **152**, 771.
Boersma, W., Betel, I., R. Daculsi, and van Der Westen, G. (1981). *Cell Tissue Kinet.* **14**, 179.
Boyse, E. A. (1973). *In* "Current Research in Oncology" (C. B. Anfinsen, M. Potter, and A. N. Schechter, eds.), pp. 57–94. Academic Press, New York.
Boyse, E. A., and Old, L. J. (1969). *Annu. Rev. Genet.* **3**, 269.
Brown, J. P., Woodbury, R. G., Hart, C. E., Hellstrom, I., and Hellstrom, K. E. (1981). *Proc. Natl. Acad. Sci. U.S.A.* **78**, 539.
Brown, J. P., Hewick, R. M., Hellstrom, I., Hellstrom, K. E., Doolittle, R. F., and Dreyer, W. J. (1982). *Nature (London)* **296**, 171.
Brown, W. R. A., Barclay, A. N., Sunderland, C. A., and Williams, A. F. (1981). *Nature (London)* **289**, 456.
Bryant, B. J. (1972). *Eur. J. Immunol.* **2**, 38.
Campbell, D. G., Gagnon, J., Reid, K. B. M., and Williams, A. F. (1981). *Biochem. J.* **195**, 13.
Cantor, H., and Boyse, E. A. (1977). *Immunol. Rev.* **33**, 105.
Chang, T. W., Kung, P. C., Gingras, S. P., and Goldstein, G. (1981). *Proc. Natl. Acad. Sci. U.S.A.* **78**, 1805.
Chapman, A., Fujimoto, K., and Kornfeld, S. (1980). *J. Biol. Chem.* **255**, 4441.
Clark, S. C., Arya, S. K., Wong-Staal, F., Matsumoto-Kobsayashi, M., Kay, R. M., Kaufman, R. J., Brown, E. L., Shoemaker, C., Copeland, T., Oroszlan, S., Smith, K., Sarngadharan, M. G., Linder, S. G., and Gallo, R. C. (1984). *Proc. Natl. Acad. Sci. U.S.A.* **8**, 2543.
Coffman, R. L., and Weissman, I. L. (1981). *Nature (London)* **289**, 681.
Cohen, F. E., Novotny, J., Sternberg, M. J. E., Campbell, D. G., and Williams, A. F. (1981). *Biochem. J.* **195**, 31.
Cotmore, S. F., Crowhurst, S. A., and Waterfield, M. D. (1981). *Eur. J. Immunol.* **11**, 597.
Coutinho, A., Larsson, E.-L., Gronvik, K.-O., and Anderson, J. (1979). *Eur. J. Immunol.* **9**, 587.
Dalchau, R., and Fabre, J. W. (1979). *J. Exp. Med.* **149**, 576.

Dalchau, R., and Fabre, J. W. (1981). *J. Exp. Med.* **153**, 753.
Dalchau, R., Kirkley, J., and Fabre, J. W. (1980). *Eur. J. Immunol.* **10**, 737.
Davignon, D., Martz, E., Reynolds, T., Kurzinger, K., and Springer, T. A. (1981a). *Proc. Natl. Acad. Sci. U.S.A.* **78**, 4535.
Davignon, D., Martz, E., Reynolds, T., Kurzinger, K., and Springer, T. A. (1981b). *J. Immunol.* **127**, 590.
Dialynas, D. P., Loken, M. R., Glasebrook, A. L., and Fitch, F. W. (1981). *J. Exp. Med.* **153**, 595.
Esselman, W. J., and Miller, H. C. (1974). *J. Exp. Med.* **139**, 445.
Gillis, S., Baker, P. E., Ruscetti, W., and Smith, K. A. (1978). *J. Exp. Med.* **148**, 1093.
Goding, J. W., and Burns, G. F. (1981). *J. Immunol.* **127**, 1256.
Gootenberg, J. E., Ruscetti, F. W., Mier, J. W., Gazdar, A., and Gallo, R. C. (1981). *J. Exp. Med.* **154**, 1403.
Gorer, P. A., and O'Gorman, P. (1956). *Transplant. Bull.* **3**, 142.
Greaves, M., and Janossy, G. (1978). *Biochim. Biophys. Acta* **516**, 193.
Haynes, B. F. (1981). *Immunol. Rev.* **57**, 127.
Haynes, B. F., Hemler, M., Cotne, T., Mann, D. L., Eisenbarth, G. S., Strominger, J. L., and Fauci, A. S. (1981). *J. Immunol.* **127**, 347.
Hedrick, S. M., Cohen, D. I., Nielsen, E. A., and Davis, M. M. (1984). *Nature (London)* **308**, 149.
Hirsch, S., Austyn, J. M., and Gordon, S. (1981). *J. Exp. Med.* **154**, 713.
Hogarth, P. M., Craig, J., and McKenzie, I. F. C. (1980). *Immunogenetics* **11**, 65.
Holbrook, N. J., Smith, K. A., Fornace Jr., A. J., Comeau, C. M., Wiskocil, R. L., and Crabtree, G. R. (1984). *Proc. Natl. Acad. Sci. U.S.A.* **81**, 1634.
Hopkins, C. R., and Trowbridge, I. (1983). *J. Cell Biol.* **97**, 508.
Hughes, E. N., and August, J. T. (1981). *J. Biol. Chem.* **256**, 664.
Hughes, E. N., Mengod, G., and August, J. T. (1981). *J. Biol. Chem.* **256**, 7023.
Hunt, S. V. (1979). *Eur. J. Immunol.* **9**, 853.
Hunt, S. V., Mason, D. W., and Williams, A. F. (1978). *Eur. J. Immunol.* **7**, 817.
Hyman, R. (1973). *J. Natl. Cancer Inst.* **50**, 415.
Hyman, R., and Stallings, V. (1977). *Immunogenetics* **4**, 171.
Hyman, R., and Stallings, V. (1978). *Immunogenetics* **6**, 447.
Hyman, R., and Trowbridge, I. S. (1978). *In* "Differentiation of Normal and Neoplastic Hematopoietic Cells" (B. Clarkson, P. Marks, and J. Till, eds.), pp. 741–754. Cold Spring Harbor Laboratory, Cold Spring Harbor, N.Y.
Hyman, R., and Trowbridge, I. (1981). *Immunogenetics* **12**, 511.
Hyman, R., Trowbridge, I., and Cunningham, K. (1980a). *J. Cell. Physiol.* **105**, 469.
Hyman, R., Cunningham, K., and Stallings, V. (1980b). *Immunogenetics* **12**, 511.
Hyman, R., Trowbridge, I., Stallings, V., and Trotter, J. (1982). *Immunogenetics* **15**, 413.
Jandl, J. H., and Katz, J. H. (1963). *J. Clin. Invest.* **42**, 314.
Kersey, J. H., LeBien, T. W., Abramson, C. S., Newman, R., Sutherland, R., and Greaves, M. (1981). *J. Exp. Med.* **153**, 726.
Kohler, G., and Milstein, C. (1975). *Nature (London)* **256**, 495.
Kolyanaraman, V. S., Sarngadharan, M. G., Nakao, Y., Ito, Y., Aoki, T., and Gallo, R. C. (1982). *Proc. Natl. Acad. Sci. U.S.A.* **79**, 1653.
Kuchel, P. W., Campbell, D. G., Barclay, A. N., and Williams, A. F. (1978). *Biochem. J.* **169**, 411.
Kuhn, L. C., McClelland, A., and Ruddle, F. H. (1984). *Cell* **37**, 95.
Kung, P. C., Goldstein, G., Reinherz, E. L., and Schlossman, S. F. (1979). *Science* **206**, 347.
Kurzinger, K., Reynolds, T., Germain, R. N., Davignon, D., Martz, E., and Springer, T. A. (1981). *J. Immunol.* **127**, 596.
Lanier, L. L., Warner, N. L., Ledbetter, J. A., and Herzenberg, L. A. (1981). *J. Exp. Med.* **153**, 998.

Larrick, J. W., and Cresswell, P. (1979). *J. Supramol. Struct.* **11**, 579.

Ledbetter, J. A., Goding, J. W., Tsu, T. T., and Herzenberg, L. A. (1979). *Immunogenetics* **8**, 347.

Ledbetter, J. A., Rouse, R. V., Micklem, H. S., and Herzenberg, L. A. (1980). *J. Exp. Med.* **152**, 280.

Ledbetter, J. A., Evans, R. L., Lipinski, M., Cunningham-Rundles, C., Good, R. A., and Herzenberg, L. A. (1981). *J. Exp. Med.* **153**, 310.

Leonard, W. J., Depper, J. M., Crabtree, G. R., Rudikoff, S., Pumphrey, J., Robb, R. J., Kranke, M., Svetlik, P. B., Peffer, N. J., Waldmann, T. A., and Greene, W. C. (1984). *Nature (London)* **311**, 626.

Lesley, J., and Trowbridge, I. W. (1982). *Immunogenetics* **15**, 313.

Levy, R., Dilley, J., Fox, R. I., and Warnke, R. (1979). *Proc. Natl. Acad. Sci. U.S.A.* **76**, 6552.

McEver, R. P., Baenziger, N. L., and Majerus, P. W. (1980). *J. Clin. Invest.* **66**, 1311.

McMaster, W. R., and Williams, A. F. (1979). *Eur. J. Immunol.* **9**, 426.

Martin, P. J., Hansen, J. A., Nowinski, R. C., and Brown, M. A. (1980). *Immunogenetics* **11**, 429.

Meuer, S. C., Fitzgerald, K. A., Hussey, R. E., Hodgdon, J. C., Schlossman, S. F., and Reinherz, E. L. (1983). *J. Exp. Med.* **157**, 705.

Metzgar, R. S., Borowitz, J. M., Jones, N. H., and Dowell, B. L. (1981). *J. Exp. Med.* **154**, 1249.

Milewicz, C., Miller, H. C., and Esselman, W. J. (1976). *J. Immunol.* **117**, 1774.

Miller, R. A., Maloney, D. G., Warnke, R., and Levy, R. (1982). *N. Engl. J. Med.* **306**, 517.

Milstein, C., and Lennox, E. (1980). *Curr. Top. Dev. Biol.* **14**, 1.

Minowada, J., Koshiba, H., Sagawa, K., Morita, M., Saito, M., Pauly, J. L., Kubonishi, I., Lok, M. S., Tatsumi, E., Han, T., Srivastava, B. I. S., Ohnuma, T., Trowbridge, I. S., Evans, R. L., Goldstein, G., Freeman, A. I., and Henderson, E. S. (1981). *In* "Leukemia Markers" (W. Knapp, ed.), pp. 179–182. Academic Press, London.

Morgan, D. A., Ruscetti, F. W., and Gallo, R. C. (1976). *Science* **193**, 1007.

Nabholz, M. (1982). *In* "Isolation, Characterization and Utilization of T Lymphocytes" (G. Fathman and F. Fitch, eds.), pp. 205–215. Academic Press, New York.

Nikaido, T., *et al.* (1984). *Nature (London)* **311**, 631.

Nussenzweig, M. C., Steinman, R. M., Unkeless, J. C., Witmer, M. D., Gutchinov, B., and Cohn, Z. A. (1981). *J. Exp. Med.* **154**, 168.

Old, L. J., and Stockert, E. (1977). *Annu. Rev. Genet.* **11**, 127.

Omary, M. B., and Trowbridge, I. S. (1980). *J. Biol. Chem.* **255**, 1662.

Omary, M. B., and Trowbridge, I. S. (1981a). *J. Biol. Chem.* **256**, 4715.

Omary, M. B., and Trowbridge, I. S. (1981b). *J. Biol. Chem.* **256**, 12888.

Omary, M. B., Trowbridge, I. S., and Battifora, H. (1980). *J. Exp. Med.* **152**, 842.

Parnes, J. R., Velan, B., Felsenfeld, A., Ramanathan, L., Ferrini, V., Appella, E., and Seidman, J. G. (1981). *Proc. Natl. Acad. Sci. U.S.A.* **78**, 2253.

Pesando, J. M., Ritz, J., Levine, H., Terhorst, C., Lazarus, H., and Schlossman, S. F. (1980). *J. Immunol.* **124**, 2794.

Pizzolo, G., Sloane, J., Beverley, P., Thomas, J. A., Bradstock, K. F.. Mattingly, S., and Janossy, G. (1980). *Cancer* **46**, 2640.

Ploegh, H., Orr, H. T., and Strominger, J. L. (1980). *Proc. Natl. Acad. Sci. U.S.A.* **77**, 6081.

Poiesz, B. J., Ruscetti, F. W., Reitz, M. J., Kolyanaraman, V. S., and Gallo, R. C. (1981). *Nature (London)* **294**, 268.

Reif, A. E., and Allen, J. M. V. (1964). *J. Exp. Med.* **120**, 413.

Reinherz, E. L., and Schlossman, S. F. (1980). *Cell* **19**, 821.

Reinherz, E. L., Kung, P. C., Goldstein, G., and Schlossman, S. F. (1979). *J. Immunol.* **123**, 1312.

Ritz, J., Pesando, J. M., Notis-McConarty, J., Lazarus, H., and Schlossman, S. F. (1980). *Nature (London)* **283**, 583.

Robb, R. J. (1984). *Immunol. Today* **5**, 203.

Robb, R. J., Munck, A., and Smith, K. A. (1981). *J. Exp. Med.* **154,** 1455.

Royston, I., Majda, J. A., Baird, S. M., Meserve, B. L., and Griffiths, J. C. (1980). *J. Immunol.* **125,** 725.

Sallan, S. E., Ritz, J., Pesando, J., Gelber, R., O'Brien, C., Hitchcock, S., Coral, F., and Schlossman, S. F. (1980). *Blood* **55,** 395.

Sarmiento, M., Loken, M. R., Trowbridge, I. S., Coffman, R., and Fitch, F. W. (1982). *J. Immunol.* **128,** 1676.

Seligman, P. A., Schleicher, R. B., and Allen, R. H. (1979). *J. Biol. Chem.* **254,** 9943.

Shaw, J., Monticone, V., and Paetkau, V. (1978). *J. Immunol.* **120,** 1967.

Springer, T., Galfre, G., Secher, D. S., and Milstein, C. (1978). *Eur. J. Immunol.* **8,** 539.

Springer, T., Galfre, G., Secher, D. S., and Milstein, C. (1979). *Eur. J. Immunol.* **9,** 301.

Stern, P. L., Willison, K. R., Lennox, E., Galfre, G., Milstein, C., Secher, D., and Ziegler, A. (1978). *Cell* **14,** 775.

Sunderland, C. A., McMaster, W. R., and Williams, A. F. (1979). *Eur. J. Immunol.* **9,** 155.

Sutherland, R., Delia, D., Schneider, C., Newman, R., Kemshead, J., and Greaves, M. (1981). *Proc. Natl. Acad. Sci. U.S.A.* **78,** 4515.

Taniguchi, T., Matsui, H., Funita, T., Takaoka, C., Kashima, N., Yoshimoto, R., and Hamuro, J. (1983). *Nature (London)* **302,** 305.

Terhorst, C., Van Agthoven, A., Reinherz, E., and Schlossman, S. (1980). *Science* **209,** 520.

Terhorst, C., Van Agthoven, A., LeClair, K., Snow, P., Reinherz, E., and Schlossman, S. (1981). *Cell* **23,** 771.

Trowbridge, I. S. (1978). *J. Exp. Med.* **148,** 313.

Trowbridge, I. S., and Domingo, D. (1981). *Nature (London)* **294,** 171.

Trowbridge, I. S., and Hyman, R. (1979). *Cell* **17,** 503.

Trowbridge, I. S., and Lopez, F. (1982). *Proc. Natl. Acad. Sci. U.S.A.* **79,** 1175.

Trowbridge, I. S., and Omary, M. B. (1981a). *Proc. Natl. Acad. Sci. U.S.A.* **78,** 3039.

Trowbridge, I. S., and Omary, M. B. (1981b). *J. Exp. Med.* **154,** 1517.

Trowbridge, I. S., Ralph, P., and Bevan, M. J. (1975). *Proc. Natl. Acad. Sci. U.S.A.* **72,** 157.

Trowbridge, I. S., Lesley, J., and Schulte, R. (1982a). *J. Cell. Physiol.* **112,** 403.

Trowbridge, I. S., Lesley, J., Schulte, R., Hyman, R., and Trotter, J. (1982b). *Immunogenetics* **15,** 299.

Ugolini, V., Nunez, G., Smith, R. G., Stasty, P., and Capra, J. D. (1980). *Proc. Natl. Acad. Sci. U.S.A.* **77,** 6764.

Van Agthoven, A., Terhorst, C., Reinherz, E., and Schlossman, S. F. (1981). *Eur. J. Immunol.* **11,** 18.

Van Wauwe, J. P., De Mey, J. R., and Goossens, J. G. (1980). *J. Immunol.* **124,** 2708.

Vitetta, E. S., Boyse, E. A., and Uhr, J. W. (1973). *Eur. J. Immunol.* **3,** 446.

Wang, C. Y., Good, R. A., Ammirata, P., Dymbort, G., and Evans, R. L. (1981). *J. Exp. Med.* **151,** 1539.

Watson, J. (1979). *J. Exp. Med.* **150,** 1510.

White, R. A. H., Mason, D. W., Williams, A. F., Galfre, G., and Milstein, C. (1978). *J. Exp. Med.* **148,** 664.

Williams, A. F., and Gagnon, J. (1982). *Science* **216,** 696.

Williams, A. F., Galfre, G., and Milstein, C. (1977). *Cell* **12,** 663.

Woodbury, R. G., Brown, J. P., Yeh, M.-Y., Hellstrom, I., and Hellstrom, E. (1980). *Proc. Natl. Acad. Sci. U.S.A.* **77,** 2183.

Yanagi, Y., Yoshikai, Y., Legget, K., Clark, S. P., Aleksander, I., and Mak, T. W. (1984). *Nature (London)* **308,** 145.

11

Some Biochemical Properties Associated with the Metastatic Potential of Tumor Cells

A. RAZ AND I. J. FIDLER

I. INTRODUCTION

Metastasis, the spread and growth of neoplastic cells, is responsible for most failures of cancer therapy. Despite the development of complex and aggressive therapeutic regimens, most deaths from cancer are probably caused by the relentless growth of metastatic cells that are resistant to the treatment. Therefore, it is of utmost importance that we understand the process of metastasis in order to develop rational and effective treatments for cancer.

Metastasis is defined as ''the transfer of disease from one organ or part, to another not directly connected with it.'' Although metastasis is poorly under-

RECEPTORS IN CELLULAR RECOGNITION
AND DEVELOPMENTAL PROCESSES

stood, results from experimental animal models and analysis of clinical data indicate that there are interrelated, sequential steps in the process that are influenced by both the unique properties of metastatic porgenitor cells and host factors. The pathogenesis of a metastasis is initiated by the detachment of tumor cells from the primary growth and their invasion into surrounding normal tissues and into the circulatory system. Once in the circulation, the tumor cells can reach distant organs where they can be arrested, extravasate, and proliferate to form secondary tumor foci (Fidler *et al.*, 1978; Nicolson *et al.*, 1978; Poste and Fidler, 1980). Because successful metastasis requires cells to survive multiple destructive events, and because different properties may be essential at different steps of the process, it is important to determine whether all the cells within a primary tumor have an equal chance to form metastases, or whether only specialized cells are capable of forming metastases.

The possibility that cells of different metastatic potential populate the parental tumor was first reported by Koch (1939), who selected a highly metastatic subline from an Ehrlich carcinoma. More recently, more direct evidence that malignant neoplasms contain cell variants with different metastatic capabilities was presented by Fidler and Kripke (1977), who performed cell cloning and fluctuation analysis with the B16 melanoma parent tumor syngeneic to the C57BL/6 mouse and numerous clones derived from it. The metastatic potential of the majority of the clones differed from that of the parent line, and there was also significant variation among the clones in their metastatic potential and the organ site of the metastases. Similar heterogeneity in the metastatic behavior of cells isolated from other tumors has now been demonstrated for murine mammary tumors (Dexter *et al.*, 1978; Tarin and Price, 1979), MCA-induced fibrosarcoma (Kerbel *et al.*, 1978), Lewis lung carcinoma (Fogel *et al.*, 1979), ultraviolet-induced fibrosarcoma, and virally induced sarcoma (Kripke *et al.*, 1978; Raz *et al.*, 1981).

A complexity to the interpretation of these data has recently come to the fore with the studies by Ling and associates (Harris *et al.*, 1982; Hill *et al.*, 1984; Ling *et al.*, 1984). These workers have made attempts to measure the rate of progression (to a metastatic phenotype) in a variety of tumor populations, including the B16 melanoma lines studied so extensively by Fidler and co-workers (Fidler, 1973a; Fidler and Nicolson, 1976, 1977). The data obtained suggest that most of the cells in clonal isolates of all of the tumors were non-metastatic (as defined by lung-nodule formation after intravenous injection into syngeneic hosts). However, effective metastatic variants were generated at unexpectedly high rates (in the order of 10^{-5} per cell per generation) even from low metastatic lines of B16 melanoma derived by Fidler and Nicolson (1976). Indeed, the difference in rate of formation of variants between "high" and "low" metastatic B16 melanoma lines was only of the order of threefold (Hill *et al.*, 1984). While the rates measured were effective rates only (uncorrected for influence of possible differences in the ability of metastatic cells to form lung colonies), the high frequencies are quite striking. Furthermore, these variants were lost at even higher rates (of the order of 10^{-1} or 10^{-2} per cell per generation in a steady-state nature) when cells were grown *in vitro* or *in vivo*. Thus a feature of metastatic disease seems to be the rapid turnover of metastatic variants, with the frequency of these in the population determined by the effective rates of their generation/loss. This model has been termed a "dynamic heterogeneity" model.

Equally striking in terms of the challenge it poses to researchers interested in making models that might help explain production of such variants at high rates in populations of malignant cells is a study by Poste *et al.* (1982). These workers found that the stability of the metastatic phenotype in a population of B16 melanoma cells was critically dependent upon the cellular environmental conditions under which the cells were cultured. Short-term passage of individual clones isolated from a heterogeneous parental population led to rapid emergence of variant subclones with different metastatic phenotypes. When several clones were mixed before culture, this instability was suppressed. Clearly this phenomenon has great implications for tumor therapy.

The above data clearly suggest that not all cells that reside within a tumor complete all the steps of the metastatic process to produce a secondary tumor growth. For the sake of convenience, we can divide the process of metastasis into five major steps: (a) dislodgement of cells from the primary tumor into surrounding tissue and penetration into the blood vessels; (b) release of single tumor cells or emboli into the circulation; (c) arrest of the emboli in the capillary beds of distant organs; (d) tumor-cell invasion of the wall of the arresting vessel, infiltration into adjacent tissue, and multiplication; and (e) growth of vascularized stroma into new tumor nodules. The ability of a tumor cell to complete successfully any of the steps required for the production of a metastasis depends to a great extent on its surface properties. However, the relationship of any specific cell-surface change to the outcome of metastasis has not yet been firmly established. This chapter is not intended to be an all-inclusive review on the topic of tumor cell-surface properties that influence metastasis. Rather, we have attempted to summarize studies of some of the biochemical properties of metastatic tumor cells.

II. TUMOR–HOST INTERACTIONS

A. Soluble or Membrane-Bound Degradative Enzymes That Might Be Associated with Metastasis

Histological examinations of many human and animal (invasive) neoplasms frequently reveal destruction of host tissues that surround the tumor, which is generally not the case for noninvasive benign tumors. This tissue destruction could be attributed to the release of degradative enzymes from tumor cells. Indeed, in general, cells of malignant neoplasms produce a higher level of hydrolytic enzymes than cells from noninvasive (benign) tumors or nontumorigenic (normal) cells (Sylven, 1968; Dresden et al., 1972; Hashimoto et al., 1973; Yaminishi et al., 1973). Necrosis of tissue could facilitate the detachment of tumor cells from the primary neoplasms and promote dissemination (Weiss, 1977, 1978b). Moreover, at the time when spontaneous metastases of the Lewis lung carcinoma are grossly evidence in the lung of syngeneic C57BL/6 mice, the activity of lysosomal degradative enzymes in the primary tumor is twofold higher than that in the metastases. This elevated enzyme activity in a primary tumor may be related to the release of cells into the circulation (Dobrossy et al., 1980).

Studies of the pathogenesis of metastasis and the host tissues with which metastatic cells must interact suggest that, in the main, two specific cellular enzymatic activities, fibrinolytic and collagenolytic, are associated with the process. Fibrin formed at the host–tumor interzone may act as a barrier to tumor-cell infiltration, and the fibrinogen–fibrin system may be a vital factor in determining events at the tumor-cell invasion zone (Strauli and Weiss, 1977). Many neoplastic transformed cells produce high levels of a serine protease that, by hydrolysis of serum plasminogen, generates an active fibrinolytic protease, plasmin (Ossowski et al., 1973). The hydrolytic products of plasminogen activator may provide contact guidance to tumor-cell locomotion to induce the vascularization of tumors and to raise the levels of plasma proteins in the extracellular milieu by increasing the permeability of blood vessels (Dvorak et al., 1981; Reich, 1973). Cell-associated fibrinolytic activities may also aid in the transplantation of circulating tumor cells. Some tumor cells can be thromboplastic and elicit fibrin formation (Chew et al., 1976), which, in turn, leads to the formation of larger tumor emboli with increased efficiency of arrest in small blood vessels. If this were always the case, however, it would be difficult to explain why fibrinolysis induced by human plasminogen in mice increased the incidence of experimental metastasis, whereas antifibrinolysis induced with transexamic acid or the presence of protease inhibitors decreased metastasis (Saito et al., 1980). Interference with coagulation by prolonged treatment of mice with heparin markedly increased the number of metastases in mice that were previously inoculated with sarcoma cells (Retik et al., 1962). An increase in the number of hepatic and pulmonary metastases was also detected in mice treated with avrin, an agent that depletes fibrinogen (Hagmar, 1972). In contrast, a decreased number of metastases was found in mice pretreated with heparin, fibrinolysin, and Warfarin (Grossi et al., 1960; Brown, 1973; Donati et al., 1978). Generalizing from such data is difficult because of the contradictory reports on the levels of plasminogen activator activity associated with B16 melanoma cell lines that have low and high metastatic potential. One group of investigators found a correlation between the production of plasminogen activator by the cells and their metastatic potential (Wang *et al.*, 1980), while others have failed to find any such correlation (Nicolson *et al.*, 1977). Clearly, the possible relationship between the production and release of plasminogen activator by tumor cells and expression of the metastatic phenotype merits further investigation.

The epithelial and endothelial basement membranes are resilient structures that present a mechanical barrier for the invasion and extravasation of tumor cells. Histological examinations of many invasive tumors revealed a degradation of the basement membrane at its area of contact with invading tumor cells (Babai, 1976; Vlaeminck *et al.*, 1972). The initial observation of Strauch (1972), who analyzed a range of human breast carcinomas and found that areas showing morphological evidence of invasion coincided with high collagenolytic activity, was further elucidated by Liotta and his colleagues (Liotta *et al.*, 1977, 1980), who found that invading tumor cells secrete specific collagenase capable of degrading the basement membrane type IV collagen. Liotta *et al.* also found a direct correlation between the presence of cell-associated type IV collagenase and metastasis.

B. Cell-Surface Components That May Be of Importance in Metastasis

The detachment of tumor cells from the primary tumor mass has been attributed, at least in part, to reduced cell cohesiveness. It was suggested that a decrease in binding sites for calcium ions on malignant cells could lead to the formation of fewer calcium "bridges" between adjacent cells (DeLong *et al.*, 1950; Coman, 1953). However, many subsequent experiments have failed to reveal any reduction in the calcium-binding ability of malignant cells (Patinkin *et al.*, 1970). More recently, reduced adhesiveness of tumor cells was attributed to a decrease in a specific cell-associated glycoprotein, the fibronectin, which may mediate cell-to-cell and cell-to-substratum adhesion (Pearlstein, 1976). Fibronectin was found to be present on the surface of certain nontransformed normal cells and to be greatly reduced or even absent in their virally transformed counterparts. At first it appeared that the absence of cell-associated fibronectin could be equated to transformation. This concept, however, is not universally accepted (Jones *et al.*, 1976). The relationship of cell-associated fibronectin to metastatic potential has also been controversial. While Chen *et al.* (1974) reported that an increase in the metastatic potential of tumor cells was associated with a decrease in the cellular content of fibronectin, other groups failed to demonstrate a correlation between the metastatic potential of, e.g., rat mammary carcinoma cells and cell-associated fibronectin.

The presence of tumor cells in the circulation is necessary, but not sufficient, to produce metastases. Only a few cells can survive the hostile circulatory system to produce metastases (Fidler, 1970; Salsbury, 1975; Der and Stanbridge, 1978; Liotta et al., 1978). Clinical observations in humans and various animal tumor systems have shown that the patterns of metastasis are not random (Parks, 1974; del Regato, 1977). In 1889, Paget asked, "What is it that decides which organ shall suffer in a case of disseminated cancer?" Studies utilizing radiolabeled tumor cells have shown that tumor cells can reach many organs, but proliferate into visible metastases in only a few (Nicolson et al., 1978; Fidler and Nicolson, 1977; Fidler et al., 1977). The preferential growth of malignant cells in particular organs was demonstrated by experiments in which mice carrying ectopically implanted organs were injected with tumor cells; metastatic foci developed in both in situ lungs and grafted lungs but not in other grafted control organs (Kinsey, 1960; Hart and Fidler, 1980). These observations, as well as the fact that tumor cells can pass through the first capillary bed encountered (Zeidman and Buss, 1952; Fisher and Fisher, 1967), indicate that tumor-cell properties per se influence the process of metastasis. The successful selection from heterogeneous parent populations of tumor cells with specific organ-homing (growth) properties also supports this conclusion (Fidler, 1973a; Brunson et al., 1978; Brunson and Nicolson, 1979; Tao et al., 1979; Raz and Hart, 1980; Shearman and Longnecker, 1981). The observation that tumor-cell aggregates produce significantly more metastases after intravenous injection (Fidler, 1973b) led to an extensive analysis of the possible relationship between the intracellular interactions and adhesion capabilities of malignant tumor cells with regard to their metastatic potential. Several parameters of tumor cell adhesiveness have been measured and will be summarized.

Comparisons of variant cells from the same tumor can facilitate the characterization of metastatic tumor cells. By sequential in vivo selection, Fidler (1973b) described the isolation of variant cells from the murine B16 melanoma that colonize in the lung after intravenous injection more efficiently than the average cell populating the parental tumor. One distinctive feature of murine B16 melanoma cell variants is the increased tendency of highly metastatic cells to aggregate in vitro with other tumor cells (homotypic aggregation; Nicolson and Winkelhake, 1975; Raz et al., 1980a) or with host cells (heterotypic aggregation; Gasic et al., 1973). Furthermore, tumor cells that home to specific organs have been shown to exhibit increased aggregation or adherence to cells derived from those receptive host organs (Nicolson and Winkelhake, 1975). Clearly, these cell-to-cell interactions are influenced to a great extent by the surface of the participating cells. Recent data substantiate this influence.

Another recent advance in this area has come from the studies that indicate that there may be a correlation between cell shape response growth control and metastatic phenotype (Ben-Ze'ev and Raz, 1981; Raz and Ben Ze'ev, 1982, 1983). Loss of cell shape-dependent growth control seems to be a key feature of the malignant process (Wittelsberger *et al.*, 1981). Thus, Raz and Ben Ze'ev (1983) used a recently developed technology to control cell shape, using growth of cells on a nonadhesive substrate [plates coated with nontoxic films of poly(2-hydroxyethyl methacrylate), poly(HEMA)]. When compared to the characteristic bipolar morphology of B16 melanoma cells grown on tissue-culture plastic, cells grown on poly(HEMA)-coated plates formed small aggregates with a gradual increase over time in culture of three-dimensional aggregates with irregular configurations. When cells from the two types of culture were injected intravenously into syngeneic C57B1/6 mice, there was an increase in lung metastatic potential in tumor cells harvested from poly(HEMA) cultures. Replating cells from poly(HEMA) plates for some 24–48 hr on regular plastic culture dishes before transplantation led to a reversion in metastatic phenotype to that of control cultures (grown throughout in plastic). Thus a reversible modulation, not a tumour selection process, was presumably in operation. Along with this change in metastatic properties there was a gradual loss of iodinatable cell-surface protein components in poly(HEMA)-cultured tumor cells. Again this loss was reversed after culture on plastic, along with the change to a flattened morphology.

Other features of potential interest in the spherical aggregates of tumor cells (compared with the normal monolayer of cells) were found. These included a decrease in synthesis of the cytoskeletal component vimentin, again a process reversed on replating to conventional plates, and a marked increase in the formation of homotypic aggregates (47% and 62% of cells at 30 or 60 min of incubation, respectively) compared to aggregate formation in control cultures (21% and 47%, respectively, at 30 and 60 min of incubation). These data are consistent with models that suggest that environmental signals could regulate the expression of genetic programs, which in turn activate/repress those cellular activities that influence metastasis (Weiss, 1978a; Schirrmacher, 1980).

III. BIOCHEMISTRY OF TUMOR CELL SURFACE IN METASTATIC GROWTH

A. Cell-Surface Negative Charge and/or Receptors

The initial sites for cell membrane-to-membrane recognition, contact, and subsequent interaction are thought to be associated with dense membrane anionic sites (Weiss and Subjeck, 1974; Hackenbrook and Miller, 1975). For this reason, we examined the distribution of cell-surface dense membrane anionic sites on B16 melanoma cell variants exhibiting different metastatic behavior. Indeed, the distribution of negative charge on the surface of B16 melanoma cells correlated with their propensity to aggregate *in vitro* and with the production of pulmonary tumor colonies after intravenous injection into syngeneic mice (Raz *et al.,* 1980a). We also observed that not all the cells in the highly metastatic B16–F10 cell population form homotypic aggregates. The frequency of dense membrane anionic sites among the nonaggregated cells was remarkably similar to that observed in cells from the low metastatic B16 lines. In *in vivo* studies, B16–F10 cells that did not participate in the aggregation process *in vitro* produced significantly fewer pulmonary lung colonies than B16–F10 cells that were recovered from cell aggregates and suspended into single cells (Raz *et al.,* 1980a). Studies with lymphosarcomas that metastasize to the liver (Guy *et al.,* 1980) also show the importance of the cell-surface negative charge.

Despite these data, the role of cell-surface charge in metastasis is still unclear. In some tumor systems, surface charge may be important; in others it may not. Cell-surface content of sialic acid has been implicated in regulating cellular adhesiveness (Kemp, 1970), the ability of intravenously injected cells to implant into various organs (Weiss *et al.,* 1974), immunogenicity (Ray, 1977), and in cell motility and invasiveness *in vitro* (Yarnell and Ambrose, 1969). All of these are properties believed to be of importance in development of metastatic growth. However studies in different tumors on the relationship between the level of cell-surface-associated sialic acid and metastasis produced different results. Direct labeling of the sialic acid present on the cell surface of three B16 melanoma cell variants that differ in their metastatic potential revealed that the quantity of labeling of the major sialoglycoprotein (ca. 78,000 molecular weight) was inversely proportional to the capacity of the cells to colonize in the lung (Raz *et al.,* 1980b). Koyama *et al.* (1979) found that less tumorigenic methylcholanthrene-induced tumor cells are characterized by a higher degree of labeling of a 150,000-dalton sialoglycoprotein than are more tumorigenic cells, whereas no difference in total and cell-surface sialic acid was found in cells obtained from seven metastatic variants of a rat mammary adenocarcinoma (Chatterjee *et al.,* 1976). A decrease in the sialyation of surface components was also found in nonmetastasizing variants of murine tumor cells (Finne *et al.,* 1980; Yogeeswaran and Tao, 1980; Yogeeswaran and Salk, 1981). B16 melanoma cells treated with colchicine, cytochalsin B, or both showed marked alterations in the pattern of tumor dissemination (Hart *et al.,* 1980). In a different set of experiments, Poste and Nicolson (1980) showed that fusing membrane vesicles shed by highly metastatic lung-colonizing cells (B16–F10) onto less metastatic cells (B16–F1) increased the lung colonizing capabilities of the modified B16–F1. However, the nature of cell-membrane molecules that control the metastatic expression of the tumor cells is not known. As discussed above, both low and high metastatic tumor cell lines exhibit differences in adhesion abilities. Adhesiveness is a cell-surface-mediated property that is altered during malignant transformation (Poste and Weiss, 1975). As yet, the relationship of altered cell adhesiveness to tumorigenicity is not completely understood.

From various investigations of intracellular recognition between normal (and tumor) cells, evidence has accumulated that proteins or glycoproteins on the surface of vertebrate cells are present that mediate specific cell-to-cell recognition and adhesion (Burger *et al.*, 1978). Similar cell-aggregating molecules were seen in tumor cells (Kudo *et al.*, 1976).

The involvement of carbohydrate residues in mediating adhesion between cells also has been reported in studies on the ability of simple sugars, glycopeptides, and glycoproteins to inhibit the aggregation of certain types of cells. One possible explanation for these findings is that sugar-binding proteins are present on the surface of the vertebrate cells and that these proteins bind multivalent glycoproteins that serve as "bridges" between adjacent cells (Simpson *et al.*, 1978; see also Part I of this volume). Nicolson and Winkelhake (1975) showed that tumor cells aggregate preferentially with cells isolated from the organ to which they metastasize. Schirrmacher *et al.* (1979, 1980) then reported that liver-colonizing tumor cells formed rosettes with hepatocytes. Neuraminidase treatment of low metastatic liver-colonizing cells promoted their aggregation with the hepatocytes. Kolb and Kolb-Bachofen (1978) speculated that tumor-cell haplotype interaction is mediated by the lectin-like receptor of asialoglycoproteins of the hepatocytes.

Most of the functions attributed to the lectins found in nonmalignant tissues pertain to the role of lectins in the regulation of physiologic contact-dependent processes as well as specific cell-to-adhesion (Harrison and Chesteron, 1980). The presence of lectins in tumor cells has not been investigated thoroughly; however, a β-galactoside-specific lectin was detected in neuroblastoma cells (Teichberg *et al.*, 1975), and a lectin specific for oligomannosyl units was observed in teratocarcinoma cells (Grabel *et al.*, 1979).

As was stated in the previous section, the homotypic aggregation of B16 cells *in vitro* requires fetal bovine serum (Raz *et al.*, 1980a). Preliminary studies indicated that fetuin alone can satisfy the requirement for fetal bovine serum in the homotypic aggregation of B16 melanoma cells (A. Raz, unpublished observations). Furthermore, fetuin or its desialated derivative, asialofetuin, has been shown to promote the homotypic aggregation of various neoplastic cells such as human melanoma, human cervical carcinoma, murine melanoma, and murine fibrosarcoma (Raz and Lotan, 1981).

Lectins present on tumor cells invading blood vessels may promote cognitive intracellular interactions that lead to embolization and eventually to arrest of distant organs. In addition, a lectin-like glycopeptide has been isolated from human glomerular basement membrane, suggesting that cell attachment to the basement membrane may be mediated via cell-surface carbohydrate residues (Gerfaux *et al.*, 1979). These observations offer circumstantial support to the suggestion that some cell-recognition processes in tumor-cell dissemination involve tumor-cell surface sugar-binding proteins and/or carbohydrate residues.

B. Surface Oligosaccharides on Metastatic and Nonmetastatic Tumor Cells

Alterations in oligosaccharide structures of glycoproteins (Takasaki *et al.*, 1980; Rachesky *et al.*, 1982) and glycolipids (Yogeeswaran and Stein, 1980) have been observed following chemical and/or viral transformation. Asparagine-linked oligosaccharides of transformed cells may be larger as a result of a decrease in terminal sialylation (see above) or due to increased branching (Takasaki *et al.*, 1980). A study by Sauter and Glick (1979) of glycopeptides in chemically transformed hamster embryo cells reported a change in size profile on *in vivo* transplantation, which occurred in parallel with a change to increased malignancy. As already noted, it may be that the best correlation between the content of larger sialylated oliosaccharides is with the metastatic potential of the cell (Yogeeswaran and Salk, 1980) rather than with transformation *per se*.

Additional evidence for a role for cell-surface oligosaccharides in the expression of the metastatic phenotype has come from analysis of lectin-resistant mutant cells. Wheat germ agglutinin (WGA)-resistant cells of the PG19 melanoma (Bramwell and Harris, 1978), of the B16 melanoma (Tao and Burger, 1977), and of the MDAY-D2 tumor (Dennis and Kerbel, 1981) are less malignant than the respective parental tumors. WGA binds avidly to oligosaccharides containing sialic acid and N-acetylglucosamine (Kronis and Carver, 1982; Bharanandan and Kattio, 1979). Mutant tumor cells selected for resistance to other lectins (e.g., phytohemagglutinin, leukoagglutinin, concanavalin A) have either no reduction or an apparent increase in metastatic potential (Tao and Burger, 1982). Decreased sialylation of tumor cells of glycoconjugates is also associated with decreased platelet aggregation and increased susceptibility to natural killer cell lysis (Pearlstein *et al.*, 1980; Yogeeswaran *et al.*, 1981).

In keeping with the above, many WGA-resistant avirulent tumour lines have unsialylated asparagine-linked oligosaccharides (e.g., a WGA^r mutant of the B16 melanoma that shows a decrease in sialic acid and an increase in fucose linked to $\alpha1{\rightarrow}3$ to N-acetylglucosamine). This may be due to an enhanced (60- to 70-fold) level of $\alpha1{\rightarrow}3$ fucosyltransferase relative to $\alpha2{\rightarrow}3$ sialyltransferase in the mutant (Finne *et al.*, 1980, 1982). The actual nature of the cell-surface oligosaccharide structural change may vary from system to system, however (Bhavanandan and Kattio, 1979).

A WGA^r tumorigenic, nonmetastatic mutant of the highly metastatic MDAY-D has been described by Dennis and Kerbel (1981). The mutant has a decreased expression of cell-surface sialic acid and attaches to fibronectin and type IV collagen-coated surfaces considerably more readily than the metastatic MDAY-D tumor (Dennis *et al.*, 1982). Neuraminidase treatment of the WGA^s lines increases their attachment to these substrates, suggesting a role for non-sialylated glycoconjugates in attachment (and hence in decreasing metastatic phenotype, perhaps by inhibiting escape from a primary subcutaneous site of growth). In all probability it is both the nature of the terminal sugar as well as the underlying backbone that is critical in these recognition phenomena, as has been described for the hepatic galactose-binding receptor (see also Chapter 3 of this volume).

In a more recent study, Dennis *et al.* (1984) reported analysis of asparagine-linked oligosaccharides in WGA^r and WGA^s parental MDAY-D tumor cell lines. [2-^{3}H]Mannose-labeled glycopeptides were fractionated by DEAE cellulose chromatography and concanavalin A–Sepharose, following which structural analysis was performed using protein nuclear magnetic resonance spectroscopy or sequential exoglycosidase digestion (and chromatography on lectin/agarose and BioGel columns). Nonmetastatic lines lacked a sialylated poly(*N*-acetyllactosamine)-containing glycopeptide and contained a unique triantennary class of oligopeptides lacking galactose and sialic acid. Both WGA^r and WGA^s cells contained the high-mannose glycopeptides that represented the major glycopeptide in both cell types, and both contained equivalent levels of a galactosyltransferase capable of adding galactose to the *N*-acetylglucosamine-terminated glycopeptide found in the mutant cells. To date therefore, the cause of the altered glycopeptide structure in the mutant is unclear, though the explanation may lie in a deficiency in UDP galactose or in the compartmentalization of enzyme/substrate *in vivo*.

In separate but perhaps related studies, Nicolson *et al.* (1981) and Butters *et al.* (1980) reported that tunicamycin, an inhibitor of asparagine-linked carbohydrate synthesis, could inhibit production of cellular components in melanoma cells and fibroblasts that were essential for adherence to fibronectin. Oligosaccharide structural changes have also been implicated in the diminished adhesiveness (to fibronectin) of ricin-resistant baby hamster kidney cell lines (Pena and Hughes, 1978).

C. Differences in Immune Recognition of Cell-Surface Antigens

Evidence that the immune response can influence metastatic spread has been obtained from several studies (Fidler and Kripke, 1980). The initial small subpopulations of metastatic cells that develop within a tumor may escape immune destruction if their cell-surface antigens differ from those of the nonmetastatic cells that are the majority population in the neoplasm. Indeed, tumor-associated antigenic differences between metastatic and nonmetastatic tumor cells isolated from the same neoplasm have been reported (Kerbel *et al.*, 1978; Fogel *et al.*, 1979; Ghosh *et al.*, 1979; Gorelik *et al.*, 1979; Schirrmacher *et al.*, 1980; Baetselier *et al.*, 1980). The modulation of the expression of cell-surface H-2 antigens and its relationship to the metastatic properties of tumor cells were also examined. Kerbel *et al.* (1978) have isolated from a methylcholanthrene-induced tumor a cell variant with high metastatic characteristics which was accompanied by the loss of its H-2 haplotype. Baetselier *et al.* (1980) examined T10 sarcoma cells that were induced in (C3Heb $\times$ C57BL/6)F$_1$ mice (H-2^b $\times$ H-2^k) and found that H-2^k-positive clones are more metastatic than H-2^k-negative clones. Such results suggest that the expression of the major H-2 complex may influence the ability of the circulating tumor cells to escape immune destruction, and this will result in increased survival to allow implantation at a distant organ. This may represent some phenomena similar to that known by immunologists in general as analogous to a major histocompatibility complex (MHC)-linked immune response genes and MHC-restricted antigen recognition. Alternatively, or in addition, it perhaps may reflect something to do with change in cell-surface glycosylatin of tumor antigens linked to (or occurring parallel with) changes in MHC antigen expression (see Chapter 3).

The implantation of circulating tumor cells in specific organs prompted a series of studies aimed at clarifying the possible existence of specific cell-surface antigens and their correlation with organ-specific metastasis. Shearman and Longnecker (1981) described the detection of cell-surface antigen by monoclonal antibodies. They found a direct correlation between the level of antigen expression and the development of liver metastasis: cells with a low level of antigen expression produced few liver metastases, and cells with a high level of antigen expression produced many liver metastases. Furthermore, monoclonal antibodies directed against the liver antigen inhibited the development of metastases only in the liver and not in other organs. By using immobilized lectins, Reading *et al.* (1980) succeeded in the selection of murine lymphosarcoma cells with an increased ability to produce liver metastases. Cell-surface analysis of the liver-colonizing cells showed a decrease in the major concanavalin A-binding glycoprotein (70,000 molecular weight), as compared to cells that do not colonize to the liver; this cell-surface glycoprotein was found to be an RNA tumor virus envelope component, the gp70. Given additional evidence that treatment of the tumor recipient in such a way as to impair macrophage activity eliminated the difference in metastatic growth between parental lymphosarcoma and variant, with the parent now producing as many metastases, Nicolson *et al.* (1981) recently postulated that the virally encoded gp70 may represent a critical target antigen for host macrophages. Inhibition of host killer T-cell activity or natural-killer-cell activity did not modify the relative metastatic potential of parent/mutant. More recently another surface antigen, distinct from gp70, has been identified that may be important in "guiding" metastatic cells to the liver site. This antigen is not present on adult liver tissue but is present on fetal liver. Antibodies to this antigen block metastasis and increase survival of tumor-bearing mice.

In a study, Fidler *et al.* (1979) examined the role of tumor-cell antigenicity in the pathogenesis of metastasis. Three C3H mouse fibrosarcomas of differing immunogenicities were tested for metastatic behavior in normal, sham-suppressed, immunosuppressed, and immunologically reconstituted syngeneic mice. Immunosuppression affected experimental metastasis of the three tumors in various ways. The highly immunogenic fibrosarcoma formed more pulmonary tumor colonies in immunosuppressed mice than in normal, sham-suppressed, or reconstituted animals. A fibrosarcoma of intermediate immunogenicity also formed more pulmonary metastases in immunosuppressed hosts, but this increase could not be reversed by immune reconstitution. In contrast, the least immunogenic tumor formed fewer pulmonary tumor clonies in immunosuppressed mice than in normal, sham-suppressed, or reconstituted mice. Therefore, it appears that generalizations about the nature of metastatic spread, tumor-cell antigenicity, and host immune status cannot be made.

IV. CONCLUSIONS

The development of metastasis is a complex and highly selective process that is dependent on the interplay of host and tumor cell properties. Characteristics of metastatic cells such as cell-surface properties, enzyme content, motility, and adhesive properties may determine the fate of blood-borne metastasis. Because of the clinical importance of metastasis, future studies must be directed toward elucidating which properties of tumor cells contribute to their metastatic phenotype and which do not. Such an understanding should aid the development of better approaches to the prevention or elimination of metastasis. However, it should be clear that we can anticipate the solution(s) to these problems will be complex, and the essential factors may vary with different tumor–host model systems studied (and perhaps even *in situ* for histologically different tumors in genetically different individuals). That this is not an unlikely possibility should be evident from an appraisal of the literature on immunological defence against malignancy. The spectrum of possible outcomes in these studies ranges from no evidence for immunological involvement, through evidence for antibody mediated resistance only, to later data that has claimed a critical role for T cells, macrophages, natural killer cells, or some combination of all of the above!

ACKNOWLEDGMENTS

This research was sponsored by the National Cancer Institute, Department of Health and Human Services, under contract NO1-CO-75380 with Litton Bionetics, Inc. The contents of this publication do not necessarily reflect the views or policies of the Department of Health and Human Services, nor does mention of trade names, commercial products, or organizations imply endorsement by the U.S. Government.

REFERENCES

Babai, F. J. (1976). *J. Ultrastruct. Res.* **56**, 287.

Baetselier, D. P., Katzar, S., Goreleck, G., Feldman, M., and Segal, S. (1980). *Nature (London)* **288**, 179.

Bharanandon, V. B., and Kattio, A. W. (1979). *J. Biol. Chem.* **254**, 4000.

Branwell, M. E., and Harris, H. H. (1978). *Proc. R. Soc. Lond. (Biol.)* **201**, 87.

Brown, J. M. (1973). *Cancer Res.* **33**, 1217.

Brunson, K. W., and Nicolson, G. L. (1979). *J. Supramol. Struct.* **11**, 517.

Brunson, K. W., Beattie, G., and Nicolson, G. L. (1978). *Nature (London)* **272**, 543.

Burger, M. M., Burkat, W., Weinbaum, G., and Jublatt, J. (1978). *Symp. Soc. Exp. Biol.* **32**, 1.

Butters, T. D., Devolia, B., Alpin, J. A., and Hughes, C. (1980). *J. Cell Sci.* **44**, 33.

Chatterjee, S. K., Kim, U., and Bielat, K. (1976). *Br. J. Cancer* **33**, 15.

Chew, E. C., Josephson, R. L., and Wallace, A. C. (1976). *In* "Fundamental Aspects of Metastasis" (L. Weiss, ed.), p. 121. North-Holland, Amsterdam.

Coman, D. R. (1953). *Cancer Res.* **13**, 397.

DeLong, R. P., Coman, D. R., and Zeidman, I. (1950). *Cancer* **3**, 718.

del Regato, J. A. (1977). *Semin. Oncol.* **4**, 33.

Dennis, J. W., and Kerbel, R. S. (1981). *Cancer Res.* **41**, 98.

Dennis, J. W., Waller, C., Timpl, R., and Schirrmacher, V. (1982). *Nature (London)* **300**, 274.

Dennis, J. W., Carver, J. P., and Schachter, H. (1984). *J. Cell Biol.* **99**, 1034.

Der, C. J., and Stanbridge, E. J. (1978). *Cell* **15**, 1241.

Dexter, D. L., Kowalski, H. M., Blazar, B. A., Fligiel, Z., Vogel, R., and Heppner, G. H. (1978). *Cancer Res.* **38**, 3174.

Dobrossy, L., Pavelic, Z. P., Vaughan, M., Porter, N. Z., and Bermacki, R. Y. (1980). *Cancer Res.* **40**, 3281.

Donati, M. B., Mussoni, L., Poggi, A., Gaetano, G. D., and Garattini, S. (1978). *Eur. J. Cancer* **14**, 343.

Dresden, M. H., Heilman, S. A., and Schmidt, J. D. (1972). *Cancer Res.* **32**, 993.

Dvorak, H. F., Quay, S. C., Orenstein, N. S., Dvorak, A. M., Hahn, P., and Bitzer, A. M. (1981). *Science* **212**, 923.

Fidler, I. J. (1970). *J. Natl. Cancer Inst.* **45**, 773.

Fidler, I. J. (1973a). *Nature (New Biol.)* **242**, 148.

Fidler, I. J. (1973b). *Eur. J. Cancer* **9**, 223.

Fidler, I. J., and Kripke, M. L. (1977). *Science* **197**, 893.

Fidler, I. J., and Kripke, M. L. (1980). *Cancer Immunol. Immunother.* **7**, 201.

Fidler, I. J., and Nicolson, G. L. (1976). *J. Natl. Cancer Inst.* **57**, 1199.

Fidler, I. J., and Nicolson, G. L. (1977). *J. Natl. Cancer Inst.* **58**, 1867.

Fidler, I. J., Gersten, D. M., and Riggs, C. W. (1977). *Cancer* **40**, 46.

Fidler, I. J., Gersten, D. M., and Hart, I. R. (1978). *Adv. Cancer Res.* **38**, 149.

Fidler, I. J., Gersten, D. M., and Kripke, M. L. (1979). *Cancer Res.* **39**, 3816.

Finne, J., Tao, J. W., and Burger, M. M. (1980). *Cancer Res.* **40**, 2580.

Finne, J., Burger, M. M., and Prieels, J. P. (1982). *J. Cell Biol.* **92**, 277.

Fisher, B., and Fisher, E. R. (1967). *Cancer Res.* **27**, 412.

Fogel, M., Gorelik, E., Segal, S., and Feldman, M. (1979). *J. Natl. Cancer Inst.* **62**, 585.

Gasic, G. J., Gasic, T. B., Galanti, N., Johnson, T., and Murphy, S. (1973). *Int. J. Cancer* **11**, 704.

Gerfaux, J., Chany-Fournier, T., Bardos, P., Muh, J. P., and Chany, C. (1979). *Proc. Natl. Acad. Sci. U.S.A.* **76**, 5129.

Ghosh, S. K., Grossberg, A. L., Kim, U., and Pressman, D. A. (1979). *J. Natl. Cancer Inst.* **62**, 1229.

Gorelik, E., Fogel, M., Segal, S., and Feldman, M. (1979). *J. Supramol. Struct.* **12**, 385.

Grabel, L. B., Rosen, S. D., and Martin, G. R. (1979). *Cell* **17**, 477.

Grossi, C. E., Agostino, D., and Cliffton, E. E. (1960). *Cancer Res.* **20**, 605.

Guy, D., Latner, A. L., Sherbet, G. V., and Turner, G. A. (1980). *Br. J. Cancer* **42**, 915.

Hackenbrock, C. R., and Miller, J. (1975). *J. Cell Biol.* **65**, 615.

Hagmar, B. (1972). *Eur. J. Cancer* **8**, 17.

Harris, J. F., Chambers, A. F., Hill, R. P., and Ling, V. (1982). *Proc. Natl. Acad. Sci. U.S.A.* **79**, 5547.

Harrison, F. L., and Chesterton, C. J. (1980). *Nature (London)* **286**, 502.

Hart, I. R., and Fidler, I. J. (1980). *Cancer Res.* **40**, 2281.

Hart, I. R., Raz, A., and Fidler, I. J. (1980). *J. Natl. Cancer Inst.* **64**, 891.

Hashimoto, K., Yaminishi, Y., Maeyens, E., Dabbous, M. K., and Kanzaki, T. (1973). *Cancer Res.* **33**, 2790.

Hill, R. P., Chambers, A. F., Ling, V., and Harris, J. F. (1984). *Science* **224**, 998.

Jones, P. A., Laug, W. E., Gardner, A., Nye, C. A., Fink, L. M., and Benedict, W. F. (1976). *Cancer Res.* **36**, 2863.

Kemp, R. B. (1970). *J. Cell Sci.* **6**, 751.

Kerbel, R. S., Twiddy, R. R., and Robertson, D. M. (1978). *Int. J. Cancer* **22**, 583.

Kinsey, D. L. (1960). *Cancer* **13**, 674.

Koch, F. E. (1939). *Z. Krebsforsch.* **48**, 495.

Kolb, H., and Kolb-Bachofen, V. (1978). *Biochem. Biophys. Res. Commun.* **85**, 678.

Koyama, K., Nudelman, E., Fukuda, M., and Hakamori, S. (1979). *Cancer Res.* **39**, 3677.

Kripke, M. L., Gruys, E., and Fidler, I.J. (1978). *Cancer Res.* **38**, 2962.

Kronis, K. A., and Carver, J. P. (1982). *Biochemistry* **21**, 3050.

Kudo, K., Hanaoka, J., and Hayashi, H. (1976). *Br. J. Cancer* **33**, 79.

Ling, V., Chambers, A. F., Harris, J. F., and Hill, R. P. (1984). *J. Cell. Physiol.* **3**, 99.

Liotta, L. A., Kleinerman, J., Catanzaso, P., and Rynbrandt, D. (1977). *J. Natl. Cancer Inst.* **58**, 1427.

Liotta, L. A., Vembu, D., Saini, R. R., and Boone, C. (1978). *Cancer Res.* **38**, 1231.

Liotta, L. A., Tryggvason, K., Garisa, S., Hart, I. R., Foltz, C. M., and Shafie, S. (1980). *Nature (London)* **286**, 67.

Nicolson, G. L., and Winkelhake, J. L. (1975). *Nature (London)* **255**, 230.

Nicolson, G. L., Birdwell, C. R., Brunson, K. W., Robbins, J. C., Beattie, G., and Fidler, I. J. (1977). *In* "Cell and Tissue Interactions" (J. W. Lask and M. M. Burger, eds.), p. 275. Raven Press, New York.

Nicolson, G. L., Brunson, K. W., and Fidler, I. J. (1978). *Cancer Res.* **38**, 4105.

Nicolson, G. L., Imura, T., Gonzalez, R., and Ruoslahth, S. (1981). *Exp. Cell Res.* **135**, 461.

Ossowski, L., Unkeless, J. C., Tobia, A., Quigley, J. P., Rifkin, D. B., and Reich, E. (1973). *J. Exp. Med.* **137**, 113.

Parks, R. C. (1974). *J. Natl. Cancer Inst.* **52**, 971.

Patinikin, D., Zanitsky, A., and Doljanski, F. (1970). *Cancer Res.* **30**, 489.

Pearlstein, E. (1976). *Nature (London)* **262**, 497.

Pearlstein, E., Salk, P. L., Yogeeswaran, G., and Karpatkin, S. (1980). *Proc. Natl. Acad. Sci.* **77**, 4336.

Pena, S. D. J., and Hughes, R. C. (1978). *Nature (London)* **176,** 80.
Poste, G., and Fidler, I. J. (1980). *Nature (London)* **283,** 139.
Poste, G., and Nicolson, G. L. (1980). *Proc. Natl. Acad. Sci.* **77,** 399.
Poste, G., and Weiss, L. (1975) In "Fundamental Aspects of Metastasis" (L. Weiss e.), p. 25. North-Holland, Amsterdam.
Poste, G., Doll, J., and Fidler, I. J. (1982). *Proc. Natl. Acad. Sci.* **78,** 6226.
Rachesky, M. H., Hard, G. C., and Glick, M. C. (1982). *Cancer Res.* **42,** 39.
Ray, P. K. (1977). *Adv. Appl. Microbiol.* **21,** 227.
Raz, A., and Ben Ze'ev, A. (1982). *Int. J. Cancer* **29,** 711.
Raz, A., and Ben Ze'ev, A. (1983). *Science* **221,** 1307.
Raz, A., and Hart, I. R. (1980). *Br. J. Cancer* **42,** 331.
Raz, A., and Lotan, R. (1981). *Cancer Res.* **41,** 3642.
Raz, A., Bucana, C., McLellan, W., and Fidler, I. J. (1980a). *Nature (London)* **284,** 363.
Raz, A., McLellan, W. L., Hart, I. R., Bucana, C. D., Hoyer, L. C., Sela, B. A., Dragsten, P., and Fidler, I. J. (1980b). *Cancer Res.* **40,** 1645.
Raz, A., Hanna, N., and Fidler, I. J. (1981). *J. Natl. Cancer Inst.* **66,** 183.
Reading, C. L., Belloni, P., and Nicolson, G. L. (1980). *J. Natl. Cancer Inst.* **64,** 1241.
Reich, E. (1973). *Fed. Proc. Fed. Am. Soc. Exp. Biol.* **32,** 2174.
Retik, A. B., Arons, M. S., Ketcham, A. S., and Mantel, N. (1962). *J. Surg. Res.* **2,** 49.
Saito, D., Sawamura, M., Umezawa, K., Kanai, Y., Furihata, C., Matsushima, T., and Sugimura, T. (1980). *Cancer Res.* **40,** 2539.
Salsbury, A. J. (1975). *Cancer Treat. Rev.* **2,** 55.
Sauter, U. V., and Glick, M. C. (1979). *Biochemistry* **18,** 2533.
Schirrmacher, V. (1980). *Immunobiology* **157,** 89.
Schirrmacher, V., Bosslet, K., Shantz, G., Clauer, K., and Habsch, D. (1979). *Int. J. Cancer* **23,** 245.
Schirrmacher, V., Cheinsong-Popov, R., and Arnheiter, H. (1980). *J. Exp. Med.* **151,** 984.
Shearman, P. J., and Longnecker, B. M. (1981). *Int. J. Cancer* **27,** 387.
Simpson, D. L., Thorne, D. R., and Loh, H. H. (1978). *Life Sci.* **22,** 727.
Strauch, P. (1972). *In* "Tissue Interaction in Carcinogenesis" (D. Tarin, ed.), p. 399. Academic Press, London.
Strauli, P., and Weiss, L. (1977). *Eur. J. Cancer* **13,** 1.
Sugarbaker, E. V., Cohen, A. M., and Ketcham, A. S. (1971). *Ann. Surg.* **174,** 161.
Sylven, B. (1968). *Eur. J. Cancer* **4,** 463.
Takasaki, T., Ikehira, H., and Kobata, A. (1980). *Biochem. Biophys. Res. Commun.* **92,** 735.
Tao, T. W., and Burger, M. M. (1977). *Nature (London)* **270,** 437.
Tao, T. W., and Burger, M. M. (1982). *Int. J. Cancer* **29,** 425.
Tao, T. W., Matter, A., Vogel, K., and Burger, M. M. (1979). *Int. J. Cancer* **23,** 854.
Tarin, D., and Price, J. E. (1979). *Br. J. Cancer* **39,** 740.
Teichberg, V. I., Silman, I., Beitsch, D. D., and Resheff, G. (1975). *Proc. Natl. Acad. Sci. U.S.A.* **72,** 1383.
Vlaeminck, M. N., Adenis, L., Mouton, Y., and Demaille, A. (1972). *Int. J. Cancer* **10,** 619.
Wang, B. S., McLoughlin, G. A., Richie, J. P., and Mannick, J. A. (1980). *Cancer Res.* **40,** 288.
Weiss, L. (1969). *Int. Rev. Cytol.* **26,** 63.
Weiss, L. (1977). *Int. J. Cancer* **20,** 87.
Weiss, L. (1978a). *Am. J. Pathol.* **97,** 601.
Weiss, L. (1978b). *Int. J. Cancer* **22,** 196.
Weiss, L., and Subjeck, J. R. (1974). *J. Cell Sci.* **14,** 215.
Weiss, L., Glaves, D., and Waite, D. A. (1974). *Int. J. Cancer* **13,** 850.
Wittelsburger, S. C., Kleene, K., and Penman, S. (1981). *Cell* **24,** 859.

Yaminishi, Y., Maeyens, E., Dabbous, M. L., Ohyama, H., and Hashimoto, K. (1973). *Cancer Res.* **33**, 2507.

Yarnell, M. M., and Ambrose, E. J. (1969). *Eur. J. Cancer* **5**, 265.

Yogeeswaran, G., and Salk, P. L. (1981). *Science* **212**, 1514.

Yogeeswaran, G., and Stein, B. S. (1980). *J. Natl. Cancer Inst.* **65**, 967.

Yogeeswaran, G., and Tao, T. (1980). *Biochem. Biophys. Res. Commun.* **95**, 1452.

Yogeeswaran, G., Granberg, A., Hanson, M., Dalionis, T., Kiessling, R., and Walsh, R. M. (1981). *Int. J. Cancer* **28**, 517.

Zeidman, I. (1957). *Cancer Res.* **17**, 157.

Zeidman, I., and Buss, J. M. (1952). *Cancer Res.* **12**, 731.

12

Cell Receptors and Cell Recognition Seen in a Sociological Perspective: Prospects for the Future

REGINALD M. GORCZYNSKI

I. INTRODUCTION

A major determining feature in the patterns of behavior expressed by animals lies in their ability to identify closely related kin. Biologically, it is known that too close a relationship between parents, leading to extensive inbreeding, is generally avoided either by dispersal of the young or by the expression of discriminatory behavior (Packer, 1979; Pusey, 1980). The mechanisms whereby

305

RECEPTORS IN CELLULAR RECOGNITION
AND DEVELOPMENTAL PROCESSES

outbreeding is favored are only recently coming to light. Odor differences among mates, with the detection of these pheromones by specific receptors on chemosensory cells, is one explanation offered for mating preferences in species as distant as the honey bee and the mouse (Getz and Smith, 1983; Berger, 1983). New evidence points to a role for the early learning (during developmental life) of, and tolerance to, the odors of close relatives as an important mechanism of kin recognition (Buckle and Greenberg, 1981). Perhaps most dramatic of all, it has now been suggested that there is a close analogy between the inherited sensory systems that identify potential sexual partners and the system (the immune system) that responds to inherited differences in cell-surface antigens and recognizes (distinguishes) self from non-self. Thus, mice of different genetic constitution (haplotypes) apparently produce strain-specific odors that allow them to identify mating partners of nonisologous strains (Lenington, 1983). In humans it has been shown that couples sharing a large number of cell-surface antigens have an above average frequency of spontaneous abortions (Beer *et al.,* 1981). The early reaction of maternal cells to foreign antigens on the trophoblast is believed to be of importance for long-term survival and growth of the fetus (Sio and Beer, 1982). Indeed, treatment of three such women with a history of spontaneous abortions with a mixture of foreign lymphocytes led to the birth of healthy children (Taylor and Faulk, 1981).

This correlation between mating preference and polymorphic cell surface antigens of the major histocompatibility complex (MHC) seems to have developed early in evolution. In *Botryllus,* a colonial ascidian, a similar somatic cell–cell recognition system has been reported to influence patterns of sexual reproduction (Scofield *et al.,* 1982). Could this mating preference of itself lead to the genetic polymorphism seen in the MHC, quite apart from the known (subsequent) influence of the latter on resistance to disease? Population genetic models of sexual selection indicate that polymorphisms will evolve in any genetic system that promotes outcrossing (O'Donald, 1983). Allegretti (1978) has written at length on the unsatisfactory way in which attempts have been made to explain the highly developed vertebrate immune system (and its intimate regulation by polymorphic MHC genes) as a driving force in the evolution of complex multicellular organisms.

I would like to examine below the case to be made for a useful analogy between the evolution and integration of multicellular organisms and, at the sociocultural level, complex stratified societies. Integration and stability in each case is dependent upon a central coordinating system using a specialized mobile scavenger group to maintain homeostasis relative to the internal and external environments. In addition the specialization per se is dependent on the development of more and more sophisticated classification schemes whereby superficially presented markers (molecules at the cell surface; readily identified and socially important class-specific markers, e.g., age/sex) are used to bring order

to the subsequent interaction between independent units. Certain of these classificatory markers are themselves a result of a dynamic interaction between the
preprogramed information and the environmental milieu. In discussing cultural
group organization I will draw freely on discussions by Fried (1960) and Service
(1962), who have proposed models based primarily upon social complexity or
economic diversity, respectively.

II. CELL-SURFACE CHANGES ASSOCIATED WITH THE DEVELOPMENT OF MULTICELLULAR ORGANISMS

A. From Single Cell to Cellular Aggregates

Most analyses of cellular development emphasize the progressive nature of the
process whereby structural and functional organization of the system, whether
unicellular or multicellular, become apparent. There are two basic schools of
thought concerning the appropriate scientific method by which to study the
mechanism for change. The one, scientific reductionism, is emphasized by
scholars of molecular biology. An alternative viewpoint holds that true com-
prehension of the means whereby the properties of the entire structure emerge
from the individual components cannot be achieved by simple integration of
isolated pieces of knowledge—development must be viewed holistically. The
general philosophy of the discussion that follows is dictated by the author's
allegiance to this latter concept.

Why do unicellular forms differentiate and develop into complex multicellular
structures? It is pertinent to note that single somatic cells or nonsexual cells of
plants are capable of such differentiation, without benefit of the internal reserves
available in vertebrate eggs, e.g., single spores of fern fronds develop in water
into gametophytes, small multicellular plants with chlorophyll-containing cells,
which produce eggs and sperm. Indeed, the primitive unicellular system represented by *Dictyostelium discoideum,* which is close to the taxonomic base of the
eukaryotic ladder, is yet another such example of an organism which can develop
into a multicellular form. Germinating spores of *Dictyostelium* liberate small
amoebae, which will continue to divide repeatedly in the presence of an adequate
food supply (the ubiquitous *Escherichia coli* is the most commonly used laboratory diet). For this organism at least, it is environmental hardship that triggers
change. Deprived of a supply of food, the amoebae undergo differentiation and
morphogenesis leading to the production of a sporocarp containing spores able to
reproduce the whole life cycle once more. The initial stage in this differentiation
process is marked by an aggregation into a "sluglike" form, in which the leading
edge of the aggregate is destined to become the stalk cells and the trailing edge the

spore cells. Attainment of the multicellular state produces two essential changes in further development. First, cell–cell contact leads to enhanced potential for intercellular communication, and for differences in the signals received by previously identical cells—the environment for cells on the outer edges of the mass is obviously quite different from that of the innermost cells. Indicative of the importance of such intercellular communication in the appropriate organization of the multicellular state in *Dictyostelium* is the evidence that mechanical disruption of the slug-like aggregate into single cells, followed by provision of those cells with an adequate food source, reverses the whole differentiation process (Newell *et al.*, 1972). Second, with the multicellular state comes the opportunity for specialization so that different cells can accomplish different tasks. The mechanism(s) whereby signals received at the interface of the organism and its environment become transduced into those messages that lead to differential gene activity in the receiving cell and/or are relayed to other cells to provoke similar changes in them is not understood. However, it seems self-evident that investigation of events at the cell surface are likely to provide at least some of the answers to these problems.

A further effect of cell specialization is to be found in the changes that subsequently occur in the mechanism(s) used for extraction of energy from the environment for food and reproduction. For the autonomous free-living amoebae, a highly mobile lifestyle, accompanied by a limited diversity in the range of foodstuffs from which energy can be extracted efficiently, is apparent. At the other extreme, the highly specialized invertebrates and vertebrates are examples of intraorganismic redistributive systems with the potential to utilize a more diverse range of nutrients from the surrounding environment, in a manner that is both more efficient and more energy productive, owing to the specialized nature of the cell systems that perform this particular task. But specialization brings along with it other essential differences. For the unicellular organism, entire unto itself, discrimination of foreign matter is a relatively elementary affair. It seems, in fact, that amoebae use simple cell-surface lectin molecules for such self–nonself discrimination, and indeed for feeding purposes and for the little cellular organization that is evident (Brown *et al.*, 1975; Frazier *et al.*, 1975). In a multicellular society the requirements for harmonious intercellular communication and interaction demand a more sophisticated process. During development the appropriate orientation of germ cells necessary for embryogenesis must be achieved. In a classic series of studies by Holtfreter (1944), analyzing the regulation of cell organization during gastrulation in amphibian embryos, it was demonstrated that relative cell-surface adhesiveness (tissue affinity) was an essential feature of the process. Separation of the independent germ-cell layers (endoderm, mesoderm, and ectoderm), followed by their subsequent mixture under artificial conditions, showed that ectoderm and endoderm would not form a stable aggregate together. In the presence of mesodermal tissue, which became interposed between these layers, a stable united aggregate was formed.

B. Organogenesis and Regulation
of Cellular Differentiation

The universality of the phenomenon of cell-surface adhesiveness in the organization of multicellular organisms is evident from work on cellular aggregation in sponges (Moscona, 1963) (which has been found to be species- and tissue-specific) and in vertebrate embryos where aggregation is in the main tissue-specific, leading to the ability to produce interspecies chimaeras (Townes and Holtfreter, 1955). Organogenesis is in itself another reflection of the hierarchical organization of cells within individual parts of a multicellular organism. If disaggregated heart and protocartilage tissue are mixed under tissue culture conditions, the heart tissue is found to the outside of the cell aggregate that forms. In contrast, mixing liver tissue with heart tissue leads to the production of an aggregate with liver tissue to the outside (Steinberg, 1970). The problem of explaining such apparently spontaneous self-assembly is not unique to cell biology—it is a major challenge of modern physical chemistry to explain in thermodynamic terms the molecular self-assembly of, e.g., collagen (Gross, 1956). Once again, an attempt has been made to explain the phenomenon (for cells at least) at the level of cell-surface adhesion. According to the so-called differential adhesion hypothesis, under certain conditions (such as when cell sorting leads to the development of a "sphere within a sphere") the adhesiveness between different cells is greater than between like cells. Presumably the molecular organization at the cell surface directs cell organization toward the thermodynamically most stable entity. What regulates this link between the three-dimensional "position" of the cell and its cell-surface molecular organization? How is this positional information transduced into the differential gene activity, which leads to the specific functions appropriate for that cellular organization within the cell hierarchy?

Once the adult multicellular complex is in place, its survival and growth depend both upon its stability and its superiority to less complex systems in its adaptability to a changing environment. Stability within the organism's lifetime is a concomitant feature of the differentiation and specialization process, since it is accepted dogma that "commitment" of immature cells to one cell path is accompanied by the loss of the ability to choose alternative modes of differentiation. A given mature cell need only "learn" its appropriate position in the cellular hierarchy at one stage in its lifespan. However, during the life of the organism a generally small (numerically) population of cells, stem cells, retains the capacity to choose among at least two paths of cell evolution, namely, differentiation or self-renewal. Thus, not only must mature cells within the cell aggregate communicate effectively with one another, but there must be also some regulatory mechanism whereby environmental signals can determine either the frequency of division of stem cells and/or the probability of these cells choosing particular lines of differentiation. In some cases, of course, the stimuli for appropriate differentiation along certain pathways are apparently expressed at only certain times during an individual cell's lifespan and are not consistent features of the adult cellular milieu. This seems to be the case for instance for limb bud formation and differentiation in vertebrates (contrast this with amphibians, where adult limb regeneration is well documented), in comparison to hemopoiesis and lymphopoiesis, which occur throughout the lifespan of the vertebrate animal.

C. Integration and Monitoring of Multicellular Societies

In general, integration of the cellular hierarchy (of mature and as yet unspecialized cells) is itself a function adopted by a specialized cell complex (the brain), while continued appraisal of appropriate cellular order within the system is a function of another specialized domain, the immune system. As the organ systems over which supervision is maintained become more differentiated from one another, the scavenger system that maintains the integrity of the individual in the face of the external milieu becomes in itself more complex, and the function of the specialized phagocytes of invertebrate organisms becomes the domain of the immune system of vertebrates, with unlimited potential to respond to any foreign determinant. Sophisticated recognition of non-self moieties by this system is achieved by unique molecular entities on the surface of individual lymphocytes. Recent evidence points to a role for cell-surface structures on the surface of the individual's own cells in the evolution (during the individual's lifetime) of recognition markers on the lymphocyte surface. In other words, the very surface markers (A) by which cells of organism P recognize cells from organism Q as non-self, themselves control the recognition molecules (B) by which P performs this action (Zinkernagel and Doherty, 1977). The plasticity of the development of this recognition potential has been demonstrated by the effect of introducing new cells into the environment during differentiation—these cells then become seen as part of the natural "self" milieu and also themselves contribute to the evolution of the new set of recognition molecules on the scavenger cells' surface. By involving a scavenger discrimination system in which learning takes place "initially" in the context of self-determinants, the vertebrate immune system thus produces a cornucopia of different cells, each with different recognition potential, and the external cellular environment secondarily selects and expands (clonal proliferation) the appropriate population recognizing any given foreign determinant. This enormous repertoire of recognition specificities is generated at the molecular level by a process of gene rearrangement in a multigene family encoding the receptor molecules (immunoglobulins on so-called B lymphocytes) as well as by subsequent somatic diversification in the rearranged gene pool (Leder, 1982). In contrast to the complexity of mammalian lymphocytes, more primitive multicellular organisms are limited in their ability to discriminate between foreign determinants; however, given their less complex cellular lifeforms and life-cycle, the recognition potential provided by lectinlike molecules (which may, nevertheless, have been the forerunner of vertebrate recognition moieties) (Marchalonis *et al.*, 1984) is sufficient.

Recent evidence is compatible with the notion that the polymorphism associated with vertebrate cell-surface markers [of the major histocompatibility complex MHC—these are analogous to those defined as (A) on P cells, which as suggested above apparently regulate the diversity of the vertebrate recognition system] evolved in multicellular organisms as a mechanism to prevent self-fertilization in hermaphrodites (Monroy and Rosati, 1979). In colonial metazoans, fusion of neighboring colonies is under strict genetic control, and it has been suggested that the regulation that enforces the heterozygous condition through control of fertilization and/or development is perhaps essential for survival. Linkage of MHC genes with elements that are responsible for recognition of all foreign determinants may thus also play a role in promoting heterozygosity in fertilization because of the adaptive advantage so conferred at the adult stage.

Finally, as indicated above, integration of the multifaceted cellular complex is a function given over to the brain of higher organisms. While it is readily accepted that conscious functions of the voluntary nervous system are under direct organismic control, and a biochemical basis for this control in terms of neurotransmitters, specific neuronal pathways, and the endocrine system has been described, the notion that the autonomic nervous system is amenable also to conscious manipulation is rather recent (Engel, 1972). Most perplexing of all, of course, is that as yet we can make only vague suggestions as to likely mechanisms to describe how conscious perception of an altered environmental milieu (either external, or, more interestingly, internal to the organism) can elicit the appropriate cellular response in a system believed to be under autonomic control. It seems unlikely to me, however, that "new rules" of intercellular communication will necessarily have been called into play during this phase of evolution, and thus the most profitable area to begin searching for mechanisms to explain the phenomenon will be in the linkage between the nervous system (the integrating device) and the immune system (the systemic "police force"). In a recent review, Blalock (1984) suggests that molecules used by lymphocytes to communicate with one another may indeed be similar to those used by cells of the nervous system for intercellular communication (and may thus serve to bridge the "communication gap" between these seemingly independent cell complexes).

III. EVOLUTION OF SOCIAL ORGANIZATION

A. Kinship

To what extent do the above statements apply to cellular groupings of the next order of magnitude, in other words sociocultural groups (and here I shall restrict myself primarily to the human species)? Exhaustive cross-cultural studies have led anthropologists to conclude that there are two principles that serve as primary modes of social organization, those of affinity (relationship by marriage) and those of descent (relationship by parentage) (Hatner, 1970). Those persons to whom one can show a relation defined in either of these terms are classificatory kin, and the shared ideology is kinship. From the kin terminology used within a group, a great deal can be inferred of the inherent sociocultural organization of the society. These "primitive" markers of the social order (and by primitive I mean to infer merely that they are "first order" markers rather than a reflection of the cultural status of the society) can be seen as the analogues of cell-surface lectins.

B. From Band to Tribal Societies

Egalitarian band-level societies follow a mobile hunting and gathering mode of existence, akin perhaps to the amoebae of *Dictyostelium*. The stochastic process, which serves as a description of lifestyle of the amoebae and the life skills of the family unit in the band, ensure that any advantage enjoyed by one unit over another is temporary. Thus economic exchange within the population is reciprocal and noninstitutionalized. Human contact with individuals outside of the band is highly infrequent, and in actuality it is rare to find social techniques within band-level societies that are capable of successfully integrating local groups into larger aggregates. Hence a more sophisticated identification system than that based on kinship is not required. All outsiders can be viewed with suspicion under the umbrella term "stranger," just as simple lectin molecules serve to define the complex extracellular world to the amoebae as "non-self."

Egalitarian band-level societies are propelled towards a more ranked tribal society in response to environmental pressure, just as food deprivation forces cellular complexity upon the autonomous amoebae of *Dictyostelium discoideum*. The change from the mobile hunting and gathering forms of subsistence to a sedentary and mobile-intensive pattern of hunting and gathering allows for a superior, more diversified adaptation to limited resources from which energy needs to be extracted more productively and more efficiently than is necessary for band-level groups (Redman, 1974). Nevertheless, there is no great change in the complexity whereby individuals categorize others in the functioning social whole, with kinship terminology still playing a preeminent role in marker termi- nology. The major technique that serves to integrate local groups into larger aggregates are so-called pan-tribal sodalities (e.g., members of a given age, belonging to the same religious order, etc.), and these additional markers may be viewed as the social equivalent of the cell-surface markers of primitive germ layers in embryonic development. Thus, it has, for instance, already been men- tioned that interspecies chimeras can be formed by virtue of the preferential adhesiveness of endoderm for mesodermal tissue, rather than for ectoderm of the same species (or individual organism).

C. Stratification and Ancillary Class-Specific Markers

Sociologically there is a major difference between chiefdom and tribal societies, in that while the latter often accords members possessing similar abilities a differential status, there is no privileged economic or political position associated with this status. In contrast, in a stratified society some members have access to important strategic resources that are categorically denied to others in the social group (Fried, 1967). By its very nature this implies that the emergence of stratified social orders cannot depend entirely on enculturation and internal (to the organism) self-administered sanctions to maintain stability but demands a more formal elaboration of rules and a mechanism to enforce those rules. This hierarchical system of social integration has a focal point in the chief, with descent from the chief being a primary determinant of rank in the social order. As Service (1962) states, "Chiefdoms are particularly distinguished from tribes by the presence of centers which coordinate economic, social and religious activities. . . ."

Specialization of labor becomes more marked, but in particular specialization occurs in the authority that holds power and the mobile group that "polices" cultural values within the group. What has also become of new importance is the recognition of territorial domain for the social aggregate. In consequence, a mobile specialized group capable of both recognizing the complexity of labels associated with members of the group, while simultaneously perceiving the presence of foreigners within this complex, develops to ensure the stability of the social order. That stability, in turn, is critically dependent upon the regulation of the specialized police force by the coordinating center.

As an example of this dependency one need only consider the chaos that occurred in the Roman Empire in 69 A.D. following Nero's suicide. During the next 12 months, with the Roman army now out of control of the Senate and Emperor, four Emperors were proclaimed in succession by various of the provincial armies, with three meeting a violent death at the hands of subsequent claimants to the throne.

Any advance in cultural complexity on the physical border of any group's territory generally acts as a trigger to induce parallel changes within that group. Failure to change inevitably results in assimilation into the more complex culture, usually by force. This phenomenon has been referred to as the "law" of Cultural Dominance.

There is no clear line of demarcation between chiefdoms and state societies (Fried, 1967). Stratification becomes continually more complex, and even where the state society apparently provides its citizens with an economic security and material prosperity greater than that of egalitarian societies, the expropriation of the individual's output, restriction of territorial access, and the demand for obedience to a numerically small elite within the society place the latter (governing) class in an intrinsically unstable position. Institutions that counter this instability run the gamut from use of an increasingly elaborate physical control to the more powerful means of psychological control, with the latter generally taking the form of some manipulation of the group ideology (Rosche and Kircheimer, 1939; Gearing and Tindale, 1973). In this regard one can correlate the purpose of modern state-supported universal education with the state-subsidized monumental architecture and elaborate priesthoods of the preindustrial Incas, Aztecs, and Egyptians.

An auxiliary marker of stratification in state societies is class, an anthropological marker that is an estimate of the individual's power (control) over the tools for, and production of, socially available energy (Adams, 1970). While the subordinate group may derive benefit from any given task, the crucial feature in a state society is that for the performance to occur at all, the superordinate group must derive (generally greater) benefit. Within our society, class markers include the more obvious ones of age, sex, and wealth, as well as the more subtle characteristics of speech, dress, nutritional habits, etc. (Dornhoff, 1970). Furthermore, classes within society differ greatly in the manner in which membership is established (where by hereditary ascription we refer to a caste system) and the rate at which it changes. In many instances groups deliberately promote cultural pluralism reducing the rate of assimilation into society, by the practice of endogamy (Berreman, 1960). Persistence of the latter is particularly evident where it confers some adaptive advantage, e.g., in competition for a restricted environmental resource. Consider here as a parallel the difference in class of the progeny of stem cells in the hemopoietic system (a variety of different offspring assimilated into the systemic circulation) with the restricted class of the progeny of stem cells in the basal layers of epithelial tissue.

IV. COMPARISON OF CELLULAR
AND SOCIAL ORGANIZATION

This brief overview of the evolution of complex societies and complex multicellular organisms will, I hope, have brought out several parallels. First, compartmentalization (or specialization) within the stratification hierarchy is itself dependent on the evolution of more sophisticated classification schemes than the primitive ones of kinship (simple lectin molecules) used in less complex societies, though one can still expect these latter markers to operate at primary levels of communication. Secondly, effective integration and stability of a multicomponent system revolves around a central coordinator (the brain/governing body), which uses a specialized mobile scavenger group (in cellular terms an immune system) to ensure the integrity of the system whether viewed in the context of either the internal or external environments. Finally, just as complex societal organization is learned during the process of enculturation, so cellular recognition apparently depends on similar learning processes during the development of the organism, i.e., the development of the lymphocyte recognition repertoire referred to above, which seems to represent an environmental selection of genetically encoded specificities. Note, too, that just as within different cultures the same kinships may imply different sets of social expectations, so at the cellular level the same molecular marker may play different functions when expressed on different tissues.

V. SUMMARY

A. Evolution of Complexity in Intercellular Recognition

We come now, of course, to the most pertinent question. Is this device of seeing the multicellular organism in the context of social anthropology of any heuristic value in furthering our understanding of intercellular communication? To quote from an article by Young (1976), analyzing medical belief systems in different sociocultural contexts,

> . . . etiological [medical] explanations [and externalizing systems] tend to dominate in simple societies . . . [and] . . . these [explanations] are frequently derogated as 'supernatural' and non-empirical . . . [yet] . . . Internalizing systems develop as the division of labor grows increasingly complex and leads to the transformation of the homogeneous and self-sufficient community and the appearance of specialized politico-jural institutions and broadly competent professonal healers. . . . These changes [in the medical belief system] include the appearance of new models for analogical thinking based on mediated exchange among functionally specialized parts [taken from the money market experience] and hierarchical dominance amongst functionally specialized parts [taken from the state structure].

What I wish to examine is the proposition that there is value not merely in considering, as has Young, that our medical beliefs are dependent on, and a construct of, our sociocultural institutions but also that there may even be an intrinsic biological ''drive'' behind these institutions and the cellular physiology and psychobiology for which our medical belief systems attempt to offer a rational explanation.

Consider initially the evolution of cell-surface markers and recognition structures, and kinship terminology. It has been suggested that the complexity of the latter and thus the ability to proscribe exogamous practices evolved in parallel with incest taboos. In actual fact, infanticide is a far more efficient mechanism to remove homozygous deleterious genes than is exogamy. One might also note, to give but one example, that since Cleopatra was a descendent of a long line of incestuous relationships in the Ptolmaic kingdom of Ancient Egypt, and from what contemporary Roman historians of this period (middle of the first century b.c.) tell us she showed little obvious evidence of genetic degeneracy attributable to this homozygosity, the biological necessity of exogamous relationships may also be overstated. Nor is a biological basis of exogamy easily reconciled with the independent endogamous practices carried out within seemingly exogamous societies (i.e., the frequently occurring marriage systems that demand union between cross cousins and not parallel cousins). Contemporary anthropology favors the view that the evolutionary pressure that favors exogamy is a social one. The ability to extend one's social family, with concomitant realization of more kinship ties, is of adaptive advantage in a more complex society where efficient utilization of limited environmental resources by a small group is not practical (Livingstone, 1969). As suggested above, it seems entirely feasible that the linkage of MHC genes with factors that promote heterozygosity at fertilization in colonial tunicates may be a reflection of the adaptive advantage conferred at the adult stage by a linkage of the extent of diversity in cellular recognition structures with polymorphism at the MHC locus.

B. Learning by Example

A major unresolved dilemma of molecular and cellular genetics is the explanation of how cell-surface interactions specify developmental change at the genetic level. Cultural complexity is learned in the early years of childhood by interaction with adult members practicing the culture of the group one becomes a part of. It seems that in vertebrate embryos at least, some cell specialization may intiailly be preprogramed into the fertilized egg by inhomogeneities in the cytoplasm. These inhomogeneities could presumably early regulate differential appearance of cell-surface markers, which in turn then leads to altered reception of environmental signals and the ensuing cascade of changes. As yet, further speculation in this area seems redundant.

C. Nature/Nurture and Chronic Disease

A Western culture view of medicine seeks an explanation for ill-health in terms of one overriding and obvious cause. In the case of infectious disease this analysis has been successful, but if we accept that the major medical problems that confront us now are of a more chronic type, we must admit to a singular failure in our ability to explain and treat these diseases in an effective manner. Contemporary anthropologists investigate a given culture by seeing the sociocultural system as a holistic entity, and an attempt is made to put this into the framework of the surrounding cultures by which it is influenced. Thus Boas (1938) has written:

> Culture may be defined as the totality of the mental and physical reactions and activities that characterize the behavior of the individuals composing a social group collectively and indi-

vidually in relation to their natural environment, to other groups, to members of the group itself and of each individual to himself. It also includes the products of those activities and their role in the life of the groups. The mere enumeration of these various aspects of life, however, does not constitute culture. It is more, for its elements are not independent: they have a structure.

Consider, as an example, the 10-year cycles of war and peace among the Tsembaga Maring of new Guinea, the ritual pig feasts that precede warring, and the development and retention of cultural taboos that prohibit aggression within this time frame. Extensive studies of the lifestyle of these peoples (Sahlins, 1958) suggests that this phenomenon can best be understood in the context of their slash and burn agricultural way of life in a relatively impacted environment. The domestic pig represents a key source of animal protein in an otherwise primarily vegetarian diet. However, pigs are killed periodically when the energy invested in keeping and raising them exceeds their social value (inherent in a culture that, perhaps for biological reasons, has attached such importance to these animals). Accordingly, pig-feast ceremonies are held on a regular basis, and these serve to cement social relationships, expand the social network (by leaving those invited to the pig-feast necessarily in social debt to the holder), and foster the importance of the "big man" in the village (the one who holds the feast). Nevertheless, it becomes essential periodically to redistribute land among the various village groups of the Tsembaga, since population growth fluctuates over time. Thus, at ritual 10-year intervals warring occurs (not infrequently with loss of life), followed by occupation of at least some of one village's land by that group generally able to have developed (via earlier pig-feasts) the numerically greater aggregate of adherents. A tree-planting ceremony follows, and not until these are at a specified height is further warfare culturally acceptable. During this time cultivation of the conquered territory does not occur, i.e., it lies fallow for cultural rules, allowing it to recover its fertility. After this time, should the group that initially sequestered the extra land have retained a sufficient number of the followers by appropriate pig-feasting, etc., to maintain control of the land, the latter now becomes formally recognized as its territory and can be used as land suitable for subsistence crops. It is clear then that the warring cycles are culturally integrated into the Tsembaga Maring's lifestyle in a way that makes good agricultural, nutritional, and social sense, and in a manner that permits flexibility of land redistribution in the jungle environment.

Pursuing the analogy with cell biology, we would thus conclude that successful understanding of chronic disease in humans depends on our viewing the process as an imbalance in both internal homeostasis (generally regulated by the nervous, endocrine, and immune system) and in the interaction with the external environment (psychosocial causes). Given that our flexibility for cultural adaptation diminishes with age (Piaget, 1965), it would thus seem most appropriate not to try to impose new long-term constraints upon this system (routine drug therapy) but to reassert those regulatory mechanisms that are an evolutionary adjunct of the system under investigation. If this sounds surprisingly like an epigenetic view of disease, one can do no better than point to the remarkable studies by Mintz and associates, who have demonstrated that under the appropriate regulatory influences a long-term passaged tumor cell can contribute to the production of specialized germ-line cells for normal mouse development (Stewart and Mintz, 1981). An alternative viewpoint would argue that a loss of the classificatory markers that describe a cellular (or social) hierarchy would also be a predisposing factor towards disease. There is evidence that genetic factors associated with changes in agglutinability and in contact inhibition of growth and/or movement may also be of importance in the development and/or growth of neoplastic cells (Dulbecco, 1969; Azarnia and Loewenstein, 1971; Ford et al., 1978). The major causal factor (genetic/epigenetic; nature/nurture) may vary from disease to disease, or even from individual to individual. Nevertheless, it seems crucial that we do not lose sight of the fact that both will likely be relevant, though in varying degrees in most (all) cases. In this light the reader is referred to a recent review by Cunningham (1985) that examines the possible role of psychological factors in the physiological processes that may regulate the onset and/or progression of cancer.

D. Adaptive Evolution

Finally, there is an interesting speculation one can make concerning the mechanisms for adaptive change in sociological and cellular systems. Physical anthropologists would argue that in the past 25,000 years it is cultural evolution rather than physical evolution that is most marked in the human species. Students of molecular evolution may soon have data which could force a rethinking of this notion. The explanation commonly offered is that culture is learned during the individual's lifetime, and is thus more rapidly changed than are those traits dictated by molecular genetics. Nevertheless, there is reason to believe that some cultural patterns are inherited in fashions predictable using Mendelian genetics. If, in complex societies, the major mechanism(s) to control culture change occur at the level of the central coordinator and the specialized police force, it seems likely that analysis of a parallel to this rapid adaptation technique at the cellular level should investigate the equivalents of these organs of change in the context of the cellular hierarchy.

There are a variety of studies in the scientific literature that point to an unusual mode of inheritance of drug-induced or behaviorally induced physiological effects that seem unlikely to be simply explained as transplacentally mediated (or colostrum-mediated) phenomena. Bakke et al. (1975) found such a peculiar inheritance after neonatal administration of L-thyroxine (T4) to rats, where F_1 offspring of such treated male rats were clearly abnormal, though to a lesser degree than their fathers. An elegant study of the effect of early weaning on susceptibility to restraint-induced gastric ulcers in rats showed, using a cross-fostering experimental design, that inheritance of this susceptibility could also not be simply explained within a seemingly conventional biological framework (Skolnick *et al.*, 1980).

Studies from our laboratory looking at a peculiar pattern of inheritance within the immune system of mice produced more such anomalous findings (Gorczynski and Steele, 1980). There may be several earlier documentations of similar phenomena (Steele, 1981a). Despite the controversy surrounding this issue (Steele, 1981b), it seems not unlikely that information acquired somatically during the breeding lifespan of males can indeed be transmitted to F_1 progeny (perhaps indirectly in the case of the immune system discussed) by immunization of females on copulation (Gorczynski *et al.*, 1983).

Hypothesizing that the cellular parallel of the cultural "police force," the specialized mobile aggressor cells, possesses some mechanism for the transfer to the next generation of information learned within one generation at high frequency, can we then consider similar possibilities for the cellular equivalent of the "cultural coordinator"? Do adaptive differentiation and inheritance of behavior occur?

REFERENCES

Adams, R. N. (1970). "Crucifixion by Power." University of Texas Press, Austin.
Allegretti, N. (1978). *Dev. Comp. Immunol.* **2**, 573.
Azarnia, R., and Loewenstein, W. R. (1971). *J. Membrane Biol.* **6**, 368.
Bakke, J. L., Lawrence, N. L., Bennet, J., and Robinson, S. (1975). *In* "Perinatal Thyroid Physiology and Disease" (D. A. Fisher and G. N. Burrows, eds.), p. 79. Raven Press, New York.
Beer, A. E., Quebbeman, J. F., Auers, J. W., and Haines, R. F. (1981). *Am. J. Obstet. Gynaecol.* **141**, 987.
Berger, S. (1983). *Nature (London)* **303**, 59.
Berreman, G. D. (1960). *Am. J. Sociol.* **66**, 120.
Blalock, J. E. (1984). *J. Immunol.* **132**, 1067.
Boas, F. (1938). "General Anthropology." D.C. Heath, Boston.
Brown, R. C., Bass, H., and Coombs, J. P. (1975). *Nature (London)* **254**, 434.
Buckle, G. R., and Greenberg, L. (1981). *Anim. Behav.* **29**, 802.
Cunningham, A. J. (1985). *Can. J. Psychol.* **26**, 1.
Dornhoff, G. W. (1970). "The Higher Circles: The Governing Class in America." Random House, New York.
Dulbecco, R. (1969). *Science* **166**, 962.
Engel, B. T. (1972). *Psychophysiology* **9**, 161.
Ford, W. L., Smith, M. E., and Andrews, P. (1978). *In* "Cell to Cell Recognition" (A. S. G. Curtis, ed.), p. 359. Cambridge Univ. Press, London and New York.
Frazier, W. A., Rosen, S. D., Reitheman, R. W., and Barondes, S. A. (1975). *J. Biol. Chem.* **250**, 7714.

Fried, M. H. (1960). *In* "Culture in History" (S. Diamond, ed.), pp. 713–731. Columbia University Press, New York.

Fried, M. H. (1967). *In* "The Evolution of Political Society: An Essay in Political Anthropology." Random House, New York.

Gearing, F., and Tindale, B. A. (1973). *Annu. Rev. Anthropol.* **2**, 95.

Getz, W. M., and Smith, K. B. (1983). *Nature (London)* **302**, 147.

Gorczynski, R. M., and Steele, E.-J. (1980). *Proc. Natl. Acad. Sci. U.S.A.* **77**, 287.

Gorczynski, R. M., Kennedy, M., MacRae, S., and Ciampi, A. (1983). *J. Immunol.* **131**, 1115.

Gross, J. (1956). *J. Biophys. Biochem. Cytol. (Suppl.)* **2**, 261.

Hatner, M. J. (1970). *Southwest. J. Anthropol.* **26**, 67.

Holtfreter, J. (1944). *J. Exp. Zool.* **95**, 171.

Leder, P. (1982). *Sci. Am.* **246**(5), 72.

Lenington, S. (1983). *Anim. Behav.* **31**, 325.

Livingstone, F. B. (1969). *Curr. Anthropol.* **10**, 45.

Marchalonis, J. J., and Barker, W. (1984). *Immunol. Today* **5**, 222.

Monroy, A., and Rosati, F. (1979). *Nature (London)* **27**, 165.

Moscona, A. A. (1963). *Proc. Natl. Acad. Sci. U.S.A.* **49**, 742.

Newell, P. C., Franke, J., and Sussman, M. (1972). *J. Mol. Biol.* **63**, 373.

O'Donald, P. (1983). *Theor. Pop. Biol.* **23**, 64.

Packer, C. (1979). *Anim. Behav.* **27**, 1.

Piaget, J. (1965). "The Moral Judgement of the Child," p. 436. New York Free Press, New York.

Pusey, A. E. (1980). *Anim. Behav.* **28**, 543.

Redman, C. L. (1974). *In* "The Explanation of Culture Change" (C. Renfrew, ed.), pp. 717–725. University of Pittsburgh Press, Pittsburgh.

Rosche, G., and Kircheimer, O. (1939). "Punishment and Social Structure." Columbia University Press, New York.

Sahlins, M. D. (1958). "Social Stratification in Polynesia." Seattle University of Washington Press, Seattle.

Scofield, V. L., Schlumpberger, J. M., West, L. A., and Weissman, I. L. (1982). *Nature (London)* **295**, 499.

Service, E. (1962). "Primitive Social Organization: An Evolutionary Perspective." Random House, New York.

Sio, F. J., and Beer, A. E. (1982). *Biol. Reprod.* **26**, 15.

Skolnick, N. J., Ackerman, S. H., Hofer, M. A., and Weiner, H. (1980). *Science* **208**, 1161.

Steele, E.-J. (1981a). *New Scientist* **90**, 360.

Steele, E.-J. (1981b). "Somatic Selection and Adaptive Evolution." University of Chicago Press, Chicago.

Steinberg, M. S. (1970). *J. Exp. Zool.* **173**, 395.

Stewart, T. A., and Mintz, B. (1981). *Proc. Natl. Acad. Sci. U.S.A.* **78**, 6314.

Taylor, C., and Faulk, W. P. (1981). *Lancet ii*, 68.

Townes, P. L., and Holtfreter, J. (1955). *J. Exp. Zool.* **128**, 53.

Young, A. (1976). *Soc. Sci. Med.* **10**, 147.

Zinkernagel, R. M., and Doherty, P. C. (1977). *Contemp. Top. Immunobiol.* **7**, 179.

Index

A

ABH-active glycoprotein, 246, 251, 252–258
  characteristics, 252–254
  identification, 252–254, 257–258
ABO blood group system, 252–258
Abortion, spontaneous, 306
Acanthamoeba, feeding recognition, 47
Acetylcholine receptor, on vertebrate skeletal
    muscle cells, 183–213
  in adult muscle, 186–188
  aggregation stimulation, 185–196
  clustering control, 196–205
  clustering induction signal, 197–200
  clustering process, 200–204
  clustering sequence of events, 206–207
  in cultured cells, 192–196
  in cultures without nerve cells, 192–194
  cytoskeletal proteins and, 202–204, 206, 207
  ectopic clusters, 188, 190–191
  extrajunctional receptors, 191–192
  in vivo, 186–192
  in innervated cells, 194–195
  during neuromuscular junction development,
      188–191
  receptor distribution, 185
  receptor synthesis, 191
  mRNA, 191
  structure, 184
Acetylcholinesterase, neuronal 228
N-Acetylgalactosamine, 246
N-Acetylgalactosaminyltransferase, 153
N-Acetylglucosamine, 78
α1 Acid glycoprotein, 52
Acquired immune deficiency syndrome, in-
    terleukin-2 therapy, 278
Acrosome reaction
  in mammals, 141
  in sea urchins, 136

Actin
  in acetylcholine receptor clustering, 203, 204,
      206
  in growth cone microfilaments, 216–217, 219
  α-Actinin, in acetylcholine receptor cluster-
      ing, 203
Adaptive differentiation, 6–7
  in chimeras, 105, 106–109
  definition, 105
  environmental restriction, 108–109, 111
  of lymphocytes, 105, 106–109
  of T cell phenotype, 110–113
Adenylate cyclase, in morphine addiction, 6
Adhesion
  of blastomeres, 133–134
  calcium-dependent, 151–153
  calcium-independent, 151–152
  carbohydrates in, 36
  differential adhesion hypothesis, 126, 171, 309
  of embryonic receptors, 150–153
  during gastrulation, 308
  interaction modulation factors in, 125
  of lymphocytes, 126, 248
  oligosaccharide regulation of, 248
  in pathogenesis, 11
  of tumor cells, 249, 291, 294–295
  of zoospores, 10
Adhesion molecule, 218–219, 230 see also Cell
    adhesion molecule
Agglutinin, see also Hemagglutinin
  as infection defense mechanism, 29
  of invertebrates, 29–33
Aggregation
  of acetylcholine receptors, 130–131, 185–
      208
  carbohydrates in, 295
  embryonic, 309
  in organogenesis, 125, 309
  of slime molds, 307–308